T0296131

ELECTRONS (+ and −), PROTONS, PHOTONS, NEUTRONS, AND COSMIC RAYS

Electrons (+ and −), Protons, Photons, Neutrons, and Cosmic Rays

by

ROBERT ANDREWS MILLIKAN

Formerly Professor of Physics, the University of Chicago
Director Norman Bridge Laboratory of Physics
California Institute of Technology

CAMBRIDGE
AT THE UNIVERSITY PRESS
1935

CAMBRIDGE
UNIVERSITY PRESS

University Printing House, Cambridge CB2 8BS, United Kingdom

Published in the United States of America by Cambridge University Press, New York

Cambridge University Press is part of the University of Cambridge.

It furthers the University's mission by disseminating knowledge in the pursuit of education, learning and research at the highest international levels of excellence.

www.cambridge.org
Information on this title: www.cambridge.org/9781107689213

© Cambridge University Press 1935

First published 1935
First paperback edition 2014

A catalogue record for this publication is available from the British Library

ISBN 978-1-107-68921-3 Paperback

PREFACE

In 1917 the University of Chicago Press published a small volume entitled *The Electron* which was intended to be a rather simple presentation of some of the newer developments in physics with which my own work had been closely associated. In 1924 a revised edition of this work appeared. A few years later I had the honor of giving "The Messenger Lectures" at Cornell University and in them still further expanded and brought up to date these "newer developments."

The present volume grows *immediately* out of these Messenger Lectures, but is, of course, an attempt to make the presentation as true a picture as I am able to give of the situation as it exists at the date of publication, January 1, 1935.

In a sense this work may be looked upon as a third revision of *The Electron*, but it differs from most revisions in two particulars. First, an effort has been made at the request of the publishers to introduce into *The Electron* portion only such changes as are demanded by correctness of presentation today, and in fact I have been pleased and somewhat surprised to find that the historical mode of presentation originally adopted has rendered radical changes even today both unnecessary and undesirable. Second, the growth of discovery and the rapidity of the advance in physics from the base occupied in 1924 has made it altogether necessary to add six entirely new chapters (xi to xvi) on "Waves and Particles," on "The Discovery and Origin of the Cosmic Rays," on "The Spinning Electron," on "The Positron," on "The Neutron and the Transmutation of the Elements," and on "The Nature of the Cosmic Rays," so that in this particular the book is not at all a revision. The illustrative material, too, has been much more than doubled, and now becomes quite a notable feature.

It is hoped that this volume, like its predecessors, may be of some interest both to the physicist and to the reader of somewhat less technical training. It has been thought desirable for the sake of both classes of readers not to break the thread of the discussion in the body of the book with the detailed analyses that the careful student demands. It is for this reason that all mathematical proofs have been thrown into appendixes. If, in spite of this, the general student finds that certain chapters, such as vii, viii, and xii, require more familiarity with the general background of physics than he possesses, it is still hoped that without them he may yet gain some idea of certain phases, at least of the fascinating progress of modern science.

ROBERT A. MILLIKAN

CALIFORNIA INSTITUTE OF TECHNOLOGY
PASADENA, CALIFORNIA
December 1, 1934

CONTENTS

		PAGE
INTRODUCTION	. . . .	1

CHAPTER

I. EARLY VIEWS OF ELECTRICITY 6

II. THE EXTENSION OF THE ELECTROLYTIC LAWS TO CONDUCTION IN GASES 25

III. EARLY ATTEMPTS AT THE DIRECT DETERMINATION OF e 45

IV. GENERAL PROOF OF THE ATOMIC NATURE OF ELECTRICITY 66

V. THE EXACT EVALUATION OF e 90

VI. THE MECHANISM OF IONIZATION OF GASES BY X-RAYS AND RADIUM RAYS 125

VII. BROWNIAN MOVEMENTS IN GASES 145

VIII. IS THE ELECTRON ITSELF DIVISIBLE? 158

IX. THE STRUCTURE OF THE ATOM 182

X. THE NATURE OF RADIANT ENERGY 232

XI. WAVES AND PARTICLES 260

XII. THE SPINNING ELECTRON 270

XIII. THE DISCOVERY AND ORIGIN OF THE COSMIC RAYS . 301

XIV. THE DIRECT MEASUREMENT OF THE ENERGY OF COSMIC RAYS AND THE DISCOVERY OF THE FREE POSITIVE ELECTRON 320

XV. THE NEUTRON AND THE TRANSMUTATION OF THE ELEMENTS 360

XVI. THE NATURE OF THE COSMIC RAYS 404

APPENDIX A. ne FROM MOBILITIES AND DIFFUSION CO-EFFICIENTS 457

APPENDIX B.   TOWNSEND'S FIRST ATTEMPT AT A DETERMI-
NATION OF e   460

APPENDIX C.   THE BROWNIAN-MOVEMENT EQUATION .   .   463

APPENDIX D.   THE INERTIA OR MASS OF AN ELECTRICAL
CHARGE ON A SPHERE OF RADIUS a   467

APPENDIX E.   MOLECULAR   CROSS-SECTION   AND   MEAN
FREE PATH   470

APPENDIX F.   NUMBER OF FREE POSITIVE ELECTRONS IN
THE   NUCLEUS   OF   AN   ATOM   BY   RUTHERFORD'S
METHOD   472

APPENDIX G.   BOHR'S THEORETICAL DERIVATION OF THE
VALUE OF THE RYDBERG CONSTANT   477

APPENDIX H.   A. H. COMPTON'S THEORETICAL DERIVA-
TION OF THE CHANGE IN THE WAVE-LENGTH OF
ETHER-WAVES BECAUSE OF SCATTERING BY FREE
ELECTRONS   479

APPENDIX I.   THE ELEMENTS, THEIR ATOMIC NUMBERS,
ATOMIC WEIGHTS, AND CHEMICAL POSITIONS .   .   .   481

APPENDIX J.   PHYSICAL CONSTANTS   482

INDEXES   483

INTRODUCTION

Perhaps it is merely a coincidence that the man who first noticed that the rubbing of amber would induce in it a new and remarkable state now known as the state of *electrification* was also the man who first gave expression to the conviction that there must be some great unifying principle which links together all phenomena and is capable of making them rationally intelligible; that behind all the apparent variety and change of things there is some primordial element, out of which all things are made and the search for which must be the ultimate aim of all natural science. Yet if this be merely a coincidence, at any rate to Thales of Miletus must belong a double honor. For he first correctly conceived and correctly stated, as far back as 600 B.C., the spirit which has actually guided the development of physics in all ages, and he also first described, though in a crude and imperfect way, the very phenomenon the study of which has already linked together several of the erstwhile isolated departments of physics, such as radiant heat, light, magnetism, and electricity, and has very recently brought us nearer to the primordial element than we have ever been before.

Whether this perpetual effort to reduce the complexities of the world to simpler terms, and to build up the infinite variety of objects which present themselves to our senses out of different arrangements or motions of the least possible number of elementary substances, is a

I

modern heritage from Greek thought, or whether it is a native instinct of the human mind may be left for the philosopher and the historian to determine. Certain it is, however, that the greatest of the Greeks aimed at nothing less than the complete banishment of caprice from nature and the ultimate reduction of all her processes to a rationally intelligible and unified system. And certain it is also that the periods of greatest progress in the history of physics have been the periods in which this effort has been most active and most successful.

Thus the first half of the nineteenth century is unquestionably a period of extraordinary fruitfulness. It is at the same time a period in which for the first time men, under Dalton's lead, began to get direct, experimental, quantitative proof that the atomic world which the Greeks had bequeathed to us, the world of Leucippus and Democritus and Lucretius, consisting as it did of an infinite number and variety of atoms, was far more complex than it needed to be, and that by introducing the idea of molecules built up out of different combinations and groupings of atoms the number of necessary elements could be reduced to but about seventy. The importance of this step is borne witness to by the fact that out of it sprang in a very few years the whole science of modern chemistry.

And now this twentieth century, though but thirty-four years old, has already attempted to take a still bigger and more significant step. By superposing upon the molecular and the atomic worlds of the nineteenth century a third electronic world, it has sought to reduce the number of primordial elements to not more than two, namely, positive and negative electrical charges. Along with this effort has come the present period of most extraordinary development and fertility—a period in which new viewpoints and indeed wholly new phenomena follow one another so rapidly across the stage of physics that the actors themselves scarcely know what is happening—a period too in which the commercial and industrial world is adopting and adapting to its own uses with a rapidity hitherto altogether unparalleled the latest products of the laboratory of the physicist and the chemist. As a consequence, the results of yesterday's researches, designed for no other purpose than to add a little more to our knowledge of the ultimate structure of matter, are today seized upon by the practical business world and made to multiply tenfold the effectiveness of the telephone or to extract six times as much light as was formerly obtained from a given amount of electric power.

It is then not merely a matter of academic interest that electricity has been proved to be atomic or granular in structure, that the elementary electrical charge has been isolated and accurately measured, and that it has been found to enter as a constitutent into the making of all the seventy-odd atoms of chemistry. These are indeed matters of fundamental and absorbing interest to the man who is seeking to unveil nature's inmost secrets, but they are also events which are pregnant with meaning for the man of commerce and for the worker in the factory. For it usually happens that when nature's inner workings have once been laid bare, man sooner or later finds a way to put his brains inside the machine and to drive it whither he wills. Every increase in man's knowledge of the way in which nature works must, in the long run, increase by just so much man's ability to control nature and to turn her hidden forces to his own account.

The purpose of this volume is to present the evidence for the atomic structure of electricity, to describe some of the most significant properties of the elementary electrical unit, the electron, and to discuss the bearing of these properties upon the two most important problems of modern physics: the structure of the atom and the nature of electromagnetic radiation. In this presentation I shall not shun the discussion of exact quantitative experiments, for it is only upon such a basis, as Pythagoras asserted more than two thousand years ago, that any real scientific treatment of physical phenomena is possible. Indeed, from the point of view of that ancient philosopher, the problem of all natural philosophy is to drive out qualitative conceptions and to replace them by quantitative relations. And this point of view has been emphasized by the farseeing throughout all the history of physics clear down to the present. One of the greatest of modern physicists, Lord Kelvin, writes:

> When you can measure what you are speaking about and express it in numbers, you know something about it, and when you cannot measure it, when you cannot express it in numbers, your knowledge is of a meagre and unsatisfactory kind. It may be the beginning of knowledge, but you have scarcely in your thought advanced to the stage of a science.

Although my purpose is to deal mostly with the researches of which I have had most direct and intimate knowledge, namely, those which have been carried on during the past thirty years in this general field, first in the Ryerson Laboratory at the University of Chicago, and later at the Norman Bridge Laboratory of Physics at the California Institute at Pasadena, I shall hope to be able to give a correct and just review of the preceding work out of which these researches grew, as well as of parallel work carried on in other laboratories. In popular writing it seems to be necessary to link every great discovery, every new theory, every important principle, with the name of a single individual. But it is an almost universal rule that developments in physics actually come about in a very different way. A science, like a plant, grows in the main by a process of infinitesimal accretion. Each research is usually a modification of a preceding one; each new theory is built like a cathedral through the addition by many builders of many different elements. This is pre-eminently true of the electron theory. It has been a growth, and I shall endeavor in every case to trace the pedigree of each research connected with it.

CHAPTER I

EARLY VIEWS OF ELECTRICITY

I. GROWTH OF THE ATOMIC THEORY OF MATTER

There is an interesting and instructive parallelism between the histories of the atomic conception of matter and the atomic theory of electricity, for in both cases the ideas themselves go back to the very beginnings of the subject. In both cases too these ideas remained absolutely sterile until the development of precise quantitative methods of measurement touched them and gave them fecundity. It took two thousand years for this to happen in the case of the theory of matter and one hundred and fifty years for it to happen in the case of electricity; and no sooner had it happened in the case of both than the two domains hitherto thought of as distinct began to move together and to appear as perhaps but different aspects of one and the same phenomenon, thus recalling again Thales' ancient belief in the essential unity of nature. How this attempt at union has come about can best be seen by a brief review of the histories of the two ideas.

The conception of a world made up of atoms which are in incessant motion was almost as clearly developed in the minds of the Greek philosophers of the School of Democritus (420 B.C.), Epicurus (370 B.C.), and Lucretius (Roman, 50 B.C.) as it is in the mind of the modern physicist, but the idea had its roots in one case in a mere speculative philosophy; in the other case, like most of our twentieth-century knowledge, it rests upon direct, exact, quantitative observations and measurement. Not that the human eye has ever seen or indeed can ever see an individual atom or molecule. This is forever impossible, and for the simple reason that the limitations on our ability to see small objects are imposed, not by the imperfections of our instruments, but by the nature of the eye itself, or by the nature of the light-wave to which the eye is sensitive. If we are to see molecules our biological friends must develop wholly new types of eyes, viz., eyes which are sensitive to waves one thousand times shorter than those to which our present optic nerves can respond.

But after all, the evidence of our eyes is about the least reliable kind of evidence which we have. We are continually seeing things which do not exist, even though our habits are unimpeachable. It is the relations which are seen by the mind's eye to be the logical consequences of exact measurement which are for the most part dependable. So far as the atomic theory of matter is concerned, these relations have all been developed since 1800, so that both the modern atomic and the modern kinetic theories of matter, in spite of their great antiquity, are in a sense less than one hundred years old. Indeed, nearly all of our definite knowledge about molecules and atoms has come since 1851, when Joule[1] in England made the first absolute determination of a molecular magnitude, namely, the average speed with which gaseous molecules of a given kind are darting hither and thither at ordinary temperatures. This

[1] *Mem. of the Manchester Lit. and Phil. Soc.* (1851; 2d series), 107; *Phil. Mag.*, XIV (1857), 211.

result was as surprising as many others which have
followed in the field of molecular physics, for it showed
that this speed, in the case of the hydrogen molecule, has
the stupendous value of about a mile a second. The
second molecular magnitude to be found was the mean
distance a molecule of a gas moves between collisions,
technically called the mean free path of a molecule.
This was computed first in 1860 by Clerk Maxwell.[1] It
was also 1860 before anyone had succeeded in making any
sort of an estimate of the number of molecules in a cubic
centimeter of a gas. When we reflect that we can now
count this number with probably greater precision than
we can attain in determining the number of people living
in New York, in spite of the fact that it has the huge
value of 27.05 billion billion, one gains some idea of how
great has been our progress in mastering some at least
of the secrets of the molecular and atomic worlds. The
wonder is that we got at it so late. Nothing is more surprising to the student brought up in the atmosphere of
the scientific thought of the present than the fact that the
relatively complex and intricate phenomena of light and
electromagnetism had been built together into moder-
ately consistent and satisfactory theories long before the
much simpler phenomena of heat and molecular physics
had begun to be correctly understood. And yet almost
all the qualitative conceptions of the atomic and kinetic
theories were developed thousands of years ago. Tyndall's statement of the principles of Democritus, whom
Bacon considered to be "a man of mightier metal than

[1] *Phil. Mag.*, XIX (1860; 4th series), 28. Clausius had discussed
some of the relations of this quantity in 1858 (*Pogg. Ann.*, CV [1858],
239), but Maxwell's magnificent work on the viscosity of gases first
made possible its evaluation.

Plato or Aristotle, though their philosophy was noised and celebrated in the schools amid the din and pomp of professors," will show how complete an atomic philosophy had arisen 400 years B.C. "That it was entirely destroyed later was not so much due to the attacks upon it of the idealistic school, whose chief representatives were Plato and Aristotle, as to the attacks upon all civilization of Genseric, Attila, and the barbarians." That the Aristotelian philosophy lasted throughout this period is explained by Bacon thus: "At a time when all human learning had suffered shipwreck these planks of Aristotelian and Platonic philosophy, as being of a lighter and more inflated substance, were preserved and came down to us, while things more solid sank and almost passed into oblivion."

Democritus' principles, as quoted by Tyndall, are as follows:

1. From nothing comes nothing. Nothing that exists can be destroyed. All changes are due to the combination and separation of molecules.

2. Nothing happens by chance. Every occurrence has its cause from which it follows by necessity.

3. The only existing things are the atoms and empty space; all else is mere opinion.

4. The atoms are infinite in number and infinitely various in form; they strike together and the lateral motions and whirlings which thus arise are the beginnings of worlds.

5. The varieties of all things depend upon the varieties of their atoms, in number, size, and aggregation.

6. The soul consists of fine, smooth, round atoms like those of fire. These are the most mobile of all. They interpenetrate the whole body and in their motions the phenomena of life arise.

These principles with a few modifications and omissions might almost pass muster today. The great advance which has been made in modern times is not so much in the conceptions themselves as in the kind of foundation upon which the conceptions rest. The principles enumerated above were simply the opinions of one man or of a school of men. There were scores of other rival opinions, and no one could say which was the better. Today there is absolutely no philosophy in the field other than the atomic philosophy, at least among physicists. Yet this statement could not have been made even as much as thirty years ago. For in spite of all the multiple relationships between combining powers of the elements, and in spite of all the other evidences of chemistry and nineteenth-century physics, a group of the foremost of modern thinkers, until quite recently, withheld their allegiance from these theories. The most distinguished of this group was the German chemist and philosopher, Wilhelm Ostwald. However, in the preface to the last edition of his *Outlines of Chemistry* he now makes the following clear and frank avowal of his changed position. He says:

I am now convinced that we have recently become possessed of experimental evidence of the discrete or grained nature of matter for which the atomic hypothesis sought in vain for hundreds and thousands of years. The isolation and counting of gaseous ions on the one hand . . . . and on the other the agreement of the Brownian movements with the requirements of the kinetic hypothesis . . . . justify the most cautious scientist in now speaking of the experimental proof of the atomic theory of matter. The atomic hypothesis is thus raised to the position of a scientifically well-founded theory.

II. GROWTH OF ELECTRICAL THEORIES

The granular theory of electricity, while unlike the atomic and kinetic theories of matter in that it can boast no great antiquity in any form, is like them in that the first man who speculated upon the nature of electricity at all conceived of it as having an atomic structure. Yet it is only within very recent years—forty at the most—that the modern electron theory has been developed. There are no electrical theories of any kind which go back of Benjamin Franklin (1750). Aside from the discovery of the Greeks that rubbed amber had the power of attracting to itself light objects, there was no knowledge at all earlier than 1600 A.D., when Gilbert, Queen Elizabeth's surgeon, and a scientist of great genius and insight, found that a glass rod and some twenty other bodies, when rubbed with silk, act like the rubbed amber of the Greeks, and he consequently decided to describe the phenomenon by saying that the glass rod had become electrified (amberized, electron being the Greek word for amber), or, as we now say, had acquired a charge of electricity. In 1733 Dufay, a French physicist, further found that sealing wax, when rubbed with cat's fur, was also electrified, but that it differed from the electrified glass rod, in that it strongly attracted any electrified body which was repelled by the glass, while it repelled any electrified body which was attracted by the glass. He was thus led to recognize two kinds of electricity, which he termed "vitreous" and "resinous." About 1747 Benjamin Franklin, also recognizing these two kinds of electrification, introduced the terms "positive" and "negative," to distinguish them. Thus, he said, we will arbitrarily call any body positively electrified if it is repelled by a glass rod which has been rubbed with silk, and we will call any body negatively electrified if it is repelled by sealing wax which has been rubbed with cat's fur. *These are today our definitions of positive and negative electrical charges.* Notice that in setting them up we propose no theory whatever of electrification, but content ourselves simply with describing the phenomena.

In the next place it was surmised by Franklin and indeed asserted by him in the very use of the terms "positive" and "negative," although the accurate proof of the relation was not made until the time of Faraday's ice-pail experiment in 1837, that when glass is positively electrified by rubbing it with silk, the silk itself takes up a negative charge of exactly the same amount as the positive charge received by the glass, and, in general, that *positive and negative electrical charges always appear simultaneously and in exactly equal amounts.*

So far, still no theory. But in order to have a rational explanation of the phenomena so far considered, particularly this last one, Franklin now made the assumption that something which he chose to call the electrical fluid or "electrical fire" exists in normal amount as a constituent of all matter in the neutral, or unelectrified state, and that more than the normal amount in any body is manifested as a positive electrical charge, and less than the normal amount as a negative charge. Aepinus, professor of physics at St. Petersburg and an admirer of Franklin's theory, pointed out that, in order to account for the repulsion of two negatively electrified bodies, it was necessary to assume that matter, when divorced from Franklin's electrical fluid, was self-repellent, i.e., that it possessed properties quite different from those which are found in ordinary unelectrified matter. In order, however, to leave matter, whose independent existence was thus threatened, endowed with its familiar old properties, and in order to get electrical phenomena into a class by themselves, other physicists of the day, led by Symmer, 1759, preferred to assume that *matter in a neutral state shows no electrical properties because it contains as constituents equal amounts of two weightless fluids which they called positive and negative electricity, respectively.* From this point of view a positively charged body is one in which there is more of the positive fluid than of the negative, and a negatively charged body is one in which the negative fluid is in excess.

Thus arose the so-called two-fluid theory—a theory which divorced again the notions of electricity and matter after Franklin had taken a step toward bringing them together. This theory, in spite of its intrinsic difficulties, dominated the development of electrical science for one hundred years and more. This was because, if one did not bother himself much with the underlying physical conception, the theory lent itself admirably to the description of electrical phenomena and also to mathematical formulation. Further, it was convenient for the purposes of classification. It made it possible to treat electrical phenomena in a category entirely by themselves, without raising any troublesome questions as to the relation, for example, between electrical and gravitational or cohesive forces. But in spite of these advantages it was obviously a makeshift. For the notion of two fluids which could exert powerful forces and yet which were absolutely without weight—the most fundamental of physical properties—and the further notion of two fluids which had no physical properties whatever, that is, which disappeared entirely when they were mixed in equal proportions—these notions were in a high degree non-physical. Indeed, Sir J. J. Thomson remarked in his Silliman Lectures in 1903 that

the physicists and mathematicians who did most to develop the fluid theories confined their attention to questions which involved only the law of forces between electrified bodies and the simultaneous production of equal quantities of plus and minus electricity, and refined and idealized their conception of the fluids themselves until any reference to their physical properties was considered almost indelicate.

From the point of view of economy in hypothesis, Franklin's one-fluid theory, as modified by Aepinus, was the better. Mathematically the two theories were identical. The differences may be summed up thus. The modified one-fluid theory required that matter, when divorced from the electrical fluid, have exactly the same properties which the two-fluid theory ascribed to negative electricity, barring only the property of fluidity. So that the most important distinction between the theories was that the two-fluid theory assumed the existence of three distinct entities, named positive electricity, negative electricity, and matter, while the one-fluid theory reduced these three entities to two, which Franklin called matter and electricity, but which might perhaps as well have been called positive electricity and negative electricity, unelectrified matter being reduced to a mere combination of these two.

Of course, the idea of a granular structure for electricity was foreign to the two-fluid theory, and since this dominated the development of electrical science, there was seldom any mention in connection with it of an electrical atom, even as a speculative entity. But with Franklin the case was different. His theory was essentially a material one, and he unquestionably believed in the existence of an electrical particle or atom, for he says: "The electrical matter consists of particles extremely subtle, since it can permeate common matter, even the densest, with such freedom and ease as not to receive any appreciable resistance." When Franklin wrote that, however, he could scarcely have dreamed that it would ever be possible to isolate and study by itself one of the ultimate particles of the electrical fluid. The atomic theory of electricity was to him what the atomic theory of matter was to Democritus, a pure speculation.

The first bit of experimental evidence which appeared in its favor came in 1833, when Faraday found that the passage of a given quantity of electricity through a solution containing a compound of hydrogen, for example, would always cause the appearance at the negative terminal of the same amount of hydrogen gas irrespective of the kind of hydrogen compound which had been dissolved, and irrespective also of the strength of the solution; that, further, the quantity of electricity required to cause the appearance of one gram of hydrogen would always deposit from a solution containing silver exactly 107.05 grams of silver.   This meant, since the weight of the silver atom is exactly 107.05 times the weight of the hydrogen atom, that the hydrogen atom and the silver atom are associated in the solution with exactly the same quantity of electricity.   When it was further found in this way that all atoms which are univalent in chemistry, that is, which combine with one atom of hydrogen, carry precisely the same quantity of electricity, and all atoms which are bivalent carry twice

this amount, and, in general, that valency, in chemistry.
is always exactly proportional to the quantity of electricity carried by the atom in question, it was obvious
that the atomic theory of electricity had been given very
strong support.

But striking and significant as were these discoveries,
they did not serve at all to establish the atomic hypothesis of the nature of electricity. They were made at the
very time when attention began to be directed strongly
away from the conception of electricity as a substance
of any kind, and it was no other than Faraday himself
who, in spite of the brilliant discoveries just mentioned,
started this second period in the development of electrical
theory, a period lasting from 1840 to about 1900. In
this period electrical phenomena are almost exclusively
thought of in terms of stresses and strains in the medium
which surrounds the electrified body. Up to this time
a more or less definite something called a charge of electricity had been thought of as existing *on* a charged body
and had been imagined to exert forces on other charged
bodies at a distance from it in quite the same way in
which the gravitational force of the earth acts on the
moon or that of the sun on the earth. This notion of
action at a distance was repugnant to Faraday, and he
found in the case of electrical forces experimental reasons
for discarding it which had not then, nor have they as yet,
been found in the case of gravitational forces. These
reasons are summed up in the statement that the electrical force between two charged bodies is found to depend
on the nature of the intervening medium, while gravitational pulls are, so far as is known, independent of intervening bodies. Faraday, therefore, pictured to himself

the intervening medium as transmitting electrical force in quite the same way in which an elastic deformation started at one end of a rod is transmitted by the rod. Further, since electrical forces act through a vacuum, Faraday had to assume that it is the ether which acts as the transmitter of these electrical stresses and strains. The properties of the ether were then conceived of as modified by the presence of matter in order to account for the fact that the same two charges attract each other with different forces according as the intervening medium is, for example, glass, or ebonite, or air, or merely ether. These views, conceived by Faraday and put into mathematical form by Maxwell, called attention away from the electrical phenomena in or on a conductor carrying electricity and focused it upon the stresses and strains taking place in the medium about the conductor. When in 1887 Heinrich Hertz in Bonn, Germany, proved by direct experiment that electrical forces are indeed transmitted in the form of electric waves, which travel through space with the speed of light exactly as the Faraday-Maxwell theory had predicted, the triumph of the ether-stress point of view was complete. Thereupon textbooks were written by enthusiastic, but none too cautious, physicists in which it was asserted that an electric charge is nothing more than a "state of strain in the ether," and an electric current, instead of representing the passage of anything definite along the wire, corresponds merely to a continuous "slip" or "breakdown of a strain" in the medium within the wire. Sir Oliver Lodge's early book, *Modern Views of Electricity*, was perhaps the most influential disseminator and expounder of this point of view.

Now what had actually been proved was not that electricity is a state of strain, but that when any electrical charge appears upon a body the medium about the body does indeed become the seat of new forces which are transmitted through the medium, like any elastic forces, with a definite speed.   Hence it is entirely proper to say that the medium about a charged body is in a state of strain.   But it is one thing to say that the electrical charge on the body *produces* a state of strain in the surrounding medium, and quite another thing to say that the electrical charge *is nothing but* a state of strain in the surrounding medium, just as it is one thing to say that when a man stands on a bridge he produces a mechanical strain in the timbers of the bridge, and another thing to say that the man is nothing more than a mechanical strain in the bridge.   The practical difference between the two points of view is that in the one case you look for other attributes of the man besides the ability to produce a strain in the bridge, and in the other case you do not look for other attributes.   So the strain theory, although not irreconcilable with the atomic hypothesis, was actually antagonistic to it, because it led men to think of the strain as distributed continuously about the surface of the charged body, rather than as radiating from definite spots or centers peppered over the surface of the body.   Between 1833 and 1900, then, the physicist was in this peculiar position: when he was thinking of the passage of electricity through a solution, he for the most part, following Faraday, pictured to himself definite specks or atoms of electricity as traveling through the solution, each atom of matter carrying an exact multiple, which might be anywhere between one and eight, of a definite elementary electrical atom, while, when he was thinking of the passage of a current through a metallic conductor, he gave up altogether the atomic hypothesis, and attempted to picture the phenomenon to himself as a continuous "slip" or "breakdown of a strain" in the material of the wire.   In other words, he recognized two types of electrical conduction which were wholly distinct in kind—electrolytic conduction and metallic conduction; and since more of the problems of the physicist dealt with metallic than with electrolytic conduction, the atomic conception, as a general hypothesis, was almost, though not quite, unheard of.   Of course it would be unjust to the thinkers of this period to say that they failed to recognize and appreciate this gulf between current views as to the nature of electrolytic and metallic conduction, and simply ignored the difficulty.   This they did not do, but they had all sorts of opinions as to the causes. Maxwell himself in his text on *Electricity and Magnetism*, published in 1873, recognizes, in the chapter on "Electrolysis,"[1] the significance of Faraday's laws, and even goes so far as to say that "for convenience in description we may call this constant molecular charge (revealed by Faraday's experiments) one molecule of electricity." Nevertheless, a little farther on he repudiates the idea that this term can have any physical significance by saying that "it is extremely improbable that when we come to understand the true nature of electrolysis we shall retain in any form the theory of molecular charges, for then we shall have obtained a secure basis on which to form a true theory of electric currents and so become independent of these provisional hypotheses."

[1] I, 375–86.

And as a matter of fact, Faraday's experiments had not shown at all that electrical charges on metallic conductors consist of specks of electricity, even though they had shown that the charges on ions in solutions have definite values which are always the same for univalent ions. It was entirely logical to assume, as Maxwell did, that an ion took into solution a definite quantity of electricity because of some property which it had of always charging up to the same amount from a charged plate. There was no reason for assuming the charge *on the electrode* to be made up of some exact number of electrical atoms.

On the other hand, Wilhelm Weber, in papers written in 1871,[1] built up his whole theory of electromagnetism on a basis which was practically identical with the modified Franklin theory and explained all the electrical phenomena exhibited by conductors, including thermoelectric and Peltier effects, on the assumption of two types of electrical constituents of atoms, one of which was very much more mobile than the other. Thus the hypothetical molecular current, which Ampere had imagined fifty years earlier to be continually flowing inside of molecules and thereby rendering these molecules little electromagnets, Weber definitely pictures to himself as the rotation of light, positive charges about heavy negative ones. His words are:

The relation of the two particles as regards their motions is determined by the ratio of their masses e and e', on the assumption that in e and e' are included the masses of the ponderable atoms which are attached to the electrical atoms. Let e be the positive electrical particle. Let the negative be exactly equal and opposite

[1] See *Werke*, IV, 281.

and therefore denoted by $-e$ (instead of e'). But let a ponderable atom be attracted to the latter so that its mass is thereby so greatly increased as to make the mass of the positive particle vanishingly small in comparison. The particle $-e$ may then be thought of as at rest and the particle $+e$ as in motion about the particle $-e$. The two unlike particles in the condition described constitute then an Amperian molecular current.

It is practically this identical point of view which has been elaborated and generalized by Lorentz and others within the past four decades in the development of the modern electron theory, with this single difference, that we now have proof that it is, in general, the negative particle whose mass or inertia is negligible in comparison with that of the positive instead of the reverse. Weber even went so far as to explain thermoelectric and Peltier effects by differences in the kinetic energies in different conductors of the electrical particles.[1] Nevertheless his explanations are here widely at variance with our modern conceptions of heat.

Again, in a paper read before the British Association at Belfast in 1874, G. Johnstone Stoney not only stated clearly the atomic theory of electricity, but actually went so far as to estimate the value of the elementary electrical charge, and he obtained a value which was about as reliable as any which had been found until within quite recent years. He got, as will be more fully explained in the next chapter, $.3 \times 10^{-10}$ absolute electrostatic units, and he got this result from the amount of electricity necessary to separate from a solution one gram of hydrogen, combined with kinetic theory estimates as to the number of atoms of hydrogen in two grams, i.e., in one

[1] *Op. cit.*, p. 294.

gram molecule of that element. This paper was entitled, "On the Physical Units of Nature," and though read in 1874 it was not published in full until 1881.[1] After showing that all physical measurements may be expressed in terms of three fundamental units, he asserts that it would be possible to replace our present purely arbitrary units (the centimeter, the gram, and the second) by three natural units, namely, the velocity of light, the coefficient of gravitation, and the elementary electrical charge. With respect to the last he says:

Finally nature presents us with a single definite quantity of electricity which is independent of the particular bodies acted on. To make this clear, I shall express Faraday's law in the following terms, which, as I shall show, will give it precision, viz.: *For each chemical bond which is ruptured within an electrolyte a certain quantity of electricity traverses the electrolyte which is the same in all cases.* This definite quantity of electricity I shall call E_1. If we make this our unit of electricity, we shall probably have made a very important step in our study of molecular phenomena.

Hence we have very good reason to suppose that in V_1, G_1, and E_2, we have three of a series of systematic units that in an eminent sense are the units of nature, and stand in an intimate relation with the work which goes on in her mighty laboratory.

Take one more illustration from prominent writers of this period. In his Faraday lecture delivered at the Royal Institution in 1881, Helmholtz spoke as follows:

Now the most startling result of Faraday's law is perhaps this, if we accept the hypothesis that the elementary substances are composed of atoms, we cannot avoid concluding that electricity also, positive as well as negative, is divided into definite elementary portions which behave like atoms of electricity.[2]

[1] *Phil. Mag.*, XI (1881; 5th series), 384.
[2] *Wissenschaftliche Abhandlungen*, III, 69.

This looks like a very direct and unequivocal statement of the atomic theory of electricity, and yet in the same lecture Helmholtz apparently thinks of metallic conduction as something quite different from electrolytic when he says:

All these facts show that electrolytic conduction is not at all limited to solutions of acids or salts. It will, however, be rather a difficult problem to find out how far the electrolytic conduction is extended, and I am not yet prepared to give a positive answer.

The context shows that he thought of extending the idea of electrolytic conduction to a great many insulators. But there is no indication that he thought of extending it to metallic conductors and imagining these electrical atoms as existing as discrete individual things on charged metals or as traveling along a wire carrying an electrical current. Nevertheless, the statement quoted above is one of the most unequivocal which can be found anywhere up to about 1899 as to the atomic nature of electricity.

The foregoing quotations are sufficient to show that the atomic theory of electricity, like the atomic theory of matter, is not at all new so far as the conception alone is concerned. In both cases there were individuals who held almost exactly the modern point of view. In both cases, too, the chief new developments have consisted in the appearance of new and exact *experimental* data which has silenced criticism and compelled the abandonment of other points of view which up to about 1900 flourished along with, and even more vigorously than, the atomic conception. Even in 1897 Lord Kelvin, with a full knowledge of all the new work which was appearing on X-rays and cathode rays, could seriously raise the question whether electricity might not be a "continuous homogeneous liquid." He does it in these words:

Varley's fundamental discovery of the cathode rays, splendidly confirmed and extended by Crookes, seems to me to necessitate the conclusion that resinous electricity, not vitreous, is *The Electric Fluid*, if we are to have a one-fluid theory of electricity. Mathematical reasons prove that if resinous electricity is a continuous homogeneous liquid it must, in order to produce the phenomena of contact electricity, which you have seen this evening, be endowed with a cohesional quality. It is just conceivable, though it does not at present seem to me very probable, that this idea may deserve careful consideration. I leave it, however, for the present and prefer to consider an atomic theory of electricity foreseen as worthy of thought by Faraday and Clerk-Maxwell, very definitely proposed by Helmholtz in his last lecture to the Royal Institution, and largely accepted by present-day workers and teachers. Indeed Faraday's laws of electrolysis seem to necessitate something atomic in electricity, ...[1]

What was the new experimental work which already in 1897 was working this change in viewpoint? Much of it was at first little if at all more convincing than that which had been available since Faraday's time. Nevertheless it set physicists to wondering whether stresses and strains in the ether had not been a bit overworked, and whether in spite of their undoubted existence electricity itself might not after all be something more definite, more material, than the all-conquering Maxwell theory had assumed it to be.

The result of the past thirty-five years has been to bring us back very close to where Franklin was in 1750, with the single difference that our modern electron theory rests upon a mass of very direct and convincing evidence, which it is the purpose of the next chapters to present.

[1] Kelvin, "Contact Electricity and Electrolysis," *Nature*, LVI (1897), 84.

CHAPTER II

THE EXTENSION OF THE ELECTROLYTIC LAWS TO CONDUCTION IN GASES

I. THE ORIGIN OF THE WORD "ELECTRON"

The word "electron" was first suggested in 1891 by Dr. G. Johnstone Stoney as a name for the "natural unit of electricity," namely, that quantity of electricity which must pass through a solution in order to liberate at one of the electrodes one atom of hydrogen or one atom of any univalent substance. In a paper published in 1891 he says:

Attention must be given to Faraday's Law of Electrolysis, which is equivalent to the statement that in electrolysis a definite quantity of electricity, the same in all cases, passes for each chemical bond that is ruptured. The author called attention to this form of the law in a communication made to the British Association in 1874 and printed in the *Scientific Proceedings of the Royal Dublin Society* of February, 1881, and in the *Philosophical Magazine* for May, 1881, pp. 385 and 386 of the latter. It is there shown that the amount of this very remarkable quantity of electricity is about the twentiethet $\left(\text{that is } \frac{1}{10^{20}}\right)$ of the usual electromagnetic unit of electricity, i.e., the unit of the Ohm series. This is the same as 3 eleventhets $\left(\frac{3}{10^{11}}\right)$ of the much smaller C.G.S. electrostatic unit of quantity. A charge of this amount is associated in the chemical atom with each bond. There may accordingly be several such charges in one chemical atom, and there appear to be at least two in each atom. These charges, which it will be convenient to call "electrons," cannot be removed from the atom, but they become disguised when atoms chemically unite. If an

electron be lodged at the point P of the molecule which undergoes
the motion described in the last chapter, the revolution of this
charge will cause an electromagnetic undulation in the surrounding
ether.[1]

It will be noticed from this quotation that the word
"electron" was introduced to denote simply a definite elementary quantity of electricity without any reference to
the mass or inertia which may be associated with it, and
Professor Stoney implies that every atom must contain
at least two electrons, one positive and one negative,
because otherwise it would be impossible that the atom
as a whole be electrically neutral. As a matter of fact
the evidence is now altogether convincing that the
hydrogen atom does indeed contain just one positive and
one negative electron.

It is unfortunate that all writers have not been more
careful to retain the original significance of the word
introduced by Professor Stoney, for it is obvious that a
word is needed which denotes merely the elementary unit
of electricity and has no necessary implication as to where
that unit is found, to what it is attached, with what
inertia it is associated, or whether it is positive or negative
in sign; and it is also apparent that the word "electron"
is the logical one to associate with this conception.
Further, there is no difficulty in retaining this original
and derivative significance of the word "electron," and at
the same time permitting its common use as a convenient
abridgment for "the free negative electron." In other
words, in view of the omnipresence of the negative electron in experimental physics and the extreme rarity of

[1] *Scientific Transactions of the Royal Dublin Society*, IV (1891; 11th
series), 563.

the isolated positive electron, it may be generally agreed that the negative is understood unless the positive is specified. The case is then in every way identical with that found in the use of the word "man," which serves admirably both to designate the genus "homo" and also to denote the male representative of that genus, the female being then differentiated by the use of a prefix. The terms "electron" and "positive electron" would then be used altogether conveniently precisely as are the terms "man" and "woman." Indeed, the most authoritative writers—Thomson, Rutherford, Campbell, Richardson, etc.—have in fact retained the original significance of the word "electron" instead of using it to denote *solely* the free negative electron, the mass of which is $1/1,835$ of that of the hydrogen atom. All of these writers in books or articles written since 1913[1] have treated of positive as well as negative electrons, although the mass associated with the former is most commonly that of the hydrogen atom. Nor is this altogether logical use confined at all to English. Perrin has approved it, and Nernst in the 1921 edition of his *Theoritische Chemie*, on pp. 197 and 456, definitely and unambiguously defines the positive and negative electrons, precisely as has been done above, as the elementary positive and negative electrical charges, respectively.

II. THE DETERMINATION OF $\dfrac{e}{m}$ AND Ne FROM THE FACTS

OF ELECTROLYSIS

Farady's experiments had of course not furnished the data for determining anything about how much

[1] See particularly Rutherford's presidential address at the 1923 Liverpool meeting of the British Association, *Science*, LVIII (1923), 213.

electricity an electron represents in terms of the standard
unit by which electrical charges are ordinarily measured
in the laboratory. This is called the coulomb, and
represents the quantity of electricity conveyed in
one second by one ampere. Faraday had merely shown
that a given current flowing in succession through
solutions containing different univalent elements like
hydrogen or silver or sodium or potassium would deposit
weights of these substances which are exactly proportional to their respective atomic weights. This enabled
him to assert that one and the same amount of electricity is associated in the process of electrolysis with
an atom of each of these substances. He thought of this
charge as carried by the atom, or in some cases by a group
of atoms, and called the group with its charge an "ion,"
that is, a "goer," or "traveler." Just how the atoms
come to be charged in a solution Faraday did not know,
nor do we know now with any certainty. Further, we
do not know how much of the solvent an ion associates
with itself and drags with it through the solution. But
we do know that when a substance like salt is dissolved
in water many of the neutral NaCl molecules are split
up by some action of the water into positively charged
sodium (Na) ions and negatively charged chlorine (Cl)
ions. The ions of opposite sign doubtless are all the
time recombining, but others are probably continually
forming, so that at each instant there are many uncom-
bined ions. Again, we know that when a water solution
of copper sulphate is formed many of the neutral $CuSO_4$
molecules are split up into positively charged Cu ions
and negatively charged SO_4 ions. In this last case too
we find that the same current which will deposit in a

given time from a silver solution a weight of silver equal to its atomic weight will deposit from the copper-sulphate solution in the same time a weight of copper equal to exactly one-half its atomic weight. Hence we know that the copper ion carries in solution twice as much electricity as does the silver ion, that is, it carries a charge of two electrons.

But though we could get from Faraday's experiments no knowledge about the quantity of electricity, e, represented by one electron, we could get very exact information about the ratio of the ionic charge E to the mass of the atom with which it is associated in a given solution.

For, if the whole current which passes through a solution is carried by the ions—and if it were not we should not always find the deposits exactly proportional to atomic weights—then the ratio of the total quantity of electricity passing to the weight of the deposit produced must be the same as the ratio of the charge E on each ion to the mass m of that ion. But by international agreement one absolute unit of electricity has been defined in the electromagnetic system of units as the amount of electricity which will deposit from a silver solution 0.01118 grams of metallic silver. Hence if m refers to the silver ion and E means the charge on the ion, we have

$$\text{for silver } \frac{E}{m} = \frac{1}{0.01118} = 89.44 \text{ electromagnetic units;}$$

or if m refers to the hydrogen ion, since the atomic weight of silver is $\dfrac{107.880}{1.00777}$ times that of hydrogen, for hydrogen $\frac{E}{m} = \frac{1}{0.01118} \times \frac{107.88}{1.008} = 9,573,$

which is about 10^4 electromagnetic units.

Thus in electrolysis $\frac{E}{m}$ varies from ion to ion, being for univalent ions, for which E is the same and equal to one electron e, inversely proportional to the atomic weight of the ion. For polivalent ions E may be 2, 3, 4, or 5 electrons, but since hydrogen is at least 7 times lighter than any other ion which is ever found in solution, and its charge is but one electron, we see that the largest value which $\frac{E}{m}$ ever has in electrolysis is its value for hydrogen, namely, about 10^4 electromagnetic units.

Although $\frac{E}{m}$ varies with the nature of the ion, there is a quantity which can be deduced from it which is a universal constant. This quantity is denoted by Ne, where e means as before an electron and N is the Avogadro constant or the number of molecules in 32 grams of oxygen, i.e., in one gram molecule. We can get this at once from the value of $\frac{E}{m}$ by letting m refer to the mass of that imaginary univalent atom which is the unit of our atomic weight system, namely, an atom which is exactly $1/16$ as heavy as oxygen or $1/107.88$ as heavy as silver. For such an atom

$$\frac{E}{m} = \frac{e}{m} = \frac{107.88}{0.01118} = 9649,4.$$

Multiplying both numerator and denominator by N and remembering that for this gas one gram molecule means 1 gram, that is $Nm = 1$, we have

$Ne = 9649.4$ international electromagnetic units,.......(1)

and since the electromagnetic unit is equivalent to 3×10^{10} electrostatic units, we have

$Ne = 28,948 \times 10^{10}$ international electrostatic units.

Further, since a gram molecule of an ideal gas under standard conditions, i.e., at $0°$ C. 76 cm. pressure, occupies 22412 c.c., if n_1 represents the number of molecules of such a gas per cubic centimeter at $0°$ C., 76 cm., we have

$$n_1 e = \frac{28,948 \times 10^{10}}{22,412} = 1.292 \times 10^{10} \text{ electrostatic units.}$$

Or if n represent the number of molecules per cubic centimeter at $15°$ C. 76 cm., we should have to multiply the last number by the ratio of absolute temperatures, i.e., by $273/288$ and should obtain then

$$ne = 1.225 \times 10^{10}.\dots\dots\dots\dots\dots(2)$$

Thus, even though the facts of electrolysis give us no information at all as to how much of a charge one electron e represents, they do tell us very exactly that if we should take e as many times as there are molecules in a gram molecule we should get exactly 9,649.4 international electromagnetic units of electricity. This is the amount of electricity conveyed by a current of 1 ampere in 10 seconds. Until quite recently we have been able to make nothing better than rough guesses as to the number of molecules in a gram molecule, but with the aid of these guesses, obtained from the kinetic theory, we have, of course, been enabled by (1) to make equally good guesses about e. Those guesses, based for the most part on quite uncertain computations as to the average radius of a molecule of air, placed N anywhere between 2×10^{23} and 20×10^{23}. It was in this way that G. Johnstone Stoney in 1874 estimated e at $.3 \times 10^{-10}$ E.S. units. In O. E. Meyer's *Kinetische Theorie der Gase* (p. 335; 1899), n, the number of molecules in a cubic centimeter, is given as 6×10^{19}. This would correspond to $e = 2 \times 10^{-10}$. In all this e is the charge carried by a univalent ion in solution and N or n is a pure number, which is a characteristic gas constant, it is true, but the analysis has nothing whatever to do with gas conduction.

III. THE NATURE OF GASEOUS CONDUCTION

The question whether gases conduct at all, and if so, whether their conduction is electrolytic or metallic or neither, was scarcely attacked until about 1895. Coulomb in 1785 had concluded that after allowing for the leakage of the supports of an electrically charged conductor, some leakage must be attributed to the air itself, and he explained this leakage by assuming that the air molecules became charged by contact and were then repelled—a wholly untenable conclusion, since, were it true, no conductor in air could hold a charge long even at low potentials, nor could a very highly charged conductor lose its charge very rapidly when charged above a certain potential and then when the potential fell below a certain critical value cease almost entirely to lose it. This is what actually occurs. Despite the erroneousness of this idea, it persisted in textbooks written as late as 1900.

Warburg in 1872 experimented anew on air leakage and was inclined to attribute it all to dust particles. The real explanation of gas conduction was not found until after the discovery of X-rays in 1895. The convincing experiments were made by J. J. Thomson, or at his instigation in the Cavendish Laboratory at Cambridge, England. The new work grew obviously and simply out of the fact that X-rays, and a year or two later radium rays, were found to discharge an electroscope, i.e., to produce conductivity in a gas. Theretofore no agencies had been known by which the electrical conductivity of a gas could be controlled at will.

Thomson and his pupils found that the conductivity induced in gases by X-rays disappeared when the gas was sucked through glass wool.[1] It was also found to be reduced when the air was drawn through narrow metal tubes. Furthermore, it was removed entirely by passing the stream of conducting gas between plates which were maintained at a sufficiently large potential difference. The first two experiments showed that the conductivity was due to something which could be removed from the gas by filtration, or by diffusion to the walls of a metal tube; the last proved that this something was electrically charged.

When it was found, further, that the electric current obtained from air existing between two plates and traversed by X-rays rose to a maximum as the P.D. between the plates increased, and then reached a value which was thereafter independent of this potential difference; and, further, that this conductivity of the air died out slowly through a period of several seconds when the X-ray no longer acted, it was evident that the qualitative proof was complete that gas conduction must be due to charged particles produced in the

[1] J. J. Thomson and E. Rutherford, *Phil. Mag.*, XLII (1896), 392.

air at a definite rate by a constant source of X-rays, and that these charged particles, evidently of both plus and minus signs, disappear by recombination when the rays are removed. The maximum or *saturation* currents which could be obtained when a given source was ionizing the air between two plates whose potential difference could be varied were obviously due to the fact that when the electric field between the plates became strong enough to sweep all the ions to the plates as fast as they were formed, none of them being lost by diffusion or recombination, the current obtained could, of course, not be increased by further increase in the field strength. Thus gas conduction was definitely shown about 1896 to be electrolytic in nature.

IV. COMPARISON OF THE GASEOUS ION AND THE ELECTROLYTIC ION

But what sort of ions were these that were thus formed? We did not know the absolute value of the charge on a univalent ion in electrolysis, but we did know accurately *ne*. Could this be found for the ions taking part in gas conduction? That this question was answered affirmatively was due to the extraordinary insight and resourcefulness of J. J. Thomson and his pupils at the Cavendish Laboratory in Cambridge, both in working out new theoretical relations and in devising new methods for attacking the new problems of gaseous conduction. These workers found first a method of expressing the quantity *ne* in terms of two measurable constants, called (1) the mobility of gaseous ions and (2) the coefficient of diffusion of these ions. Secondly, they devised new methods of measuring these two constants—constants which had never before been determined. The theory of the relation between these constants and the quantity *ne* will be found in Appendix A. The result is

$$ne = \frac{v_0}{D} P \dots\dots\dots\dots\dots\dots(3)$$

in which P is the pressure existing in the gas and v_0 and D are the mobility and the diffusion coefficients respectively of the ions at this pressure.

If then we can find a way of measuring the mobilities v_0 of atmospheric ions and also the diffusion coefficients D, we can find the quantity *ne*, in which *n* is a mere number, viz., the number of molecules of air per cubic centimeter at 15° C., 76 cm. pressure, and *e* is the average charge on an atmosphere ion. We shall then be in position to compare this with the product we found in (2) on p. 31, in which *n* had precisely the same significance as here, but *e* meant the average charge carried by a univalent ion in electrolysis.

The methods devised in the Cavendish Laboratory between 1897 and 1903 for measuring the mobilities and the diffusion coefficients of gaseous ions have been used in most later work upon these constants. The mobilities were first determined by Rutherford in 1897,[1] then more accurately by another method in 1898.[2] Zeleny devised a quite distinct method in 1900,[3] and Langevin still another method in 1903.[4] These observers all agree closely in finding the average mobility (velocity in unit

[1] *Phil. Mag.*, XLIV (1898), 422.

[2] *Proc. Camb. Phil. Soc.*, IX, 401.

[3] *Phil. Trans.*, A 195, p. 193.

[4] *Annale de Chimie et de Physique*, XXVIII, 289.

field) of the negative ion in dry air about 1.83 cm. per second, while that of the positive ion was found but 1.35 cm. per second. In hydrogen these mobilities were about 7.8 cm. per second and 6.1 cm. per second, respectively, and in general the mobilities in different gases, though not in vapors, seem to be roughly in the inverse ratio of the square roots of the molecular weights.

The diffusion coefficients of ions were first measured in 1900 by Townsend, now professor of physics in Oxford, England,[1] by a method devised by him and since then used by other observers in such measurements. If we denote the diffusion coefficient of the positive ion by $D+$ and that of the negative by $D-$, Townsend's results in dry air may be stated thus:

$$D+ = 0.028$$
$$D- = 0.043.$$

These results are interesting in two respects. In the first place, they seem to show that for some reason the positive ion in air is more sluggish than the negative, since it travels but about 0.7 $(= 1.35/1.81)$ as fast in a given electrical field and since it diffuses through air but about 0.7 $(= 28/43)$ as rapidly. In the second place, the results of Townsend show that an ion is very much more sluggish than is a molecule of air, for the coefficient of diffusion of oxygen through air is 0.178, which is four times the rate of diffusion of the negative ion through air and five times that of the positive ion. This sluggishness of ions as compared with molecules was at first universally considered to mean that the gaseous ion is not a single molecule with an attached electrical charge, but

[1] *Phil. Trans.*, A 193, p. 129.

a cluster of perhaps from three to twenty molecules held together by such a charge. If this is the correct interpretation, then for some reason the positive ion in air is a larger cluster than is the negative ion.

It has been since shown by a number of observers that the ratio of the mobilities of the positive and negative ions is not at all the same in other gases as it is in air. In carbon dioxide the two mobilities have very nearly the same value, while in chlorine, water vapor, and the vapor of alcohol the positive ion apparently has a slightly larger mobility than the negative. There seems to be some evidence that the negative ion has the larger mobility in gases which are electro-positive, while the positive has the larger mobility in the gases which are strongly electro-negative. This dependence of the ratio of mobilities upon the electro-positive or electro-negative character of the gas has usually been considered strong evidence in favor of the large cluster-ion theory as developed especially by J. J. Thomson.

More recently, however, Loeb,[1] who has worked for years on mobilities in both strong and weak electric fields, and Wellish,[2] who, at Yale, measured mobilities at very low pressures, concluded that their results were not consistent with this form of cluster-ion theory. They preferred to interpret them in terms of the so-called Atom-ion Theory. This theory seeks to explain the relative sluggishness of ions, as compared with molecules, by the additional resistance which the gaseous medium offers to the motion of a molecule through it when that

[1] Leonard B. Loeb, *Proc. Nat. Acad.*, II (1916), 345, and *Phys. Rev.*, 1917. See especially *Phys. Rev.*, XXXVIII (1931), 549.

[2] Wellish, *Amer. Jour. of Science*, XXXIX (1915), 583.

molecule is electrically charged. *According to this hypothesis, the ion would be simply an electrically charged molecule.*

This second way of accounting for the sluggishness of ions is probably in the main correct, though the atom-ion theory was too extreme in reducing the ion to one single molecule. Loeb himself[1] now explains the difference in the mobilities of positive and negative ions by the assumption that the positive charge forms a different "small ion" group from the negative by attaching itself to a different kind of molecular impurity.

Furthermore, Erikson,[2] Wahlin,[3] and Loeb[1] have apparently shown quite conclusively that if the mobility of the positive ion in air is measured within .03 second of the time of its formation, its value is identical with that of the negative, namely, 1.8 cm. per second, while a short time thereafter it has sunk to about 1.4 cm. per second because of the addition of one more molecule, thus forming a very stable two-molecule-ion-group.

Fortunately, the quantitative evidence for the electrolytic nature of gas conduction is in no way dependent upon the correctness of either one of the theories as to the nature of the ion. It depends simply upon the comparison of the values of ne obtained from electrolytic measurements, and those obtained from the substitution in equation (3) of the measured values of v_0 and D for gaseous ions.

As for these measurements, results obtained by Franck and Westphal,[4] who in 1908 repeated in Berlin

[1] L. B. Loeb, *Phys. Rev.*, XXXVIII (1931), 1716.

[2] H. A. Erikson, *ibid.*, XX (1922), 118.

[3] H. B. Wahlin, *ibid.*, p. 267.

[4] *Verh. der deutsch. phys. Ges.*, XI (1909), 146 and 276.

both measurements on diffusion coefficients and mobility coefficients, agree within 4 or 5 per cent with the results published by Townsend in 1900. According to both of these observers, the value of ne for the negative ions produced in gases by X-rays, radium rays, and ultra-violet light came out, within the limits of experimental error, which were presumably 5 or 6 per cent, the same as the value found for univalent ions in solutions, namely, 1.23×10^{10} absolute electrostatic units. This result seems to show with considerable certainty that the negative ions in gases ionized by X-rays or similar agencies carry on the average the same charge as that borne by the univalent ion in electrolysis. When we consider the work on the positive ion, our confidence in the inevitableness of the conclusions reached by the methods under consideration is perhaps somewhat shaken. For Townsend found that the value of ne for the positive ion came out about 14 per cent higher than the value of this quantity for the univalent ion in electrolysis, a result which he does not seem at first to have regarded as inexplicable on the basis of experimental uncertainties in his method. In 1908, however,[1] he devised a second method of measuring the ratio of the mobility and the diffusion coefficient and obtained this time, as before, for the negative ion, $ne = 1.23 \times 10^{10}$, but for the positive ion twice that amount, namely, 2.46×10^{10}. From these last experiments he concluded that the positive ions in gases ionized by X-rays carried on the average twice the charge carried by the univalent ion in electrolysis. Franck and Westphal, however, found in their work that Townsend's original value for ne for the

[1] *Proc. Roy. Soc.*, LXXX (1908), 207.

positive ions was about right, and hence concluded that only about 9 per cent of the positive ions could carry a charge of value $2e$. Work which will be described later indicates that neither Townsend's nor Franck and Westphal's conclusions are correct, and hence point to errors of some sort in both methods. But despite these difficulties with the work on positive ions, it should nevertheless be emphasized that Townsend was the first to bring forward strong quantitative evidence (1) that the mean charge carried by the negative ions in ionized gases is the same as the mean charge carried by univalent ions in solutions, and (2) that the mean charge carried by the positive ions in gases has not far from the same value.

But there is one other advance of fundamental importance which came with the study of the properties of gases ionized by X-rays. For up to this time the only type of ionization known was that observed in solution and here it is always some compound molecule like sodium chloride (NaCl) which splits up spontaneously into a positively charged sodium ion and a negatively charged chlorine ion. But the ionization produced in gases by X-rays was of a wholly different sort, for it was observable in pure gases like nitrogen or oxygen, or even in monatomic gases like argon and helium   Plainly, then, the neutral atom even of a monatomic substance must possess minute electrical charges as constituents. Here we had the first direct evidence (1) that an atom is a complex structure, and (2) that electrical charges enter into its make-up. *With this discovery, due directly to the use of the new agency, X-rays, the atom as an ultimate, indivisible thing was gone, and the era of the study of the constituents of the atom began.* And with astonishing rapidity during the past thirty-five years the properties of the subatomic world have been revealed.

Physicists began at once to seek diligently and to find at least partial answers to questions like these:

1. What are the masses of the constituents of the atoms torn asunder by X-rays and similar agencies?

2. What are the values of the charges carried by these constituents?

3. How many of these constituents are there?

4. How large are they, i.e., what volumes do they occupy?

5. What are their relations to the emission and absorption of light and heat waves, i.e., of electromagnetic radiation?

6. Do all atoms possess similar constituents? In other words, is there a primordial subatom out of which atoms are made?

The partial answer to the first of these questions came with the study of the electrical behavior of rarefied gases in so-called vacuum tubes.

This field had been entered and qualitatively explored with amazing insight as early as 1879 by Sir William Crookes, who in describing in that year some of his experiments said:

The phenomena in these exhausted tubes reveal to physical science a new world—a world where matter exists in a fourth state. . . . . In studying this fourth state of matter we seem at length to have within our grasp and obedient to our control the little indivisible particles which with good warrant are supposed to constitute the physical basis of the universe.[1]

[1] Fournier d'Albe, *Life of Sir William Crookes,* 1924.

Further, by 1890 Sir Arthur Schuster[1] had gone a step farther and shown how the ratio of the charge to the mass $\left[\dfrac{e}{m}\right]$ of these same hypothetical particles might be determined. Indeed he had experimentally evaluated this ratio, obtaining, however, a value very much too small, namely, 1.1×10^{-6} electromagnetic units.

But it was J. J. Thomson[2] who in 1897 first introduced a more reliable method of determining this ratio, namely, one which combines a measurement of the magnetic deflectability of a beam of cathode rays with the electrostatic deflectability of the same beam. The value which he obtained, namely, 7×10^{6} electromagnetic units, was nearly a thousand times the value of $\dfrac{e}{m}$ for the hydrogen ion in solutions. Also since the approximate equality of ne in gases and solutions meant that e was at least of the same order in both, the only possible conclusion was that the negative ion which appears in discharges in exhausted tubes has a mass, i.e., an inertia, only one-thousandth of the mass of the lightest-known atom, namely, the atom of hydrogen. Later more accurate experiments have fixed the correct value of $\dfrac{e}{m}$ for cathode rays at 1.7573×10^{7} electromagnetic units.[3,4]

Furthermore, J. J. Thomson and after him other experimenters showed that $\dfrac{e}{m}$ for the negative carrier is always the same whatever be the nature of the residual gas in the discharge tube. This was an indication of an affirmative answer to the sixth question above—an

---

[1] *Proc. Roy. Soc.*, XL (1890), 526. [2] *Phil. Mag.*, XLIV (1897), 298.

[3] Houston, *Phys. Rev*, XLV (1934), 104.

[4] Dunnington, *ibid.*, XLIII (1933), 404.

indication which was strengthened by Zeeman's discovery in 1897 of the splitting by a magnetic field of a single spectral line into two or three lines; for this, when worked out quantitatively, pointed to the existence *within* the atom of a negatively charged particle which had approximately the same value of $\frac{e}{m}$.

The study of $\frac{e}{m}$ for the *positive* ions in exhausted tubes was first carried out quantitatively by Wien,[1] and was later most elaborately and most successfully dealt with by J. J. Thomson[2] and his pupils at the Cavendish Laboratory. The results of the work of all observers up to date shows that with the exceptions considered in chapter xiv $\frac{e}{m}$ for a positive ion in gases is never larger than its value for the hydrogen ion in electrolysis, and that it varies with different sorts of residual gases just as it is found to do in electrolysis.

In a word, then, the act of ionization in gases appears to consist in the detachment from a neutral atom of one or more negatively charged particles, called by Thomson corpuscles. The residuum of the atom is of course positively charged, and it always carries practically the whole mass of the original atom. The detached corpuscle must soon attach itself, in a gas at ordinary pressure, to a neutral atom, since otherwise we could not account for the fact that the mobilities and the diffusion coefficients of negative ions are usually of the same order of magnitude as those of the positive ions. It is because of this tendency of the parts of the dissociated atom to

[1] W. Wien, *Wied. Ann.*, LXV (1898), 440.

[2] *Rays of Positive Electricity.* London: Longmans, 1913.

form new attachments in gases at ordinary pressure that the inertias of these parts had to be worked out in the rarefied gases of exhausted tubes.

The foregoing conclusions as to the masses of the positive and negative constituents of atoms had all been reached before 1900, mostly by the workers in the Cavendish Laboratory, and subsequent investigation has not modified them in any essential particulars.

The history of the development of our present knowledge of the charges carried by the constituents will be detailed in the next chapters.

CHAPTER III

EARLY ATTEMPTS AT THE DIRECT DETER-
MINATION OF e

Although the methods sketched in the preceding chapters had been sufficient to show that the mean charges carried by ions in gases are the same or nearly the same as the mean charges carried by univalent ions in solution, in neither case had we any way of determining what the absolute value of that mean charge is, nor, indeed, had we any proof even that all the ions of a given kind, e.g., silver or hydrogen, carry the same charge. Of course, the absolute value of e could be found from the measured value of ne if only n, the number of molecules in 1 c.c. of gas under standard conditions, were known. But we had only rough guesses as to this number. These guesses varied tenfold, and none of them were based upon considerations of recognized accuracy or even validity.

I. TOWNSEND'S WORK ON e

The first attempt at a direct determination of e was published by Townsend in a paper read before the Cambridge Philosophical Society on February 8, 1897.[1] Townsend's method was one of much novelty and of no little ingenuity. It is also of great interest because it contains all the essential elements of some of the subsequent determinations.

[1] *Proceedings*, IX (1897), 244.

It had been known, even to Laplace and Lavoisier a hundred years before, that the hydrogen gas evolved when a metal dissolves in an acid carries with it an electrical charge. This "natural method" of obtaining a charge on a gas was scarcely studied at all, however, until after the impulse to the study of the electrical properties of gases had been given by the discovery in 1896 that electrical properties can be artificially imparted to gases by X-rays. Townsend's paper appeared within a year of that time. Enright[1] had indeed found that the hydrogen given off when iron is dissolving in sulphuric acid carries with it a positive charge, but Sir Oliver Lodge[2] had urged that it was not the gas itself which carries the charge but merely the spray, for the frictional electrification of spray was a well-known phenomenon. Indeed, it has always been assumed that the gas molecules which rise from the electrodes in electrolysis are themselves neutral. Townsend, however, first showed that some of these molecules are charged, although there are indeed a million million neutral ones for every one carrying a charge. He found that both the oxygen and the hydrogen which appear at the opposite electrodes when sulphuric acid is electrolyzed are positively charged, while when the electrolyte is caustic potash both the oxygen and the hydrogen given off are negative. Townsend's electrolyzing currents were from 12 to 14 amperes. He got in this way many more ions per cubic centimeter than he could produce with X-rays, the total charge per cubic centimeter being as large as 5×10^{-3} electrostatic units.

[1] *Phil. Mag.*, XXIX (1890; 5th series), 56.
[2] *Ibid.*, p. 292; *Nature*, XXXVI, 412.

When these charged gases were bubbled through water they formed a cloud. This cloud could be completely removed by bubbling through concentrated sulphuric acid or any drying agent, but when the gas came out again into the atmosphere of the room it again condensed moisture and formed a stable cloud. Townsend says that "the process of forming the cloud in positive or negative oxygen by bubbling through water, and removing it again by bubbling through sulphuric acid, can be gone through without losing more than 20 or 25 per cent of the original charge on the gas." This means simply that the ions condense the water about them when there is an abundance of moisture in the air, but when the cloud is carried into a perfectly dry atmosphere, such as that existing in a bubble surrounded on all sides by concentrated sulphuric acid, the droplets of water evaporate and leave the charge on a molecule of air as it was at first. The 20 or 25 per cent loss of charge represents the fraction of the droplets with their charges which actually got into contact with and remained in the liquids through which the gas was being bubbled.

In order to find the charge on each ion, Townsend took the following five steps:

1. He assumed that in saturated water vapor each ion condensed moisture about it, so that the number of ions was the same as the number of droplets.

2. He determined with the aid of a quadrant electrometer the total electrical charge per cubic centimeter carried by the gas.

3. He found the total weight of the cloud by passing it through drying tubes and determining the increase in weight of these tubes.

4. He found the average weight of the water droplets constituting the cloud by observing their rate of fall under gravity and computing their mean radius with the aid of a purely theoretical law known as Stokes's Law.

5. He divided the weight of the cloud by the average weight of the droplets of water to obtain the number of droplets which, if assumption 1 is correct, was the number of ions, and he then divided the total charge per cubic centimeter in the gas by the number of ions to find the average charge carried by each ion, that is, to find e.

A brief description of the way in which these experiments were carried out is contained in Appendix B.

One of the interesting side results of this work was the observation that clouds from negative oxygen fall faster than those from positive oxygen, thus indicating that the negative ions in oxygen act more readily than do the positive ions as nuclei for the condensation of water vapor. This observation was made at about the same time in another way by C. T. R. Wilson,[1] also in the Cavendish Laboratory, and it has played a rather important rôle in subsequent work. Wilson's discovery was that when air saturated with water vapor is ionized by X-rays from radioactive substances and then cooled by a sudden expansion, a smaller expansion is required to make a cloud form about the negative than about the positive ions. Thus when the expansion increased the volume in a ratio between 1.25 and 1.3, only negative ions acted as nuclei for cloudy condensation, while with expansions greater than 1.3 both negatives and positives were brought down.

[1] *Proc. Camb. Phil. Soc.*, IX (1897), 333.

Townsend first obtained by the foregoing method when he worked with positive oxygen,

$$e = 2.8 \times 10^{-10} \text{ electrostatic units,}$$

and when he worked with negative oxygen,

$$e = 3.1 \times 10^{-10} \text{ electrostatic units.}$$

In later experiments[1] he obtained 2.4 and 2.9, respectively, in place of the numbers given above, but in view of the unavoidable errors, he concluded that the two charges might be considered equal and approximately 3×10^{-10} electrostatic units. Thus he arrived at about the same value for e as that which was then current because of the kinetic theory estimates of n, the number of molecules in a cubic centimeter of a gas.

The weak points in this first attempt at a direct determination of e consisted in: (1) the assumption that the number of ions is the same as the number of drops; (2) the assumption of Stokes's Law of Fall which had never been tested experimentally, and which from a theoretical standpoint might be expected to be in error when the droplets were small enough; (3) the assumption that the droplets were all alike and fell at a uniform rate wholly uninfluenced by evaporation or other causes of change; (4) the assumption of no convection currents in the gas when the rate of fall of the cloud was being measured.

II. SIR JOSEPH THOMSON'S WORK ON e

This first attempt to measure e was carried out in Professor J. J. Thomson's laboratory. The second attempt was made by Professor Thomson himself[2] by a method

[1] *Ibid.*, p. 345.

[2] *Phil. Mag.*, XLVI (1898), 528.

which resembled Townsend's very closely in all its essential particulars. Indeed, we may set down for Professor Thomson's experiment precisely the same five elements which are set down on p. 47 for Townsend's. The differences lay wholly in step 2, that is, in the way in which the electrical charge per cubic centimeter carried by the gas was determined, and in step 3, that is, in the way in which the total weight of the cloud was obtained. Thomson produced ions in the space A (Fig. 1) by an X-ray bulb which ran at a constant rate, and measured first the current which, under the influence of a very weak electromotive force E, flows through A between the surface of the water and the aluminum plate which closes the top of the vessel. Then if n' is the whole number of ions of one sign per cubic centimeter, u the velocity of the positive and v that of the negative ion under unit electric force, i.e., if u and v are the mobilities of the positive and negative ions, respectively, then the current I per unit area is evidently given by

$$I = n'e(u+v)\,E \dots\dots\dots\dots\dots(4)$$

I and E were easily measured in any experiment; $u+v$ was already known from Rutherford's previous work, so that $n'e$, the charge of one sign per cubic centimeter of gas under the ionizing action of a constant source of X-rays, could be obtained at once from (4). This then simply replaces Townsend's method of obtaining the charge per cubic centimeter on the gas, and in principle the two methods are quite the same, the difference in experimental arrangements being due to the fact that Townsend's ions are of but one sign while Thomson's are of both signs.

Having thus obtained $n'e$ of equation (4), Thomson
had only to find n' and then solve for e.   To obtain n'
he proceeded exactly as Townsend had done in letting
the ions condense droplets of water about them and
weighing the cloud thus formed.   But in order to form
the cloud, Thomson utilized C. T. R. Wilson's discovery

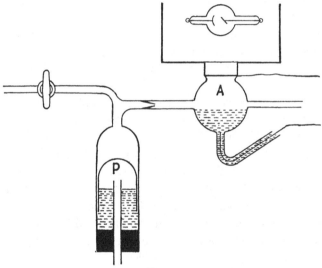

FIG. 1

just touched upon above, that a sudden expansion and
consequent cooling of the air in A (Fig. 1) would cause
the ions in A to act as nuclei for the formation of water
droplets.   To produce this expansion the piston P is
suddenly pulled down so as to increase the volume of the
space above it.   A cloud is thus formed about the ions
in A.   Instead of measuring the weight of this cloud
directly, as Townsend had done, Thomson computed it
by a theoretical consideration of the amount of cooling

produced by the expansion and the known difference
between the densities of saturated water vapor at the
temperature of the room and the temperature resulting
from the expansion. This method of obtaining the
weight of the cloud was less direct and less reliable than
that used by Townsend, but it was the only one available with Thomson's method of obtaining an ionized gas
and of measuring the charge per cubic centimeter on that
gas. The average size of the droplets was obtained precisely as in Townsend's work by applying Stokes's Law
to the observed rate of fall of the top of the cloud in
chamber A.

The careful consideration of Thomson's experiment
shows that it contains the theoretical uncertainties
involved in Townsend's work, while it adds some very
considerable experimental uncertainties. The most serious of the theoretical uncertainties arise from (1) the
assumption of Stokes's Law, and (2) the assumption that
the number of ions is equal to the number of droplets.
Both observers sought for some experimental justification
for the second and most serious of these assumptions, but
subsequent work by H. A. Wilson, by Quincke, and
by myself has shown that clouds formed by C. T. R.
Wilson's method consist in general of droplets some of
which may carry one, some two, some ten, or almost any
number of unit charges, and I have never been able,
despite quite careful experimenting, to obtain conditions
in which it was even approximately true that each
droplet carried but a single unit charge. Quincke has
also published results from which he arrives at the
same conclusion.

[1] *Verh. der deutsch. phys. Ges.*, XVI (1914), 422.

Again, when we compare the *experimental* uncertainties in Townsend's and Thomson's methods, it is at once obvious that the assumption that the clouds are not evaporating while the rate of fall is being determined is even more serious in Thomson's experiment than in Townsend's, for the reason that in the former case the clouds are formed by a sudden expansion and a consequent fall in temperature, and it is certain that during the process of the return of the temperature to initial conditions the droplets must be evaporating. Furthermore, this sudden expansion makes the likelihood of the existence of convection currents, which would falsify the computations of the radius of the drop from the observed rate of fall, more serious in Thomson's work than in Townsend's. The results which Thomson attained in different experiments gave values ranging from 5.5×10^{-10} to 8.4×10^{-10}. He published as his final value 6.5×10^{-10}. In 1903, however,[1] he published some new work on e in which he had repeated the determination, using the radiation from radium in place of that from X-rays as his ionizing agent and obtained the result $e = 3.4 \times 10^{-10}$. He explained the difference by the assumption that in his preceding work the more active negative ions had monopolized the aqueous vapor available and that the positive ions had not been brought down with the cloud as he had before assumed was the case. He now used more sudden expansions than he had used before, and concluded that the assumption made in the earlier experiments that the number of ions was equal to the number of particles, although shown to be incorrect for the former case, was correct for these

[1] *Phil. Mag.*, V (1903; 6th series), 354.

second experiments. As a matter of fact, if he had obtained only half the ions in the first experiments and all of them in the second, his second result should have come out approximately one-half as great as the first, which it actually did. Although Thomson's experiment was an interesting and important modification of Townsend's, it can scarcely be said to have added greatly to the accuracy of our knowledge of e.

The next step in advance in the attempt at the determination of e was made in 1903 by H. A. Wilson,[1] also in the Cavendish Laboratory.

III. H. A. WILSON'S METHOD

Wilson's modification of Thomson's work consisted in placing inside the chamber A two horizontal brass plates $3\frac{1}{2}$ cm. in diameter and from 4 to 10 mm. apart and connecting to these plates the terminals of a 2,000-volt battery. He then formed a negative cloud by a sudden expansion of amount between 1.25 and 1.3, and observed first the rate of fall of the top surface of this cloud between the plates when no electrical field was on; then he repeated the expansion and observed the rate of fall of the cloud when the electrical field as well as gravity was driving the droplets downward. If mg represents the force of gravity acting on the droplets in the top surface of the cloud and $mg+Fe$ the force of gravity plus the electrical force arising from the action of the field F on the charge e, and if v_1 is the velocity of fall under the action of gravity alone, and v_2 the velocity when both gravity and the electrical field are acting, then, if the ratio between the force acting and

[1] *Op. cit.*, p. 429.

the velocity produced is the same when the particle is charged as when it is uncharged, we have

$$\frac{mg}{mg+Fe}=\frac{v_1}{v_2}\ldots\ldots\ldots\ldots\ldots\ldots(5)$$

Combining this with the Stokes's Law equation which runs

$$v_1=\frac{2}{9}\frac{ga^2\sigma}{\eta}\ldots\ldots\ldots\ldots\ldots\ldots(6)$$

in which a is the radius, σ the density, v_1 the velocity of the drop under gravity g, and η is the viscosity of the air, and then eliminating m by means of

$$m=\tfrac{4}{3}\pi a^3\sigma\ldots\ldots\ldots\ldots\ldots\ldots\ldots(7)$$

Wilson obtained after substituting for η and σ the appropriate values (not accurately known, it is true, for saturated air at the temperature existing immediately after the expansion),

$$e=3.1\times10^{-9}\frac{g}{F}(v_2-v_1)v_1^{\frac{1}{2}}\ldots\ldots\ldots\ldots(8)$$

Wilson's method constitutes a real advance in that it eliminates the necessity of making the very awkward assumption that the number of droplets is equal to the number of negative ions, for since he observes only the rate of fall of the *top* of the cloud, and since the more heavily charged droplets will be driven down more rapidly by the field than the less heavily charged ones, his actual measurements would always be made upon *the least heavily charged droplets*. All of the other difficulties and assumptions contained in either Townsend's or Thomson's experiments inhere also in Wilson's, and in addition one fresh and rather serious assumption

is introduced, namely, that the clouds formed in succes
sive expansions are identical as to size of droplets. Foɪ
we wrote down the first equation of Wilson's method as
though the v_1 and v_2 were measurements made upon
the same droplet, when as a matter of fact the measurements are actually made on wholly different droplets.
I have myself found the duplication of cloud conditions
in successive expansions a very uncertain matter.
Furthermore, Wilson's method assumes uniformity in the
field between the plates, an assumption which might be
quite wide of the truth.

Although the elimination of the assumption of
equality of the number of droplets and the number of
ions makes Wilson's determination of e more reliable as
to method than its predecessors, the accuracy actually
attained was not great, as can best be seen from his own
final summary of results. He made eleven different
determinations which varied from $e = 2 \times 10^{-10}$ to
$e = 4.4 \times 10^{-10}$. His eleven results are:

TABLE I

e	e
2.3×10^{-10}	3.8×10^{-10}
2.6 "	3.0 "
4.4 "	3.5 "
2.7 "	2.0 "
3.4 "	2.3 "
3.8 "	

Mean 3.1×10^{-10}

In 1906, being dissatisfied with the variability of these
results, the author repeated Wilson's experiment without
obtaining any greater consistency than that which the
latter had found. Indeed, the instability, distortion, and
indefiniteness of the top surface of the cloud were some-

what disappointing, and the results were not considered
worth publishing. Nevertheless, it was concluded from
these observations that the accuracy might be improved
by using radium instead of X-rays for the ionizing agent,
by employing stronger electrical fields, and thus increasing the difference between v_1 and v_2, which in Wilson's
experiment had been quite small, and by observing the
fall of the cloud through smaller distances and shorter
times in order to reduce the error due to the evaporation of the cloud during the time of observation.
Accordingly, a 4,000-volt storage battery was built and
in the summer of 1908 Mr. Begeman and the author,
using radium as the ionizing agent, again repeated the
experiment and published some results which were somewhat more consistent than those reported by Wilson.[1]
We gave as the mean of ten observations which varied
from 3.66 to 4.37 the value $e = 4.06 \times 10^{-10}$. We stated
at the time that although we had not eliminated altogether the error due to evaporation, we thought that we
had rendered it relatively harmless, and that our final
result, although considerably larger than either Wilson's
or Thomson's (3.1 and 3.4, respectively), must be considered an approach at least toward the correct value.

IV. THE BALANCED-DROP METHOD

Feeling, however, that the amount of evaporation of
the cloud was still a quite unknown quantity, I next
endeavored to devise a way of eliminating it entirely.
The plan now was to use an electrical field which was
strong enough, not merely to increase or decrease slightly
the speed of fall under gravity of the top surface of the

[1] *Phys. Rev.*, XXVI (1908), 198.

cloud, as had been done in all the preceding experiments, but also sufficiently strong to hold the top surface of the cloud stationary, so that the rate of its evaporation could be accurately observed and allowed for in the computations.

This attempt, while not successful in the form in which it had been planned, led to a modification of the cloud method which seemed at the time, and which has actually proved since, to be of far-reaching importance. *It made it for the first time possible to make all the measurements on individual droplets,* and thus not merely to eliminate ultimately all of the questionable assumptions and experimental uncertainties involved in the cloud method of determining e, but, more important still, it made it possible to examine the properties of individual isolated electrons and to determine whether different ions actually carry one and the same charge. That is to say, it now became possible to determine whether electricity in gases and solutions is actually built up out of electrical atoms, each of which has exactly the same value, or whether the electron which had first made its appearance in Faraday's experiments on solutions and then in Townsend's and Thomson's experiments on gases is after all only a *statistical mean* of charges which are themselves greatly divergent. This latter view had been strongly urged up to and even after the appearance of the work which is now under consideration. It will be given further discussion presently.

The first determination which was made upon the charges carried by individual droplets was carried out in the spring of 1909. A report of it was placed upon the program of the British Association meeting at Winnipeg in August, 1909, as an additional paper, was printed in abstract in the *Physical Review* for December, 1909, and in full in the *Philosophical Magazine* for February, 1910, under the title "A New Modification of the Cloud Method of Determining the Elementary Electrical Charge and the Most Probable Value of That Charge."[1] The following extracts from that paper show clearly what was accomplished in this first determination of the charges carried by individual droplets.

THE BALANCING OF INDIVIDUAL CHARGED DROPS BY AN ELECTROSTATIC FIELD

My original plan for eliminating the evaporation error was to obtain, if possible, an electric field strong enough exactly to balance the force of gravity upon the cloud and then by means of a sliding contact to vary the strength of this field so as to hold the cloud balanced throughout its entire life. In this way it was thought that the whole evaporation-history of the cloud might be recorded, and that suitable allowances might then be made in the observations on the rate of fall to eliminate entirely the error due to evaporation. It was not found possible to balance the cloud, as had been originally planned, but it was found possible to do something much better: namely, to hold individual charged drops suspended by the field for periods varying from 30 to 60 seconds. I have never actually timed drops which lasted more than 45 seconds, although I have several times observed drops which in my judgment lasted considerably longer than this. The drops which it was found possible to balance by an electrical field always carried multiple charges, and the difficulty experienced in balancing such drops was less than had been anticipated.

The procedure is simply to form a cloud and throw on the field immediately thereafter. The drops which have charges of the same sign as that of the upper plate or too weak charges of the opposite sign rapidly fall, while those which are charged with too many multiples of the sign opposite to that of the upper plate are

[1] *Phil. Mag.*, XIX (1910), 209.

jerked up against gravity to this plate. The result is that after a lapse of 7 or 8 seconds the field of view has become quite clear save for a relatively small number of drops which have just the right ratio of charge to mass to be held suspended by the electric field. These appear as perfectly distinct bright points. I have on several occasions obtained but one single such "star" in the whole field and held it there for nearly a minute. For the most part, however, the observations recorded below were made with a considerable number of such points in view. Thin, flocculent clouds, the production of which seemed to be facilitated by keeping the water-jackets J_1 and J_2 (Fig. 2) a degree or two above the temperature of the room, were found to be particularly favorable to observations of this kind.

Furthermore, it was found possible so to vary the mass of a drop by varying the ionization, that drops carrying in some cases two, in some three, in some four, in some five, and in some six, multiples could be held suspended by nearly the same field. The means of gradually varying the field which had been planned were therefore found to be unnecessary. If a given field would not hold any drops suspended it was varied by steps of 100 or 200 volts until drops were held stationary, or nearly stationary. When the P.D. was thrown off it was often possible to see different drops move down under gravity with greatly different speeds, thus showing that these drops had different masses and correspondingly different charges.

The life-history of these drops is as follows: If they are a little too heavy to be held quite stationary by the field they begin to move slowly down under gravity. Since, however, they slowly evaporate, their downward motion presently ceases, and they become stationary for a considerable period of time. Then the field gets the better of gravity and they move slowly upward. Toward the end of their life in the space between the plates, this upward motion becomes quite rapidly accelerated and they are drawn with considerable speed to the upper plate. This, taken in connection with the fact that their whole life between plates only 4 or 5 mm. apart is from 35 to 60 seconds, will make it obvious that during a very considerable fraction of this time their motion must be exceedingly slow. I have often held drops through a

period of from 10 to 15 seconds, during which it was impossible to
see that they were moving at all.   Shortly after an expansion I
have seen drops which at first seemed stationary, but which then
began to move slowly down in the direction of gravity, then become
stationary again, then finally began to move slowly up.   This is
probably due to the fact that large multiply charged drops are not
in equilibrium with smaller singly charged drops near them, and
hence, instead of evaporating, actually grow for a time at the
expense of their small neighbors.   Be this as it may, however, it
is by utilizing the experimental fact that there is a considerable
period during which the drops are essentially stationary that it
becomes possible to make measurements upon the rate of fall in
which the error due to evaporation is wholly negligible in comparison with the other errors of the experiment.   Furthermore, in
making measurements of this kind the observer is just as likely
to time a drop which has not quite reached its stationary point as
one which has just passed through that point, so that the mean of
a considerable number of observations would, even from a theoretical standpoint, be quite free from an error due to evaporation.

THE METHOD OF OBSERVATION

The observations on the rate of fall were made with a short-
focus telescope *T* (see Fig. 2) placed about 2 feet away from the
plates.   In the eyepiece of this telescope were placed three equally
spaced cross-hairs, the distance between those at the extremes corresponding to about one-third of the distance between the plates.
A small section of the space between the plates was illuminated by
a narrow beam from an arc light, the heat of the arc being absorbed
by three water cells in series.   The air between the plates was
ionized by 200 mg. of radium, of activity 20,000, placed from
3 to 10 cm. away from the plates.   A second or so after expansion
the radium was removed, or screened off with a lead screen, and the
field thrown on by hand by means of a double-throw switch.   If
drops were not found to be held suspended by the field, the P.D.
was changed or the expansion varied until they were so held.   The
cross-hairs were set near the lower plate, and as soon as a stationary
drop was found somewhere above the upper cross-hair, it was
watched for a few seconds to make sure that it was not moving,

and then the field was thrown off and the plates short-circuited by means of the double-throw switch, so as to make sure that they retained no charge. The drop was then timed by means of an accurate stop watch as it passed across the three cross-hairs, one of the two hands of the watch being stopped at the instant of

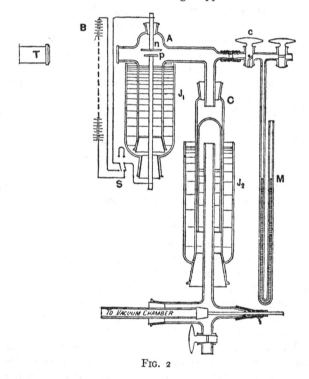

FIG. 2

passage across the middle cross-hair, the other at the instant of passage across the lower one. It will be seen that this method of observation furnishes a double check upon evaporation; for if the drop is stationary at first, it is not evaporating sufficiently to influence the reading of the rate of fall, and if it begins to evaporate appreciably before the reading is completed, the time required to pass through the second space should be greater than that required to pass through the first space. It will be seen from the observations which follow that this was not, in general, the case.

It is an exceedingly interesting and instructive experiment to watch one of these drops start and stop, or even reverse its direction of motion, as the field is thrown off and on. I have often caught a drop which was just too light to remain stationary and moved it back and forth in this way four or five times between the same two cross-hairs, watching it first fall under gravity when the field was thrown off and then rise against gravity when the field was thrown on. The accuracy and certainty with which the instants of passage of the drops across the cross-hairs can be determined are precisely the same as that obtainable in timing the passage of a star across the cross-hairs of a transit instrument.

Furthermore, since the observations upon the quantities occurring in equation (4) [see (8) p. 55 of this volume] are all made upon the same drop, all uncertainties as to whether conditions can be exactly duplicated in the formation of successive clouds obviously disappear. There is no theoretical uncertainty whatever left in the method unless it be an uncertainty as to whether or not Stokes's Law applies to the rate of fall of these drops under gravity. The experimental uncertainties are reduced to the uncertainty in a time determination of from 3 to 5 seconds, when the object being timed is a single moving bright point. This means that when the time interval is say 5 seconds, as it is in some of the observations given below, the error which a practiced observer will make with an accurate stop watch in any particular observation will never exceed 2 parts in 50. The error in the mean of a considerable number of concordant observations will obviously be very much less than this.

Since in this form of observation the v_2 of equation (5) [(8) of this volume] is zero, and since F is negative in sign, equation (5) reduces to the simple form:

$$e = 3.422 \times 10^{-9} \times \frac{g}{X} (v_1)^{\frac{3}{2}} \dots\dots\dots\dots\dots (6)^1$$

[1] I had changed the constant in Wilson's equation from 3.1 to 3.422 because of careful measurements on the temperature existing in the cloud chamber about 10 seconds after expansion and because of new measurements on the viscosity of the saturated air.

It will perhaps be of some interest to introduce two
tables from this paper to show the exact nature of these

TABLE II

SERIES 1 (BALANCED POSITIVE WATER DROPS)		

Distance between plates .545 cm.
Measured distance of fall .155 cm.

Volts	Time 1 Space	Time 2 Spaces
2,285......	2.4 sec.	4.8 sec.
2,285......	2.4	4.8
2,275......	2.4	4.8
2,325......	2.4	4.8
2,325......	2.6	4.8
2,325......	2.2	4.8
2,365......	2.4	4.8
2,312......	2.4	4.8

Mean time for .155 cm.=4.8 sec.

$$e_3 = 3.422 \times 10^{-9} \times \frac{980.3}{14.14} \times \left(\frac{.155}{4.8}\right)^{\frac{3}{2}}$$
$$= 13.77 \times 10^{-10}$$

Therefore $e = 13.85 \times 10^{-10} \div 3$
$$= 4.59 \times 10^{-10}.$$

SERIES 2 (BALANCED POSITIVE WATER DROPS)		

Distance between plates .545 cm.
Measured distance of fall .155 cm.

Volts	Time 1 Space	Time 2 Spaces
2,365......	1.8 sec.	4.0 sec.
2,365......	1.8	4.0
2,365......	2.2	3.8
2,365......	1.8	4.0
2,395......	2.0	4.0
2,395......	2.0	4.0
2,395......	2.0	3.8
2,365......	1.8	4.0
2,365......	1.8	4.0
2,365......	1.8	4.0
2,374......	1.90	3.96

Mean time for .155 cm.=3.91 sec.

$$e_4 = 3.422 \times 10^{-9} \times \frac{980.3}{14.52} \times \left(\frac{.155}{3.91}\right)^{\frac{3}{2}}$$
$$= 18.25 \times 10^{-10}$$

Therefore $e = 18.25 \div 4$
$$= 4.56 \times 10^{-10}.$$

TABLE III

Series	Charge	Value of e	Weight Assigned
1............	3e	4.59	7
2............	4e	4.56	7
3............	2e	4.64	6
4............	5e	4.83	4
5............	2e	4.87	1
6............	6e	4.69	3

Simple mean $e=4.70 \times 10^{-10}$
Weighted mean $e=4.65 \times 10^{-10}$

earliest measurements on the charges carried by individual particles.

In connection with these experiments I chanced to observe a phenomenon which interested me very much at the time and suggested quite new possibilities. While working with these "balanced drops" I noticed on several occasions on which I had failed to screen off the rays from the radium that now and then one of them would suddenly change its charge and begin to move up or down in the field, evidently because it had captured in the one case a positive, in the other a negative, ion. This opened up the possibility of measuring with certainty, not merely the charges on individual droplets as I had been doing, but the charge carried by a single atmospheric ion. *For by taking two speed measurements on the same drop, one before and one after it had caught an ion, I could obviously eliminate entirely the properties of the drop and of the medium and deal with a quantity which was proportional merely to the charge on the captured ion itself.*

Accordingly, in the fall of 1909 there was started the series of experiments described in the succeeding chapter.

The problem had already been so nearly solved by the work with the water droplets that there seemed no possibility of failure. It was only necessary to get a charged droplet entirely free from evaporation into the space between the plates of a horizontal air condenser and then, by alternately throwing on and off an electrical field, to keep this droplet pacing its beat up and down between the plates until it could catch an atmospheric ion in just the way I had already seen the water droplets do. The change in the speed in the field would then be exactly proportional to the charge on the ion captured.

CHAPTER IV

GENERAL PROOF OF THE ATOMIC NATURE OF ELECTRICITY

Although the "balanced-droplet method" just described had eliminated the chief sources of uncertainty which inhered in preceding work on e and had made it possible to assert with much confidence that the unit charge was a real physical entity and not merely a "statistical mean," it was yet very far from an exact method of studying the properties of gaseous ions. The sources of error or uncertainty which still inhered in it arose from (1) the lack of stagnancy in the air through which the drop moved; (2) the lack of perfect uniformity of the electrical field used; (3) the gradual evaporation of the drops, rendering it impossible to hold a given drop under observation for more than a minute or to time a drop as it fell under gravity alone through a period of more than five or six seconds; and (4) the assumption of the validity of Stokes's Law.

The method which was devised to replace it was not only entirely free from all of these limitations, but it constituted an entirely new way of studying ionization and one which at once yielded important results in a considerable number of directions. This chapter deals with some of these by-products of the determination of e which are of even more fundamental interest and importance than the mere discovery of the exact size of the electron.

I.   ISOLATION OF INDIVIDUAL IONS AND MEASUREMENT
OF THEIR RELATIVE CHARGES

In order to compare the charges on different ions, the procedure adopted was to blow with an ordinary commercial atomizer an oil spray into the chamber C (Fig. 3).

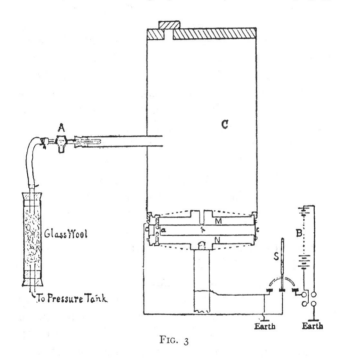

FIG. 3

The air with which this spray was blown was first rendered dust-free by passage through a tube containing glass wool.   The minute droplets of oil constituting the spray, most of them having a radius of the order of a one-thousandth of a millimeter, slowly fell in the chamber C, and occasionally one of them would find its way through the minute pinhole p in the middle of the circular brass plate M, 22 cm. in diameter, which formed one of the plates of the air condenser. The other plate, N, was held 16 mm. beneath it by three ebonite posts a. By means of the switch S these plates could be charged, the one positively and the other negatively, by making them the terminals of a 10,000-volt storage battery B, while throwing the switch the other way (to the left) short-circuited them and reduced the field between them to zero. The oil droplets which entered at p were illuminated by a powerful beam of light which passed through diametrically opposite windows in the encircling ebonite strip c. As viewed through a third window in c on the side toward the reader, it appeared as a bright star on a black background. These droplets which entered p were found in general to have been strongly charged by the frictional process involved in blowing the spray, so that when the field was thrown on in the proper direction they would be pulled up toward M. Just before the drop under observation could strike M the plates would be short-circuited and the drop allowed to fall under gravity until it was close to N, when the direction of motion would be again reversed by throwing on the field. In this way the drop would be kept traveling back and forth between the plates. The first time the experiment was tried an ion was caught within a few minutes, and the fact of its capture was signaled to the observer by the change in the speed with which it moved up when the field was on. The significance of the experiment can best be appreciated by examination of the complete record of one of the early experiments when the timing was done merely with a stop watch.

The column headed t_g gives the successive times which the droplet required to fall between two fixed cross-hairs in the observing telescope whose distance apart corresponded in this case to an actual distance of fall of .5222 cm. It will be seen that these numbers are all the same within the limits of error of a stop-watch measurement. The column marked t_F gives the successive times

TABLE IV

t_g	t_F
13.6	12.5
13.8	12.4
13.4	21.8
13.4	34.8
13.6	84.5
13.6	85.5
13.7	34.6
13.5	34.8
13.5	16.0
13.8	34.8
13.7	34.6
13.8	21.9
13.6	
13.5	
13.4	
13.8	
13.4	

Mean 13.595

which the droplet required to rise under the influence of the electrical field produced by applying in this case 5,051 volts of potential difference to the plates M and N. It will be seen that after the second trip up, the time changed from 12.4 to 21.8, indicating, since in this case the drop was positive, that a negative ion had been caught from the air. The next time recorded under t_F, namely, 34.8, indicates that another negative ion had been caught. The next time, 84.5, indicates the capture of still another negative ion. This charge was held for two trips, when the speed changed back again to 34.6, showing that a positive ion had now been caught which carried precisely the same charge as the negative ion which before caused the inverse change in time, i.e., that from 34.8 to 84.5.

In order to obtain some of the most important consequences of this and other similar experiments we need make no assumption further than this, that the velocity with which the drop moves is proportional to the force acting upon it and is independent of the electrical charge which it carries. Fortunately this assumption can be put to very delicate experimental test, as will presently be shown, but introducing it for the time being as a mere assumption, as Townsend, Thomson, and Wilson had done before, we get

$$\frac{v_1}{v_2} = \frac{mg}{Fe_n - mg} \text{ or } e_n = \frac{mg}{Fv_1}(v_1 + v_2) \ \ldots\ldots \ \ldots\ (9)$$

The negative sign is used in the denominator because v_2 will for convenience be taken as positive when the drop is going up in the direction of F, while v_1 will be taken as positive when it is going down in the direction of g. e_n denotes the charge on the drop, and must not be confused with the charge on an ion. If now by the capture of an ion the drop changes its charge from e_n to $e_{n'}$, then the value of the captured charge e_i is

$$e_i = e_{n'} - e_n = \frac{mg}{Fv_1}(v_2' - v_2) \ldots \ \ldots\ldots\ \ldots (10)$$

and since $\frac{mg}{Fv_1}$ is a constant for this drop, any charge which it may capture will always be proportional to

$(v'_2 - v_2)$, that is, to the change produced in the velocity in the field F by the captured ion. The successive values of v_2 and of $(v'_2 - v_2)$, these latter being obtained by subtracting successive values of the velocities given under v_2, are shown in Table V.

TABLE V

$$v_2 \qquad (v'_2 - v_2)$$

$\dfrac{.5222}{12.45} = .04196$

$\qquad\qquad .01806 \div 2 = .00903$

$\dfrac{.5222}{21.5} = .02390$

$\qquad\qquad .00885 \div 1 = .00885$

$\dfrac{.5222}{34.7} = .01505$

$\qquad\qquad .00891 \div 1 = .00891$

$\dfrac{.5222}{85.0} = .006144$

$\qquad\qquad .00891 \div 1 = .00891$

$\dfrac{.5222}{34.7} = .01505$

$\qquad\qquad .01759 \div 2 = .00880$

$\dfrac{.5222}{16.0} = .03264$

$\qquad\qquad .01759 \div 2 = .00880$

$\dfrac{.5222}{34.7} = .01505$

$\qquad\qquad .00891 \div 1 = .00891$

$\dfrac{.5222}{21.85} = .02390$

It will be seen from the last column that within the limits of error of a stop-watch measurement, all the charges captured have exactly the same value save in three cases. In all of these three the captured charges were just twice as large as those appearing in the other changes. Relationships of exactly this sort have been found to hold absolutely without exception, no matter in what gas the drops have been suspended or what sort of droplets were used upon which to catch the ions. In many cases a given drop has been held under observation for five or six hours at a time and has been seen to catch not eight or ten ions, as in the experiment above, but hundreds of them. Indeed, I have observed, all told, the capture of many thousands of ions in this way, and in no case have I ever found one the charge of which, when tested as above, did not have either exactly the value of the smallest charge ever captured or else a very small multiple of that value. *Here, then, is direct, unimpeachable proof that the electron is not a "statistical mean," but that rather the electrical charges found on ions all have either exactly the same value or else small exact multiples of that value.*

II. PROOF THAT ALL STATIC CHARGES BOTH ON CONDUCTORS AND INSULATORS ARE BUILT UP OF ELECTRONS

The foregoing experiment leads, however, to results of much more fundamental importance than that mentioned in the preceding section. The charge which the droplet had when it first came under observation had been acquired, not by the capture of ions from the air, but by the ordinary frictional process involved in blowing the spray. If then ordinary static charges are built up of electrons, this charge should be found to be an exact multiple of the ionic charge which had been found from the most reliable measurement shown in Table V to be proportional to the velocity .00891. This initial charge e_n on the drop is seen from equations (9) and (10) to bear the same relation to (v_1+v_2) which the ionic charge $e_n'-e_n$ bears to $(v_2'-v_2)$. Now, $v_1 = .5222/13.595 = .03842$, hence $v_1+v_2 = .03842+.04196 = .08038$. Dividing this by 9 we obtain .008931, which is within about

one-fifth of 1 per cent of the value found in the last
column of Table V as the smallest charge carried by an
ion. *Our experiment has then given us for the first time
a means of comparing a frictional charge with the ionic
charge, and the frictional charge has in this instance been
found to contain exactly 9 electrons.* A more exact means
of making this comparison will be given presently, but
suffice it to say here that experiments like the foregoing
have now been tried on thousands of drops in different
media, some of the drops being made of non-conductors
like oil, some of semi-conductors like glycerin, some of
excellent metallic conductors like mercury. In every
case, without a single exception, the initial charge placed
upon the drop by the frictional process, and all of the
dozen or more charges which have resulted from the
capture by the drop of a larger or smaller number of
ions, have been found to be exact multiples of the smallest charge caught from the air. Some of these drops
have started with no charge at all, and one, two, three,
four, five, and six elementary charges or electrons have
been picked up. Others have started with seven or
eight units, others with twenty, others with fifty, others
with a hundred, others with a hundred and fifty elementary units, and have picked up in each case a dozen or
two of elementary charges on either side of the starting-
point, so that, in all, drops containing every possible number of electrons between one and one hundred and fifty
have been observed and the number of electrons which
each drop carried has been accurately counted by the
method described. When the number is less than fifty
there is not a whit more uncertainty about this count
than there is in counting one's own fingers and toes. It

is not found possible to determine with certainty the number of electrons in a charge containing more than one hundred or two hundred of them, for the simple reason that the method of measurement used fails to detect the difference between 200 and 201, that is, we cannot measure $v_2' - v_2$ with an accuracy greater than one-half of 1 per cent. But it is quite inconceivable that large charges such as are dealt with in commercial applications of electricity can be built up in an essentially different way from that in which the small charges whose electrons we are able to count are found to be. Furthermore, since it has been definitely proved that an electrical current is nothing but the motion of an electrical charge over or through a conductor, it is evident that the experiments under consideration furnish not only the most direct and convincing of evidence that all electrical charges are built up out of these very units which we have been dealing with as individuals in these experiments, but that all electrical currents consist merely in the transport of these electrons through the conducting bodies.

In order to show the beauty and precision with which these multiple relationships stand out in all experiments of this kind, a table corresponding to much more precise measurements than those given heretofore is here introduced (Table VI). The time of fall and rise shown in the first and second columns were taken with a Hipp chronoscope reading to one-thousandth of a second. The third column gives the reciprocals of these times. These are used in place of the velocities v_2 in the field, since distance of fall and rise is always the same. The fourth column gives the successive changes in speed due to the capture of ions. These also are expressed merely as time reciprocals. For reasons which will be explained in the next section, each one of these changes may correspond to the capture of not merely one but of several distinct ions. The numbers in the fifth column represent

TABLE VI

t_g Sec.	t_F Sec.	$\dfrac{1}{t_F}$	$\left(\dfrac{1}{t'_F}-\dfrac{1}{t_F}\right)$	n'	$\dfrac{1}{n'}\left(\dfrac{1}{t'_F}-\dfrac{1}{t_F}\right)$	$\left(\dfrac{1}{t_g}+\dfrac{1}{t_F}\right)$	n	$\dfrac{1}{n}\left(\dfrac{1}{t_g}+\dfrac{1}{t_F}\right)$
11.848	80.708	.01236				.09655	18	.005366
11.890	22.366		.03234	6	.005390			
11.908	22.390	.04470				.12887	24	.005371
11.904	22.368		.03751	7	.005358			
11.882	140.565	.007192	.005348	1	.005348	.09138	17	.005375
11.906	79.600	.01254				.09673	18	.005374
11.838	34.748		.01616	3	.005387			
11.816	34.762	.02870				.11289	21	.005376
11.776	34.846							
11.840	29.286		.026872	5	.005375	.11833	22	.005379
11.904	29.236	.03414						
11.870	137.308	.007268	.021572	4	.005393	.09146	17	.005380
11.952	34.638	.02884				.11303	21	.005382
11.860			.01623	3	.005410			
11.846	22.104		.04307	8	.005384	.12926	24	.005386
11.912	22.268	.04507						
11.910	500.1	.002000				.08619	16	.005387
11.918	19.704		.04879	9	.005421			
11.870	19.668	.05079	.03794	7	**.005420**	.13498	25	.005399
11.888	77.630					.09704	18	.005390
11.894	77.806	.01285	.01079	2	.005395	.10783	20	.005392
11.878	42.302	.02364						
11.880			Means		.005386			.005384

Duration of exp.	= 45 min.		Pressure	= 75.62 cm.	
Plate distance	= 16 mm.		Oil density	= .9199	
Fall distance	= 10.21 mm.		Air viscosity	= 1,824×10⁻⁷	
Initial volts	= 5,088.8		Radius (a)	= .000276 cm.	
Final volts	= 5,081.2		$\dfrac{l}{a}$	= .034	
Temperature	= 22.82° C.		Speed of fall	= .08584 cm./sec.	

$$e_1 = 4.991 \times 10^{-10}$$

simply the small integer by which it is found that the numbers in the fourth column must be divided in order to obtain the numbers in the sixth column. These will be seen to be exactly alike within the limits of error of the experiment. The mean value at the bottom of the sixth column represents, then, the smallest charge ever caught from the air, that is, it is the elementary *ionic* charge. The seventh column gives the successive values of $v_1 + v_2$ expressed as reciprocal times. These numbers, then, represent the successive values of the *total* charge carried by the droplet. The eighth column gives the integers by which the numbers in the seventh column must be divided to obtain the numbers in the last column. These also will be seen to be invariable. The mean at the bottom of the last column represents, then, *the electrical unit out of which the frictional charge on the droplet was built up, and it is seen to be identical with the ionic charge represented by the number at the bottom of the sixth column.*

It may be of interest to introduce one further table (Table VII) arranged in a slightly different way to show

TABLE VII

n	$4.917 \times n$	Observed Charge	n	$4.917 \times n$	Observed Charge
1............	4.917	....	10.........	49.17	49.41
2............	9.834	....	11.........	54.09	53.91
3............	14.75	....	12.........	59.00	59.12
4............	19.66	19.66	13.........	63.92	63.68
5............	24.59	24.60	14.........	68.84	68.65
6............	29.50	29.62	15.........	73.75	....
7............	34.42	34.47	16.........	78.67	78.34
8............	39.34	39.38	17.........	83.59	83.22
9............	44.25	44.42	18.........	88.51	... .

how infallibly the atomic structure of electricity follows from experiments like those under consideration.

In this table 4.917 is merely a number obtained precisely as above from the change in speed due to the capture of ions and one which is proportional in this experiment to the ionic charge. The column headed $4.917 \times n$ contains simply the whole series of exact mul-

tiples of this number from 1 to 18. The column headed
"Observed Charge" gives the successive observed values
of (v_1+v_2). It will be seen that during the time of observation, about four hours, this drop carried all possible
multiples of the elementary charge from 4 to 18, save only
15. *No more exact or more consistent multiple relationship
is found in the data which chemists have amassed on the
combining powers of the elements and on which the atomic
theory of matter rests than is found in the foregoing numbers.*

Such tables as these—and scores of them could be
given—place beyond all question the view that an
electrical charge wherever it is found, whether on an
insulator or a conductor, whether in electrolytes or in
metals, has a definite granular structure, that it consists
of an exact number of specks of electricity (electrons) all
exactly alike, which in static phenomena are scattered over the surface of the charged body and in current
phenomena are drifting along the conductor. Instead
of giving up, as Maxwell thought we should some day do,
the "provisional hypothesis of molecular charges," we
find ourselves obliged to make all our interpretations of
electrical phenomena, *metallic as well as electrolytic*, in
terms of it.

III. MECHANISM OF CHANGE OF CHARGE OF A DROP

All of the changes of charge shown in Table IV were
spontaneous changes, and it has been assumed that all
of these changes were produced by the capture of ions
from the air. When a negative drop suddenly increases
its speed in the field, that is, takes on a larger charge of
its own kind than it has been carrying, there seems to be
no other conceivable way in which the change can be

produced. But when the charge suddenly *decreases* there is no a priori reason for thinking that the change may not be due as well to the direct loss of a portion of the charge as to the neutralization of this same amount of electricity by the capture of a charge of opposite sign. That, however, the changes do actually occur, when no X-rays or radioactive rays are passing between the plates, only by the capture of ions from the air, was rendered probable by the fact that drops not too heavily charged showed the same tendency on the whole to increase as to decrease in charge. This should not have been the case if there were two causes tending to decrease the charge, namely, direct loss and the capture of opposite ions, as against one tending to increase it, namely, capture of like ions. The matter was very convincingly settled, however, by making observations when the gas pressures were as low as 2 or 3 mm. of mercury. Since the number of ions present in a gas is in general directly proportional to the pressure, spontaneous changes in charge should almost never occur at these low pressures; in fact, it was found that drops could be held for hours at a time without changing. The frequency with which the changes occur decreases regularly with the pressure, as it should if the changes are due to the capture of ions. For the number of ions formed by a given ionizing agent must vary directly as the pressure.

Again, the changes do not, in general, occur when the electrical field is on, for then the ions are driven instantly to the plates as soon as formed, at a speed of, say, 10,000 cm. per second, and so do not have any opportunity to accumulate in the space between them. When the field is off, however, they do so accumulate, until, in ordinary air, they reach the number of, say, 20,000 per cubic centimeter. These ions, being endowed with the kinetic energy of agitation characteristic of the temperature, wander rapidly through the gas and become a part of the drop as soon as they impinge upon it. It was thus that all the changes recorded in Table IV took place.

It is possible, however, so to control the changes as to place electrons of just such sign as one wishes, and of just such number as one wishes, within limits, upon a given drop. If, for example, it is desired to place a positive electron upon a given drop the latter is held with the aid of the field fairly close to the negative plate, say the upper plate; then an ionizing agent—X-rays or radium—is arranged to produce uniform ionization in the gas between the plates. Since now all the positive ions move up while the negatives move down, the drop is in a shower of positive ions, and if the ionization is intense enough the drop is sure to be hit. In this way a positive charge of almost any desired strength may be placed upon the drop.

Similarly, in order to throw a negative ion or ions upon the drop it is held by the field close to the lower, i.e., to the positive, plate in a shower of negative ions produced by the X-rays. It was in this way that most of the changes shown in Table VI were brought about. This accounts for the fact that they correspond in some instances to the capture of as many as six electrons.

When X-rays are allowed to fall directly upon the drop itself the change in charge may occur, not merely because of the capture of ions, but also because the rays eject beta particles, i.e., negative electrons, from the molecules of the drop. That changes in charge were actually produced in this way in our experiments was proved conclusively in 1910 by the fact that when the pressure was reduced to a very low value and X-rays were allowed to pass through the air containing the drop, the latter would change readily in the direction of increasing positive or decreasing negative charge, but it could almost never be made to change in the opposite direction. This is because at these low pressures the rays can find very few gas molecules to ionize, while they detach negative electrons from the drop as easily as at atmospheric pressure. *This experiment proved directly that the charge carried by an ion in gases is the same as the charge on the beta or cathode-ray particle.*

When it was desired to avoid the direct loss of negative electrons by the drop, we arranged lead screens so that the drop itself would not be illuminated by the rays, although the gas underneath it was ionized by them.[1]

IV. DIRECT OBSERVATION OF THE KINETIC ENERGY OF AGITATION OF A MOLECULE

I have already remarked that when a drop carries but a small number of electrons it appears to catch ions of its own sign as rapidly as those of opposite signs—a result which seems strange at first, since the ions of opposite sign must be attracted, while those of like sign must be repelled. Whence, then, does the ion obtain the energy which enables it to push itself up against this electrostatic repulsion and attach itself to a drop already strongly charged with its own kind of electricity? It cannot obtain it from the field, since the phenomenon of capture occurs when the field is not on. It cannot

[1] See *Phil. Mag.*, XXI (1911), 757.

obtain it from any explosive process which frees the ion from the molecule at the instant of ionization, since in this case, too, ions would be caught as well, or nearly as well, when the field is on as when it is off. Here, then, is an absolutely direct proof that the ion must be endowed with a kinetic energy of agitation which is sufficient to push it up to the surface of the drop against the electrostatic repulsion of the charge on the drop.

This energy may easily be computed as follows: Let us take a drop, such as was used in one of these experiments, of radius .000197 cm. The potential at the surface of a charged sphere can be shown to be the charge divided by the radius. The value of the elementary electrical charge obtained from the best observations of this type, is 4.770×10^{-10} absolute electrostatic units. Hence the energy required to drive an ion carrying the elementary charge e up to the surface of a charged sphere of radius r, carrying 16 elementary charges, is

$$\frac{16e^2}{r} = \frac{16 \times (4.770 \times 10^{-10})^2}{.000197} = 1.95 \times 10^{-14} \text{ ergs}$$

Now, the kinetic energy of agitation of a molecule as deduced from the value of e herewith obtained, and the kinetic theory equation, $p = \frac{1}{3}nm\bar{c^2}$, is 5.75×10^{-14} ergs. According to the Maxwell-Boltzmann Law of the partition of energy, which certainly holds in gases, this should also be the kinetic energy of agitation of an ion. It will be seen that the value of this energy is approximately three times that required to push a single ion up to the surface of the drop in question. Hence the electrostatic forces due to 16 electrons on the drop are too weak to exert much influence upon the motion of an approaching ion. But if it were possible to load up a drop with negative electricity until the potential energy of its charge were about three times as great as that computed above for this drop, then the phenomenon here observed of the catching of new negative ions by such a negatively charged drop should not take place, save in the exceptional case in which an ion might acquire an energy of agitation considerably larger than the mean value. Now, as a matter of fact, it was regularly observed that the heavily charged drops had a very much smaller tendency to pick up new negative ions than the more lightly charged drops, and, in one instance, we watched for four hours another negatively charged drop of radius .000658 cm., which carried charges varying from 126 to 150 elementary units, and which therefore had a potential energy of charge (computed as above on the assumption of uniform distribution) varying from 4.6×10^{-14} to 5.47×10^{-14}. In all that time this drop picked up but one single negative ion when the field was off, and that despite the fact that the ionization was several times more intense than in the case of the drop of Table I. Positive ions too were being caught at almost every trip down under gravity. (The strong negative charge on the drop was maintained by forcing on negative ions by the field as explained above.)

V. POSITIVE AND NEGATIVE ELECTRONS EXACTLY EQUAL

The idea has at various times been put forth in connection with attempts to explain chemical and cohesive forces from the standpoint of electrostatic attractions that the positive and negative charges in a so-called neutral atom may not after all be exactly equal, in other words, that there is really no such thing as an entirely neutral atom or molecule. As a matter of fact, it is difficult to find decisive tests of this hypothesis. The present experiments, however, make possible the following sort of test. I loaded a given drop first with negative electrons and took ten or twelve observations of rise and fall, then with the aid of X-rays, by the method indicated in the last section, I reversed the sign of the charge on the drop and took a corresponding number of observations of rise and fall, and so continued observing first the value of the negative electron and then that of the positive. Table VIII shows a set of such observations taken in air with a view to subjecting this point to as rigorous a test as possible. Similar, though not quite so elaborate, observations have been made in hydrogen with the same result. The table shows in the first column the sign of the charge; in the second the successive values of the time of fall under gravity; in the third the successive times of rise in the field F; in the fourth the number of electrons carried by the drop for each value of t_F; and in the fifth the number, characteristic of this drop, which is proportional to the charge of one electron. This number is obtained precisely as in the two preceding tables by finding the greatest common divisor of the successive values of $(v_1 + v_2)$ and then multiplying this by an arbitrary constant which has nothing to do with the present experiment and hence need not concern us here (see chap. v).

It will be seen that though the times of fall and of rise, even when the same number of electrons is carried by the drop, change a trifle because of a very slight evaporation and also because of the fall in the potential

TABLE VIII

Sign of Drop	t_g Sec.	t_F Sec.	n	e
−	63.118 63.050 63.186 63.332 62.328	41.728 41.590	8	$e_1 = 6.713$
	62.728 62.926 62.900 63.214	25.740 25.798 25.510 25.806	11	
	Mean = 62.976			
+	63.538 63.244	22.694 22.830	12	$e_1 = 6.692$
	63.114 63.242 63.362	25.870 25.876 25.484	11	
	63.136 63.226 63.764 63.280 63.530 63.268	10.830 10.682 10.756 10.778 10.672 10.646	22	
	Mean = 63.325			
+	63.642 63.020 62.820	71.664 71.248	6	$e_1 = 6.702$
	63.514 63.312 63.776 63.300	52.668 52.800 52.496 52.860	7	
	63.156 63.126	71.708	6	
	Mean = 63.407			

TABLE VIII—*Continued*

Sign of Drop	t_g Sec.	t_F Sec.	n	e
—	63.228 63.294 63.184	42.006 41.920 42.108 }	8	
	63.260 63.478 63.074 63.306	53.210 52.922 53.034 53.438 }	7	$e_1 = 6.686$
	63.414 63.450 63.446 63.556	12.888 12.812 12.748 12.824 }	19	
	Mean = 63.335			

Duration of experiment 1 hr. 40 min.   Mean $e+ = 6.697$
Initial volts = 1723.5   Mean $e- = 6.700$
Final volts = 1702.1
Pressure   =   53.48 cm.

of the battery, yet the mean value of the positive electron, namely, 6.697, agrees with the mean value of the negative electron, namely, 6.700, to within less than 1 part in 2,000. Since this is about the limit of the experimental error (the probable error by least squares is 1 part in 1,500), *we may with certainty conclude that there are no differences of more than this amount between the values of the positive and negative electrons.* This is the best evidence I am aware of for the exact neutrality of the ordinary molecules of gases. Such neutrality, if it is actually exact, would seem to preclude the possibility of explaining gravitation as a result of electrostatic forces of any kind. The electromagnetic effect of moving charges might, however, still be called upon for this purpose.

VI. RESISTANCE OF MEDIUM TO MOTION OF DROP THROUGH IT THE SAME WHEN DROP IS CHARGED AS WHEN UNCHARGED

A second and equally important conclusion can be drawn from Table VIII.   It will be seen from the column headed "n" that during the whole of the time corresponding to the observations in the third group from the top the drop carried either 6 or 7 electrons, while, during the last half of the time corresponding to the observations in the second group from the top, it carried three times as many, namely, 22 electrons. Yet the mean times of fall under gravity in the two groups agree to within about one part in one thousand. The time of fall corresponding to the heavier charge happens in this case to be the smaller of the two. We may conclude, therefore, that *in these experiments the resistance which the medium offers to the motion of a body through it is not sensibly increased when the body becomes electrically charged.   This demonstrates experimentally the exact validity for this work of the assumption made on p. 70 that the velocity of the drop is strictly proportional to the force acting upon it, whether it is charged or uncharged.*

The result is at first somewhat surprising since, according to Sutherland's theory of the small ion, the small mobility or diffusivity of charged molecules, as compared with uncharged, is due to the additional resistance which the medium offers to the motion through it of a charged molecule.   This additional resistance is due to the fact that the charge on a molecule drags into collision with it more molecules than would otherwise hit it.   But with oil drops of the sizes here used

$(a = 50 \times 10^{-6})$ the total number of molecular collisions against the surface of the drop is so huge that even though the small number of charges on it might produce a few more collisions, their number would be negligible in comparison with the total number. At any rate the experiment demonstrates conclusively that the charges on our oil drops do not influence the resistance of the medium to the motion of the drop. This conclusion might also have been drawn from the data contained in Table VI. The evidence for its absolute correctness has been made more convincing still by a comparison of drops which carried but 1 charge and those which carried as many as 68 unit charges. Further, I have observed the rate of fall under gravity of droplets which were completely discharged, and in every case that I have ever tried I have found this rate precisely the same, within the limits of error of the time measurements, as when it carried 8 or 10 unit charges.

VII.   DROPS ACT LIKE RIGID SPHERES

It was of very great importance for the work, an account of which will be given in the next chapter to determine whether the drops ever suffer—either because of their motion through a resisting medium, or because of the electrical field in which they are placed—any appreciable distortion from the spherical form which a freely suspended liquid drop must assume. The complete experimental answer to this query is contained in the agreement of the means at the bottom of the last and the third from the last columns in Table VI and in similar agreements shown in many other tables, which may be found in the original articles.[1]   Since $\frac{1}{t_g}$ is in this experiment large compared to $\frac{1}{t_F}$, the value of the greatest common divisor at the bottom of the last column of Table VI is determined almost wholly by the rate of fall of the particle under gravity when there is no field at all between the plates, while the velocity at the bottom of the third from the last column is a difference between two velocities in a strong electrical field.   If, therefore, the drop were distorted by the electrical field, so that it exposed a larger surface to the resistance of the medium than when it had the spherical form, the velocity due to a given force, that is, the velocity given at the bottom of the third from the last column, would be less than that found at the bottom of the last column, which corresponds to motions when the drop certainly was spherical.

Furthermore, if the drops were distorted by their motion through the medium, then this distortion would be greater for high speeds than for low, and consequently the numbers in the third from the last column would be consistently larger for high speeds than for low.   No such variation of these numbers with speed is apparent either in Table VI or in other similar tables.

We have then in the exactness and invariableness of the multiple relations shown by successive differences in speed and the successive sums of the speeds in the third from the last and the last columns of Table VI complete experimental proof that in this work the droplets act under all circumstances like undeformed spheres.   It is of interest that Professor Hadamard,[2] of the University of

[1] *Phys. Rev.*, Series 1, XXXII (1911), 349; Series 2, II (1913), 109.
[2] *Comptes rendus* (1911), 1735.

Paris, and Professor Lunn,[1] of the University of Chicago, have both shown from theoretical considerations that this would be the case with oil drops as minute as those with which these experiments deal, so that the conclusion may now be considered as very firmly established both by the experimentalist and the theorist.

[1] *Phys. Rev.*, XXXV (1912), 227.

CHAPTER V

THE EXACT EVALUATION OF e

I. DISCOVERY OF THE FAILURE OF STOKES'S LAW

Although complete evidence for the atomic nature of electricity is found in the fact that all of the charges which can be placed upon a body as measured by the sum of speeds $v_1 + v_2$, and all the changes of charge which this body can undergo as measured by the differences of speed $(v_2' - v_2)$ are invariably found to be exact multiples of a particular speed, yet there is something still to be desired if we must express this greatest common divisor of all the observed series of speeds merely as a velocity which is a characteristic constant of each particular drop but which varies from drop to drop. We ought rather to be able to reduce this greatest common divisor to electrical terms by finding the proportionality factor between speed and charge, and, that done, we should, of course, expect to find that the charge came out a universal constant independent of the size or kind of drop experimented upon. The attempt to do this by the method which I had used in the case of the water drops (p. 55), namely, by the assumption of Stokes's Law, heretofore taken for granted by all observers, led to the interesting discovery that this law is not valid.[1]   Accord-

[1] Cunningham (*Proc. Roy. Soc.*, LXXXIII [1910], 357) and the author came independently to the conclusion as to the invalidity of Stokes's Law, he from theoretical considerations developed at about the same time, I from my experimental work.

ing to this law the rate of fall of a spherical drop under gravity, namely, v_1, is given by

$$v_1 = \frac{2ga^2}{9\eta}(\sigma - \rho) \quad \ldots\ldots\ldots\ldots \quad \ldots(11)$$

in which η is the viscosity of the medium, a the radius and σ the density of the drop, and ρ the density of the medium. This last quantity was neglected in (6), p. 55, because, with the rough measurements there possible, it was useless to take it into account, but with our oil drops in dry air all the other factors could be found with great precision.

When we assume the foregoing equation of Stokes and combine it with equation (5) on p. 55, an equation whose exact validity was proved experimentally in the last chapter, we obtain, after substitution of the purely geometrical relation $m = \frac{4\pi}{3}a^3 \ (\sigma - \rho)$, the following expression for the charge e_n carried by a drop loaded with n electrons which we will assume to have been counted by the method described:

$$e_n = \frac{4\pi}{3}\left(\frac{9\eta}{2}\right)^{\frac{3}{2}}\left(\frac{1}{g(\sigma-\rho)}\right)^{\frac{1}{2}}\frac{(v_1+v_2)v_1^{\frac{1}{2}}}{F} \quad \ldots\ldots\ldots(12)$$

According to this equation the elementary charge e_1 should be obtained by substituting in this the greatest common divisor of all the observed series of values of (v_1+v_2) or of $(v_2'-v_2)$. Thus, if we call this $(v_1+v_2)_0$, we have

$$e_1 = \frac{4\pi}{3}\left(\frac{9\eta}{2}\right)^{\frac{3}{2}}\left(\frac{1}{g(\sigma-\rho)}\right)^{\frac{1}{2}}\frac{(v_1+v_2)_0 v_1^{\frac{1}{2}}}{F} \quad \ldots\ldots\ldots(13)$$

But when this equation was tested out upon different drops, although it yielded perfectly concordant results so long as the different drops all fell with about the same speed, when drops of different speeds, and, therefore, of different sizes, were used, the values of e_1 obtained were consistently larger the smaller the velocity under gravity. For example, e_1 for one drop for which $v_1 = .01085$ cm. per second came out 5.49×10^{-10}, while for another of almost the same speed, namely, $v_1 = .01176$, it came out 5.482; but for two drops whose speeds were five times as large, namely, $.0536$ and $.0553$, e_1 came out 5.143 and 5.145, respectively. This could mean nothing save that Stokes's Law did not hold for drops of the order of magnitude here used, something like $a = .0002$ cm. (see Section IV below), and it was surmised that the reason for its failure lay in the fact that the drops were so small that they could no longer be thought of as moving through the air as they would through a continuous homogeneous medium, which was the situation contemplated in the deduction of Stokes's Law. This law ought to begin to fail as soon as the inhomogeneities in the medium—i.e., the distances between the molecules—began to be at all comparable with the dimensions of the drop. Furthermore, it is easy to see that as soon as the holes in the medium begin to be comparable with the size of the drop, the latter must begin to increase its speed, for it may then be thought of as beginning to reach the stage in which it can fall freely through the holes in the medium. This would mean that the observed speed of fall would be more and more in excess of that given by Stokes's Law the smaller the drop became. But the apparent value of the electronic charge, namely, e_1, is seen from equation (13) to vary directly with the speed $(v_1 + v_2)_0$ imparted by a given force. Hence e_1 should come out larger and larger the smaller the radius of the drop, that is, the smaller its velocity under gravity. Now, this was exactly the behavior shown consistently by all the oil drops studied. Hence it looked as though we had discovered, not merely the failure of Stokes's Law, but also the line of approach by means of which it might be corrected.

In order to be certain of our ground, however, we were obliged to initiate a whole series of new and somewhat elaborate experiments.

These consisted, first, in finding very exactly what is the coefficient of viscosity of air under conditions in which it may be treated as a homogeneous medium, and, second, in finding the limits within which Stokes's Law may be considered valid.

II. THE COEFFICIENT OF VISCOSITY OF AIR

The experiments on the coefficient of viscosity of air were carried out in the Ryerson Laboratory by Dr. Lachen Gilchrist,[1] and Dr. I. M. Rapp.[2] Dr. Gilchrist used a method which was in many respects new and which may fairly be said to be freer from theoretical uncertainties than any method which has ever been used. He estimated that his results should not be in error by more than .1 or .2 of 1 per cent. Dr. Rapp used a form of the familiar capillary-tube method, but under conditions which seemed to adapt it better to an absolute evaluation of η for air than capillary-tube arrangements have ordinarily been.

[1] *Phys. Rev.*, N.S., I, (1913), 124.
[2] *Phys. Rev.*, N.S., II (1913), 363.

These two men, as the result of measurements which were in progress for more than two years, obtained final means which were in very close agreement with one another as well as with the most careful of preceding determinations. It will be seen from Table IX that

TABLE IX

η_{23} for Air

.00018227	Rapp, Capillary-tube method, 1913 (*Phys. Rev.*, II, 363).
.00018257	Gilchrist, Constant deflection method, 1913 (*Phys. Rev.*, I, 124).
.00018229	Hogg, Damping of oscillating cylinders, 1905 (*Proc. Am. Acad.*, XL, 611).
.00018258	Tomlinson, Damping of Swinging Pendulum, 1886 (*Phil. Trans.*, CLXXVII, 767).
.00018232	Grindley and Gibson, Flow through pipe, 1908 (*Proc. Roy. Soc.*, LXXX, 114).

Mean... .00018240

every one of the five different methods which have been used for the absolute determination of η for air leads to a value that differs by less than one part in one thousand from the following mean value, $\eta_{23} = .00018240$. It was concluded, therefore, that we could depend upon the value of η for the viscosity of air under the conditions of our experiment to at least one part in one thousand. Late in 1917 Dr. E. Harrington[1] improved still further the apparatus designed by Dr. Gilchrist and the author and made with it in the Ryerson Laboratory a determination of η which is, I think, altogether unique in its reliability and precision. I give to it alone greater

[1] *Phys. Rev.*, December, 1916.

weight than to all the other work of the past fifty years in this field taken together.   The final value is

$$\eta_{23} = .00018226$$

and the error can scarcely be more than one part in two thousand.

III.   LIMITS OF VALIDITY OF STOKES'S LAW

In the theoretical derivation of Stokes's Law the following five assumptions are made: (1) that the inhomogeneities in the medium are small in comparison with the size of the sphere; (2) that the sphere falls as it would in a medium of unlimited extent; (3) that the sphere is smooth and rigid; (4) that there is no slipping of the medium over the surface of the sphere; (5) that the velocity with which the sphere is moving is so small that the resistance to the motion is all due to the viscosity of the medium and not at all due to the inertia of such portion of the media as is being pushed forward by the motion of the sphere through it.

If these conditions were all realized then Stokes's Law ought to hold.   Nevertheless, there existed up to the year 1910 no experimental work which showed that actual experimental results may be accurately predicted by means of the unmodified law, and Dr. H. D. Arnold accordingly undertook in the Ryerson Laboratory to test how accurately the rates of fall of minute spheres through water and alcohol might be predicted by means of it.

His success in these experiments was largely due to the ingenuity which he displayed in producing accurately spherical droplets of rose-metal.   This metal melts at about 82° C. and is quite fluid at the temperature of boiling water. Dr. Arnold placed some of this metal in a glass tube drawn to form a capillary at one end and suspended the whole of the capillary tube in a glass tube some 70 cm. long and 3 cm. in diameter. He then filled the large tube with water and applied heat in such a way that the upper end was kept at about 100° C., while the lower end was at about 60°. He then forced the molten metal, by means of compressed air, out through the capillary into the hot water. It settled in the form of spray, the drops being sufficiently cooled by the time they reached the bottom to retain their spherical shape. This method depends for its success on the relatively slow motion of the spheres and on the small temperature gradient of the water through which they fall. The slow and uniform cooling tends to produce homogeneity of structure, while the low velocities allow the retention of very accurately spherical shape. In this way Dr. Arnold obtained spheres of radii from .002 cm. to .1 cm., which, when examined under the microscope, were found perfectly spherical and practically free from surface irregularities. He found that the slowest of these drops fell in liquids with a speed which could be computed from Stokes's Law with an accuracy of a few tenths of 1 per cent, and he determined experimentally the limits of speed through which Stokes's Law was valid.

Of the five assumptions underlying Stokes's Law, the first, third, and fourth were altogether satisfied in Dr. Arnold's experiment. The second assumption he found sufficiently realized in the case of the very smallest drops which he used, but not in the larger ones. The question, however, of the effect of the walls of the vessel upon the motion of drops through the liquid contained in the vessel had been previously studied with great ability by Ladenburg,[1] who, in working with an exceedingly viscous oil, namely Venice turpentine, obtained a formula by which the effects of the wall on the motion might be eliminated. If the medium is contained in a cylinder of circular cross-section of radius R and of length L, then, according to Ladenburg, the simple Stokes formula should be modified to read

$$V = \frac{2}{9} \frac{ga^2(\sigma - \rho)}{\eta\left(1 + 2.4\frac{a}{R}\right)\left(1 + 3.1\frac{a}{L}\right)} .$$

Arnold found that this formula held accurately in all of his experiments in which the walls had any influence on the motion. Thus he worked under conditions under which all of the first four assumptions underlying Stokes's Law were taken care of. This made it possible for him to show that the law held rigorously when the fifth assumption was realized, and also to find by experiment the limits within which this last assumption might be considered as valid. Stokes had already found from theoretical considerations[2] that the law would not hold unless the radius of the sphere were small in comparison with $\frac{\eta}{v\rho}$, in which ρ is the density of the medium, η its viscosity, and v the velocity of the sphere. This radius is called the critical radius. But it was not known how near it was possible to approach to the critical radius. Arnold's experiments showed that the inertia of the medium has no appreciable effect upon the rate of

[1] *Ann. der Phys.*, XXII (1907), 287; XXIII (1908), 447.

[2] *Math. and Phys. Papers*, III, 59.

motion of a sphere so long as the radius of that sphere is less than .6 of the critical radius.

Application of this result to the motion of our oil drops established the fact that even the very fastest drops which we ever observed fell so slowly that not even a minute error could arise because of the inertia of the medium. This meant that the fifth condition necessary to the application of Stokes's Law was fulfilled. Furthermore, our drops were so small that the second condition was also fulfilled, as was shown by the work of both Ladenburg and Arnold. The third condition was proved in the last chapter to be satisfied in our experiments. Since, therefore, Arnold's work had shown very accurately that Stokes's Law does hold when all of the five conditions are fulfilled, the problem of finding a formula for replacing Stokes's Law in the case of our oil-drop experiments resolved itself into finding in just what way the failure of assumptions 1 and 4 affected the motion of these drops.

IV. CORRECTION OF STOKES'S LAW FOR INHOMOGENEITIES IN THE MEDIUM

The first procedure was to find how badly Stokes's Law failed in the case of our drops. This was done by plotting the apparent value of the electron e_1 against the observed speed under gravity. This gave the curve shown in Fig. 4, which shows that though for very small speeds e_1 varies rapidly with the change in speed, for speeds larger than that corresponding to the abscissa marked 1,000 there is but a slight dependence of e_1 on speed. This abscissa corresponds to a speed of .1 cm. per second. We may then conclude that for drops which are large enough to fall at a rate of 1 cm. in ten seconds or faster, Stokes's Law needs but a small correction, because of the inhomogeneity of the air.

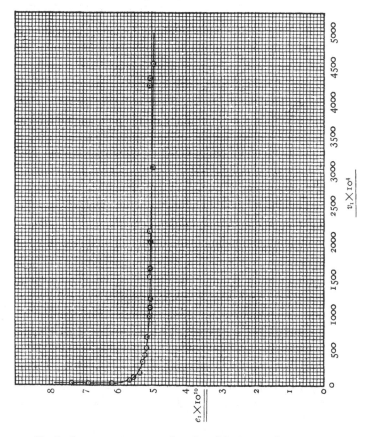

To find an exact expression for this correction we may proceed as follows: The average distance which a gas molecule goes between two collisions with its neighbors, a quantity well known and measured with some approach to precision in physics and called "the mean free path" of a gas molecule, is obviously a measure of the size of the holes in a gaseous medium. When Stokes's Law begins to fail as the size of the drops diminish, it must be because the medium ceases to be homogeneous, as looked at from the standpoint of the drop, and this means simply that the radius of the drop has begun to be comparable with the mean size of the holes—a quantity which we have decided to take as measured by the mean free path l. The increase in the speed of fall over that given by Stokes's Law, when this point is reached, must then be some function of $\dfrac{l}{a}$. In other words, the correct expression for the speed v_1 of a drop falling through a gas, instead of being

$$v_1 = \frac{2}{9}\frac{ga^2}{\eta}(\sigma - \rho) ,$$

as Arnold showed that it was when the holes were negligibly small—as the latter are when the drop falls through a liquid—should be of the form

$$v_1 = \frac{2}{9}\frac{ga^2}{\eta}(\sigma - \rho)\left(1 + f\frac{l}{a}\right)\dots\dots\dots\dots(14)$$

If we were in complete ignorance of the form of the function f we could still express it in terms of a series of undetermined constants A, B, C, etc., thus

$$f = A\frac{l}{a} + B\left(\frac{l}{a}\right)^2 + C\left(\frac{l}{a}\right)^3 , \text{ etc.,}$$

and so long as the departures from Stokes's Law were small as Fig. 4 showed them to be for most of our drops, we could neglect the second-order terms in $\dfrac{l}{a}$ and have therefore

$$v_1 = \frac{2}{9}\frac{ga^2}{\eta}(\sigma-\rho)\left(1+A\frac{l}{a}\right) \ldots \ldots \ldots \ldots (15)$$

Using this corrected form of Stokes's Law to combine with (9) (p. 20), we should obviously get the charge e_n in just the form in which it is given in (13), save that wherever a velocity appears in (13) we should now have to insert in place of this velocity $\dfrac{v}{1+A\dfrac{l}{a}}$. And since the velocity of the drop appears in the $3/2$ power in (13), if we denote now by e the absolute value of the electron and by e_1, as heretofore, the apparent value obtained from the assumption of Stokes's Law, that is, from the use of (13), we obtain at once

$$e = \frac{e_1}{\left(1+A\dfrac{l}{a}\right)^{\frac{3}{2}}} \ldots \ldots \ldots \ldots \ldots (16)$$

In this equation e_1 can always be obtained from (13), while l is a known constant, but e, A, and a are all unknown. If a can be found our observations permit at once of the determination of both e and A, as will be shown in detail under Section VI (see p. 105).

However, the possibility of determining e if we know a can be seen in a general way without detailed analysis. For the determination of the radius of the drop is equivalent to finding its weight, since its density is known. That we can find the charge on the drop as soon as we can determine its weight is clear from the simple consideration that the velocity under gravity is proportional to its weight, while the velocity in a given electrical field is proportional to the charge which it carries. Since we measure these two velocities directly, we can obtain either the weight, if we know the charge, or the charge, if we know the weight. (See equation 9, p. 70.)

V. WEIGHING THE DROPLET

The way which was first used for finding the weight of the drop was simply to solve Stokes's uncorrected equation (11) (p. 91) for a in the case of each drop. Since the curve of Fig. 4 shows that the departures from Stokes's Law are small except for the extremely slow drops, and since a appears in the second power in (11), it is clear that, if we leave out of consideration the very slowest drops, (11) must give us very nearly the correct values of a. We can then find the approximate value of A by the method of the next section, and after it is found we can solve (15) for the correct value of a. This is a method of successive approximations which theoretically yields a and A with any desired degree of precision. As a matter of fact the whole correction term, $A \frac{l}{a}$ is a small one, so that it is never necessary to make more than two approximations to obtain a with much more precision than is needed for the exact evaluation of e.

As soon as e was fairly accurately known it became possible, as indicated above, to make a direct weighing of extraordinarily minute bodies with great certainty and with a very high degree of precision. For we have already shown experimentally that the equation

$$\frac{v_1}{v_2} = \frac{mg}{Fe_n - mg} \quad \dots\dots\dots\dots\dots (17)$$

is a correct one and it involves no assumption whatever as to the shape, or size, or material of the particle. If we solve this equation for the weight mg of the particle we get

$$mg = Fe_n\frac{v_1}{v_1+v_2}\dots\dots\dots\dots\dots(18)$$

In this equation e_n is known with the same precision as e, for we have learned how to count n. It will presently be shown that e is probably now known with an accuracy of one part in a thousand, hence mg can now be determined with the same accuracy for any body which can be charged up with a counted number n of electrons and then pulled up against gravity by a known electrical field, or, if preferred, simply balanced against gravity after the manner used in the water-drop experiment and also in part of the oil-drop work.[1] *This device is simply an electrical balance in place of a mechanical one, and it will weigh accurately and easily to one ten-billionth of a milligram.*

Fifty years ago it was considered the triumph of the instrument-maker's art that a balance had been made so sensitive that one could weigh a piece of paper, then write his name with a hard pencil on the paper and determine the difference between the new weight and the old—that is, the weight of the name. This meant determining a weight as small as one-tenth or possibly one-hundredth of a milligram (a milligram is about 1/30,000 of an ounce). In about the year 1912 Ramsay and Spencer, in London, by constructing a balance entirely out of very fine quartz fibers and placing it in a vacuum, succeeded in weighing objects as small

[1] See *Phil. Mag.*, XIX (1910), 216; XXI (1911), 757.

as one-millionth of a milligram, that is, they pushed the limit of the weighable down about ten thousand times. The work which we are now considering pushed it down at least ten thousand times farther and made it possible to weigh accurately bodies so small as not to be visible at all to the naked eye. For it is only necessary to float such a body in the air, render it visible by reflected light in an ultra-microscope arrangement of the sort we were using, charge it electrically by the capture of ions, count the number of electrons in its charge by the method described, and then vary the potential applied to the plates or the charge on the body until its weight is just balanced by the upward pull of the field. The weight of the body is then exactly equal to the product of the known charge by the strength of the electric field. We made all of our weighings of our drops and the determination of their radii in this way as soon as we had located e with a sufficient degree of precision to warrant it.[1] Indeed, even before e is very accurately known it is possible to use such a balance for a fairly accurate evaluation of the radius of a spherical drop. For when we replace m in (18) by $4/3\pi a^3(\sigma-\rho)$ and solve for a we obtain

$$a = \sqrt[3]{\frac{3Fe_n}{4\pi g(\sigma-\rho)}\frac{v_1}{v_1+v_2}} \dots\dots\dots\dots(19)$$

The substitution in this equation of an approximately correct value of e yields a with an error but one-third as great as that contained in the assumed value of e, for a is seen from this equation to vary as the cube root of e. This is the method which, in view of the accurate evalua-

[1] *Phys. Rev.*, II (1913), 117. This paper was read before the Deutsche physikalische Gesellschaft in Berlin in June, 1912.

tion of e, it is now desirable to use for the determination of the weight or dimensions of any minute body, for the method is quite independent of the nature of the body or of the medium in which it is immersed. Indeed, it constitutes as direct and certain a weighing of the body as though it were weighed on a mechanical balance.

VI.   THE EVALUATION OF e AND A

With e_1 and $\dfrac{l}{a}$ known, we can easily determine e and A from the equation

$$e=\frac{e_1}{\left(1+A\dfrac{l}{a}\right)^{\frac{2}{3}}}$$

for if we write this equation in the form

$$e^{\frac{2}{3}}\left(1+A\frac{l}{a}\right)=e_1^{\frac{2}{3}} \ldots\ldots\ldots\ldots\ldots(20)$$

and then plot the observed values of e_1 as ordinates and the corresponding values of $\dfrac{l}{a}$ as abscissae we should get a straight line, provided our corrected form of Stokes's Law (15) (p. 101) is adequate for the correct representation of the phenomena of fall of the droplets within the range of values of $\dfrac{l}{a}$ in which the experiments lie. If no such linear relation is found, then an equation of the form of (15) is not adequate for the description of the phenomena within this range. As a matter of fact, a linear relation was found to exist for a much wider range of values of $\dfrac{l}{a}$ than was anticipated would be the case. The

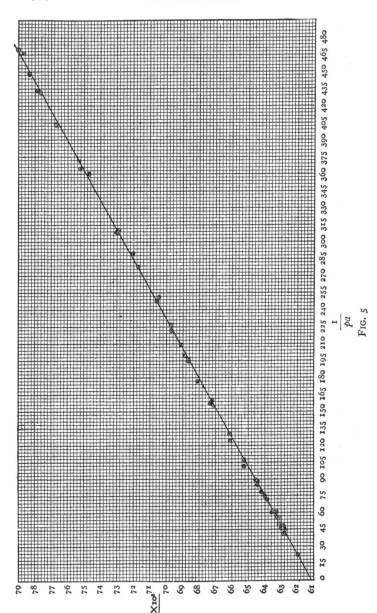

FIG. 5

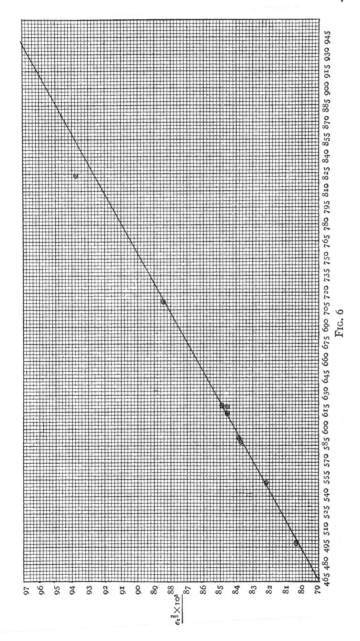

Fig. 6

intercept of this line on the axis of ordinates, that is, the value of e_1 when $\dfrac{l}{a} = 0$ is seen from (20) to be $e^{\frac{2}{3}}$, and we have but to raise this to the 3/2 power to obtain the absolute value of e. Again, A is seen from (20) to be merely the slope of this line divided by the intercept on the $e_1^{\frac{2}{3}}$ axis.

In order to carry this work out experimentally it is necessary to vary $\dfrac{l}{a}$ and find the corresponding values of e_1. This can be done in two ways. First, we may hold the pressure constant and choose smaller and smaller drops with which to work, or we may work with drops of much the same size but vary the pressure of the gas in which our drops are suspended, for the mean free path l is evidently inversely proportional to the pressure.

Both procedures were adopted, and it was found that a given value of e_1 always corresponded to a given value of $\dfrac{l}{a}$, no matter whether l was kept constant and a reduced to, say, one-tenth of its first value, or a kept about the same and l multiplied tenfold. The result of one somewhat elaborate series of observations which was first presented before the Deutsche physikalische Gesellschaft in June, 1912, and again before the British Association at Dundee in September, 1912,[1] is shown in Figs. 5 and 6. The numerical data from which these curves are plotted are given fairly fully in Table IX. It will be seen that this series of observations embraces a study of 58 drops. These drops represent all of those studied for 60 consecutive days, no single one being

[1] *Phys. Rev.*, II (1913), 136.

TABLE IX

No.	Tem. °C.	P.D. (Volts)	t_g (Sec.)	v_1 cm./sec.	$(v_1+v_2)_e$	n	$a \times 10^5$ cm.	p (cm. Hg)	$\dfrac{1}{pa}$	$\dfrac{i}{a}$	$e_1 \times 10^{10}$	$e_1^3 \times 10^3$	$e^3 \times 10^3$
1	23.00	5.168	4.363	.2357	.003293	77–102	58.56	75.80	22.52	.01615	4.877	61.90	61.14
2	22.80	5.120	8.492	.1202	.004670	27–36	32.64	75.00	40.85	.02933	4.981	62.82	61.20
3	23.46	5.100	9.995	.1032	.004006	22–27	30.29	73.71	44.88	.03122	4.971	62.75	61.04
4	22.85	5.163	10.758	.09489	.005111	18–36	28.94	75.20	45.92	.03288	5.001	63.00	61.24
5	23.08	5.072	10.663	.09575	.005176	20–30	29.14	73.25	40.85	.03353	4.962	62.82	61.13
6	22.82	5.085	11.880	.08584	.005497	17–24	27.54	75.62	48.11	.03437	4.991	62.93	61.09
7	23.79	5.090	11.950	.08368	.005480	16–19	27.57	75.10	48.44	.03466	4.981	62.82	61.07
8	23.50	5.158	12.540	.08141	.005623	16–19	26.90	75.30	49.52	.03544	5.016	63.12	61.09
9	22.87	5.139	13.562	.07375	.005962	19–23	25.71	75.00	51.73	.03702	5.016	63.13	61.23
10	23.25	5.015	13.380	.06641	.006174	13–22	24.31	76.27	54.09	.03831	5.010	63.08	61.02
11	23.01	5.066	15.193	.06720	.006416	11–14	24.36	73.90	55.52	.03973	5.015	63.12	61.10
12	23.00	5.080	15.985	.06375	.006416	12–16	23.70	75.14	56.15	.04018	5.028	63.24	61.06
13	23.00	5.024	15.695	.05463	.006873	9–15	21.91	76.06	59.94	.04290	5.043	63.35	61.21
14	23.09	5.077	18.730	.05451	.006888	8–16	21.85	75.28	60.78	.04348	5.064	63.53	61.07
15	23.85	5.078	18.959	.05274	.006966	8–18	21.78	75.24	61.03	.04368	5.040	63.33	61.21
16	23.70	5.103	18.738	.05449	.007005	9–16	21.87	74.68	61.33	.04390	5.065	63.54	61.21
17	23.06	5.060	18.415	.05545	.006890	9–18	22.06	73.47	61.69	.04411	5.054	63.43	61.00
18	22.95	5.093	26.130	.03907	.008339	5–13	18.45	75.54	71.74	.05134	5.098	63.82	61.08
19	22.91	5.033	28.568	.03570	.008651	5–9	17.63	75.87	74.77	.05350	5.120	64.00	61.12
20	23.00	5.094	9.480	.10772	.005058	23–32	30.54	41.77	78.40	.05612	5.145	64.22	61.23
21	23.08	5.018	35.253	.02803	.005660	4–11	15.80	74.32	85.08	.06089	5.166	64.36	61.11
22	23.22	5.005	40.542	.02515	.010332	3–9	14.75	76.42	88.70	.06350	5.168	64.40	61.01
23	22.76	5.008	39.900	.02554	.005810	3–8	14.85	75.40	89.35	.06395	5.190	64.59	61.18
24	23.16	5.050	12.466	.08189	.005596	15–28	26.41	37.19	101.8	.07283	5.209	64.54	61.35
25	22.98	5.066	15.157	.06737	.005399	12–17	24.01	38.95	107.2	.07660	5.278	65.28	61.20
26	23.20	4.572	7.875	.12908	.004324	33–40	33.07	24.33	124.4	.08892	5.379	66.06	61.31
27	23.18	4.570	9.408	.1085	.004730	23–29	30.23	25.37	130.4	.09330	5.381	66.16	61.18
28	23.00	5.145	84.270	.01211	.01595	1–4	4.69	75.83	156.8	.09332	5.379	66.14	61.16
29	22.99	5.973	23.223	.04303	.008488	6–12	19.06	33.47		.1117	5.529	67.36	61.37

Mean = 61.120

Mean = 61.138

TABLE IX—Continued

No.	Tem. °C.	P.D. (Volts)	t_g (Sec.)	v_1 cm./sec.	$(v_1+v_2)_a$	n	$a\times10^5$ cm.	ρ (cm. Hg)	$\dfrac{1}{\rho a}$	$\dfrac{l}{a}$	$e_1\times10^{10}$	$e_1^{\frac{2}{3}}\times10^5$	$e^{\frac{2}{3}}\times10^4$ Mean=61.138
30	23.19	5,090	26.830	.03801	.009111	5–10	17.77	35.18	160.2	.1147	5.507	67.18	61.06
31	22.89	5,098	38.479	.02649	.011180	3–5	14.71	36.51	176.5	.1263	5.621	68.12	61.38
32	23.06	5,070	14.060	.07246	.006762	12–17	24.29	21.12	195.0	.1394	5.692	68.07	61.22
33	23.07	4,582	18.229	.05601	.006981	10–13	21.33	23.86	195.6	.1405	5.687	68.64	61.13
34	23.06	5,061	38.010	.02682	.011205	3–8	14.72	34.01	199.8	.1429	5.714	68.84	61.20
35	23.00	4,246	0.265	.11032	.004653	27–34	29.84	16.00	209.5	.1499	5.739	69.07	61.07
36	22.91	4,236	9.879	.10340	.004863	24–28	28.74	15.67	222.0	.1589	5.820	69.71	61.23
37	23.06	4,236	12.040	.08496	.005362	18–24	26.27	16.75	227.5	.1625	5.821	69.72	61.03
38	22.94	2,556	10.657	.09581	.003109	32–43	27.49	14.70	247.5	.1771	5.935	70.61	61.16
39	23.00	5,054	19.950	.05115	.008370	8–15	20.12	19.73	251.8	.1802	5.910	70.41	60.79
40	23.09	5,058	21.130	.04830	.008865	7–9	18.38	18.54	278.3	.1993	6.076	71.72	61.09
41	23.05	5,062	24.008	.04254	.009490	6–8	18.16	19.01	289.6	.2073	6.110	72.03	60.97
42	22.94	4,238	18.347	.05564	.007110	9–17	20.60	15.72	308.8	.2210	6.224	73.04	61.24
43	23.18	3,254	13.909	.07340	.004729	16–28	23.70	13.55	311.0	.2227	6.214	72.83	61.00
44	23.04	4,231	29.114	.03503	.009273	5–9	16.16	17.17	360.0	.2579	6.466	74.77	60.95
45	22.97	3,377	29.776	.03425	.007430	5–12	15.90	17.27	304.2	.2006	6.537	75.30	61.39
46	22.81	3,401	25.909	.03937	.007311	6–19	16.90	14.68	403.3	.2886	6.719	76.71	61.30
47	22.80	2,550	12.891	.07921	.003935	18–42	23.80	9.70	432.8	.3097	6.841	77.66	61.13
18	23.02	2,559	32.326	.03150	.006286	7–14	15.01	15.35	433.8	.3104	6.866	77.85	61.28
49	23.45	3,370	14.983	.06815	.011353	8–9	22.00	10.10	448.8	.3221	6.936	78.36	61.22
50	23.48	2,535	11.059	.08757	.003783	25–30	24.88	8.60	466.7	.3340	6.978	78.07	60.85
51	22.98	2,539	10.924	.09346	.003615	27–34	25.69	8.26	470.7	.3368	7.024	79.02	61.36
52	23.16	3,351	50.400	.02021	.010775	2–6	11.83	16.95	498.5	.3568	7.210	80.40	61.13
53	23.46	2,451	33.379	.03055	.006623	5–10	14.39	12.61	551.8	.3945	7.470	82.19	61.18
54	22.90	2,533	19.227	.05347	.005314	11–17	18.87	9.03	587.8	.4112	7.661	83.73	61.22
55	23.27	2,546	24.254	.04206	.006041	9–18	16.72	10.11	591.5	.4233	7.672	83.82	61.11
56	23.12	1,700	5.058	.20256	.001861	117–136	30.53	4.46	614.2	.4396	7.777	84.57	61.11
57	23.12	2,321	15.473	.06509	.004360	18–24	20.85	7.74	619.7	.4435	7.774	84.54	60.87
58	23.03	3,388.5	24.33	.04196	.005183	6–10	16.62	9.070	620.2	.4439	7.810	84.83	61.14

Mean of all numbers in last column = 61.138
Mean of first 23 numbers = 61.120

omitted. They represent a thirty-fold variation in $\dfrac{l}{a}$ (from .016, drop No. 1, to .444, drop No. 58), a seventeen-fold variation in the pressure p (from 4.46 cm., drop No. 56, to 76.27 cm., drop No. 10), a twelvefold variation in a (from 4.69×10^{-5} cm., drop No. 28, to 58.56×10^{-5} cm., drop No. 1), and a variation in the

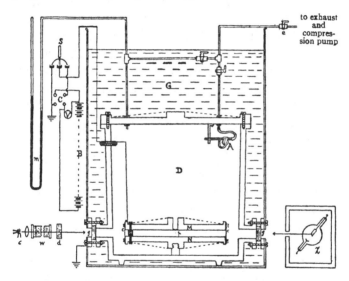

FIG. 7

number of free electrons carried by the drop from 1 on drop No. 28 to 136 on drop No. 56.

The experimental arrangements are shown in Fig. 7. The brass vessel D was built for work at all pressures up to 15 atmospheres, but since the present observations have to do only with pressures from 76 cm. down, these were measured with a very carefully made mercury manometer m, which at atmospheric pressure gave precisely the same reading as a standard barometer. Complete stagnancy of the air between the condenser plates M and N was attained, first, by absorbing all of the heat rays from the arc A by means of a water cell w, 80 cm. long, and a cupric chloride cell d, and, secondly, by immersing the whole vessel D in a constant temperature bath G of gas-engine oil (40 liters), which permitted, in general, fluctuations of not more than .02° C. during an observation. This constant-temperature bath was found essential if such consistency of measurement as is shown here was to be obtained. A long search for causes of slight irregularity revealed nothing so important as this, and after the bath was installed all of the irregularities vanished. The atomizer A was blown by means of a puff of carefully dried and dust-free air introduced through cock e. The air about the drop p was ionized when desired, or electrons discharged directly from the drop, by means of Röntgen rays from X, which readily passed through the glass window g. To the three windows g (two only are shown) in the brass vessel D correspond, of course, three windows in the ebonite strip c, which encircles the condenser plates M and N. Through the third of these windows, set at an angle of about 28° from the line Xpa and in the same horizontal plane, the oil drop is observed through a short-focus telescope having a scale in the eyepiece to make possible the exact measurement of the speeds of the droplet-star.

In plotting the actual observations I have used the reciprocal of the pressure $\frac{1}{p}$ in place of l, for the reason that l is a theoretical quantity which is necessarily proportional to $\frac{1}{p}$, while p is the quantity actually measured.

This amounts to writing the correction-term to Stokes's Law in the form $\left(1+\dfrac{b}{pa}\right)$ instead of in the form $1+A\dfrac{l}{a}$ and considering b the undetermined constant which is to be evaluated, as was A before, by dividing the slope of our line by its y-intercept.

Nevertheless, in view of the greater ease of visualization of $\dfrac{l}{a}$ all the values of this quantity corresponding to successive values of $\dfrac{1}{pa}$ are given in Table IX. Fig. 5 shows the graph obtained by plotting the values of e_1 against $\dfrac{1}{pa}$ for the first 51 drops of Table IX, and Fig. 6 shows the extension of this graph to twice as large values of $\dfrac{1}{pa}$ and e_1. It will be seen that there is not the slightest indication of a departure from a linear relation between e_1 and $\dfrac{1}{pa}$ up to the value $\dfrac{1}{pa}=620.2$, which corresponds to a value of $\dfrac{l}{a}$ of .4439 (see drop No. 58, Table IX). Furthermore, the scale used in the plotting is such that a point which is one division above or below the line in Fig. 5 represents in the mean an error of 2 in 700. *It will be seen from Figs. 5 and 6 that there is but one drop in the 58 whose departure from the line amounts to as much as 0.5 per cent. It is to be remarked, too, that this is not a selected group of drops, but represents all of the drops experimented upon during 60 consecutive days,* during which time the apparatus was taken down several times and set up anew. It is certain, then, that an equation of the form (15) holds very accurately up to

$\dfrac{l}{a} = .4$.   The last drop of Fig. 6 seems to indicate the
beginning of a departure from this linear relationship.
Since such departure has no bearing upon the evaluation
of e, discussion of it will not be entered into here, although
it is a matter of great interest for the molecular theory.

Attention may also be called to the completeness of
the answers furnished by Figs. 5 and 6 to the question
raised in chap. iv as to a possible dependence of the drag
which the medium exerts on the drop upon the amount
of the latter's charge; also, as to a possible variation of
the density of the drop with its radius.   Thus drops
Nos. 27 and 28 have practically identical values of $\dfrac{1}{pa}$,
but while No. 28 carries, during part of the time, but
1 unit of charge (see Table IX), drop No. 27 carries
29 times as much and it has about 7 times as large a
diameter.   Now, if the small drop were denser than the
large one, or if the drag of the medium upon the heavily
charged drop were greater than its drag upon the one
lightly charged, then for both these reasons drop No. 27
would move more slowly relatively to drop No. 28
than would otherwise be the case, and hence e_1 for drop
No. 27 would fall below e_1 for drop No. 28.   Instead of
this the two e_1 fall so nearly together that it is impossible
to represent them on the present scale by two separate
dots.   Drops Nos. 52 and 56 furnish an even more
striking confirmation of the same conclusion, for both
drops have about the same value for $\dfrac{l}{a}$ and both are
exactly on the line, though drop No. 56 carries at one
time 68 times as heavy a charge as drop No. 52 and has
three times as large a radius.   In general, the fact that

Figs. 5 and 6 show no tendency whatever on the part of either the very small or the very large drops to fall above or below the line is experimental proof of the joint correctness of the assumptions of constancy of drop-density and independence of drag of the medium on the charge on the drop.

The values of $e^{\frac{2}{3}}$ and b obtained graphically from the y-intercept and the slope in Fig. 5 are $e^{\frac{2}{3}} = 61.13 \times 10^{-8}$ and $b = .000625$, p being measured, for the purposes of Fig. 5 and of this computation in centimeters of Hg at $23°$ C. and a being measured in centimeters. The value of A in equations 15 and 16 (p. 101) corresponding to this value of b is $.874$.

Instead, however, of taking the result of this graphical evaluation of e, it is more accurate to reduce each of the observations on e_1 to e by means of the foregoing value of b and the equation

$$e^{\frac{2}{3}}\left(1 + \frac{b}{pa}\right) = e_1^{\frac{2}{3}}$$

The results of this reduction are contained in the last column of Table IX. These results illustrate very clearly the sort of consistency obtained in these observations. *The largest departure from the mean value found anywhere in the table amounts to 0.5 per cent and "the probable error" of the final mean value computed in the usual way is 16 in 61,000.*

Instead, however, of using this final mean value as the most reliable evaluation of e, it was thought preferable to make a considerable number of observations at atmospheric pressure on drops small enough to make t_s determinable with great accuracy and yet large enough so that the whole correction term to Stokes's Law amounted to but a small percentage, since in this case, even though there might be a considerable error in the correction-term constant b, such error would influence the final value of e by an inappreciable amount. The first 23 drops of Table IX represent such observations. It will be seen that they show slightly greater consistency than do the remaining drops in the table and that the correction-term reductions for these drops all lie between 1.3 per cent (drop No. 1) and 5.6 per cent (drop No. 23), so that even though b were in error by as much as 3 per cent (its error is actually not more than 1.5 per cent), e would be influenced by that fact to the extent of but 0.1 per cent. The mean value of $e^{\frac{2}{3}}$ obtained from the first 23 drops is 61.12×10^{-8}, a number which differs by 1 part in 3,400 from the mean obtained from all the drops.

When correction is made for the fact that the numbers in Table IX were obtained on the basis of the assumption $\eta = .0001825$, instead of $\eta = .0001824$ (see Section II), which was the value of η_{23} chosen in 1913 when this work was first published, the final mean value of $e^{\frac{2}{3}}$ obtained from the first 23 drops is 61.085×10^{-8}. This corresponds to

$$e = 4.774 \times 10^{-10} \text{ electrostatic units.}$$

I have already indicated that as soon as e is known it becomes possible to find with the same precision which has been attained in its determination the exact number of molecules in a given weight of any substance, the absolute weight of any atom or molecule, the average kinetic energy of agitation of an atom or molecule at any temperature, and a considerable number of other important molecular and radioactive constants. In addition, it has recently been found that practically all of the important radiation constants like the wave-lengths of X-rays, Planck's *h*, the Stefan-Boltzmann constant σ, the Wien constant c_2, etc., depend for their most reliable evaluation upon the value of *e*. In a word, *e* is increasingly coming to be regarded, *not only as the most fundamental of physical or chemical constants, but also the one of most supreme importance for the solution of the numerical problems of modern physics.* It seemed worth while, therefore, to drive the method herewith developed for its determination to the limit of its possible precision. Accordingly, in 1914 I built a new condenser having surfaces which were polished optically and made flat to within two wave-lengths of sodium light. These were 22 cm. in diameter and were separated by 3 pieces of echelon plates, 14.9174 mm. thick, and all having optically perfect plane-parallel surfaces. The dimensions of the condenser, therefore, no longer introduced an uncer-tainty of more than about 1 part in 10,000. The volts were determined after each reading in terms of a Weston standard cell and are uncertain by no more than 1 part in 3,000. The times were obtained from an exception-ally fine printing chronograph built by William Gaertner & Co. It is controlled by a standard astronomical clock and prints directly the time to hundredths of a second. All the other elements of the problem were looked to with a care which was the outgrowth of five years of experience with measurements of this kind. The present form of the apparatus is shown in diagram in Fig. 8, and in Fig. 9 is shown a photograph taken before the enclosing oil tank had been added. This work

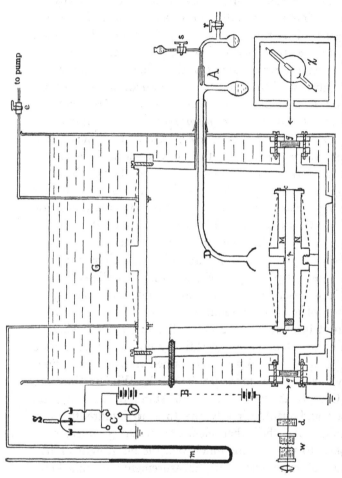

Fig. 8.—A, atomizer through which the oil spray is blown into the cylindrical vessel D. G, oil tank to keep the temperature constant. M and N, circular brass plates, electrical field produced by throwing on 10,000-volt battery B. Light from arc lamp a after heat rays are removed by passage through w and d, enters chamber through glass window g and illuminates droplet p between plates M and N through the pinhole in M. Additional ions are produced about p by X-rays from the bulb X.

FIG. 9

was concluded in August, 1916, and occupied the better part of two years of time. The final table of results and the corresponding graph are given in Table X and in Fig. 10. The final value of e^1 computed on the basis $\eta_{23} = .0001824$ is seen to be now 61.126×10^{-8} instead of 61.085, or .07 per cent higher than the value found in 1913. But Dr. Harrington's new value of η_{23}, namely, .00018226, is more reliable than the old value and is lower than it by .07 per cent. Since η appears in the first power in e^1, it will be seen that the new value[1] of e, determined with new apparatus and with a completely new determination of all the factors involved, comes out to the fourth place exactly the same as the value published in 1913, namely, 4.774×10^{-10} old international electrostatic units. In view of new determinations both c and of the ohm this now corresponds to $e = 4.770$ absolute electrostatic units. The corresponding values of b and A are now .000617 and .863, respectively.

Since in 1934 steps are on foot to again fix the value of the Faraday constant by international agreement,[2] this time at 9,648.9 absolute electromagnetic units, and since this is the number N of molecules in a gram molecule times the elementary electrical charge, we have

$$N \times 4.770 \times 10^{-10} = 9,648.9 \times 2.99778 \times 10^{-10},$$
$$N = 6\ 064 \times 10^{20}$$

Although the probable error in this number computed by the method of least squares from Table X is but one part in 4,000, it would be erroneous to infer that e and N are now known with that degree of precision, for there are

[1] See Millikan, *Phil. Mag.*, June, 1917; *Phys. Rev.*, XXXV (1930), 1231.

[2] At. wt. of Ag. = 107.880; electrochem. eq't. of Ag. = 0.0118805.

four constant factors entering into all of the results in Table X and introducing uncertainties as follows: The coefficient of viscosity η which appears in the $3/2$ power introduces into *e* and *N* a maximum possible uncertainty of less than 0.1 per cent, say 0.07 per cent. The cross-hair distance which is uniformly duplicatable to one part in two thousand appears in the $3/2$ power and introduces an uncertainty of no more than 0.07 per cent. All the other factors, such as the volts and the distance between the condenser plates, introduce errors which are negligible in comparison. The uncertainty in *e* and *N* is then that due to two factors, each of which introduces a maximum possible uncertainty of about 0.07 per cent. Following the usual procedure, we may estimate the uncertainty in *e* and *N* as the square root of the sum of the squares of these two uncertainties, that is, as about one part in 1000. We have then:

$$e = 4\ 770 \pm .005 \times 10^{-10}$$
$$N = 6.064 \pm .006 \times 10^{23}$$

Perhaps these numbers have little significance to the general reader who is familiar with no electrical units save those in which his monthly light bills are rendered. If these latter seem excessive, it may be cheering to reflect that the number of electrons contained in the quantity of electricity which courses every second through a common sixteen-candle-power electric-lamp filament, and for which we pay $1/100,000$ of 1 cent, is so large that if all the two and one-half million inhabitants of Chicago were to begin to count out these electrons and were to keep on counting them out each at the rate of two a second, and if no one of them were ever to stop to eat, sleep, or die,

TABLE X

No.	Tem. °C.	P.D. (Volts)	t_g (Sec.)	v_1 cm./Sec.	n	$a \times 10^5$ cm.	p (cm. Hg)	$\dfrac{1}{pa}$	$\dfrac{l}{a}$	$e_1^{2/3} \times 10^8$	$e^{2/3} \times 10^8$
1	23.07	6,650	16.50	.06194	7–13	23.40	74.49	57.45	.04111	63.21	61.03
2	23.00	6,100	16.76	.06099	8–11	23.22	75.00	57.5	.04115	63.204	61.03
3	23.05	5,308	19.73	.05180	7–15	21.34	74.49	63.0	.04509	63.54	61.16
4	23.08	4,132	37.82	.02703	4–6	15.33	75.37	86.7	.06205	64.27	60.97
5	23.06	4,661	40.09	.02521	3–6	14.84	75.00	90.6	.06484	64.63	61.21
6	23.12	4,111	51.53	.01983	3–4	13.05	75.77	101.3	.07250	65.02	61.19
7	23.08	5,299	51.48	.01985	2–5	13.05	74.98	102.4	.07329	65.07	61.20
8	23.00	6,661	56.06	.01823	2–3	12.50	75.40	106.3	.07608	65.13	61.11
9	23.01	6,082	59.14	.01728	1–4	12.17	75.04	109.7	.07850	65.19	61.05
10	23.00	4,077	57.46	.01779	1–8	12.34	75.67	107.3	.07680	65.21	61.16
11	23.10	4,663	16.58	.06165	10–12	22.72	29.26	150.6	.1078	66.72	61.01
12	23.13	4,661	29.18	.03502	5–7	17.08	36.61	160.1	.1146	67.12	61.07
13	23.11	4,687	18.81	.05432	8–10	21.26	30.27	155.6	.1114	67.14	61.26
14	22.98	4,651	47.65	.02145	2–7	13.20	36.80	205.4	.1477	68.90	61.11
15	23.12	4,648	32.72	.03129	4–6	15.92	31.35	200.7	.1437	68.97	61.39
16	23.10	3,393	18.34	.05572	12–16	21.11	20.58	227.8	.1630	69.88	61.27
17	23.15	3,669	46.82	.02204	2–4	13.12	29.10	262.4	.1878	70.85	60.94
18	23.12	4,091	14.10	.07249	5–7	17.32	20.54	281.4	.2014	71.60	60.98
19	23.10	3,339	30.24	.02605	15–19	23.00	13.24	321.4	.2297	73.34	61.20
20	23.14	4,682	18.30	.05585	3–5	14.00	20.72	345.4	.2472	74.27	61.22
21	23.14	3,350	43.88	.02329	10–13	20.47	13.62	359.1	.2570	74.54	60.97
22	23.13	3,381	46.90	.02179	3–6	13.17	20.47	371.5	.2659	75.00	60.97
23	23.00	3,345	19.65	.05201	0–12	12.69	20.74	380.6	.2724	75.62	61.24
24	23.09	3,345	19.65	.05201	0–12	19.65	13.12	388.5	.2781	75.92	61.18
25	23.15	3,344	26.76	.03819	6–9	16.57	13.80	438.3	.3137	77.74	61.18

Mean = 61.126.

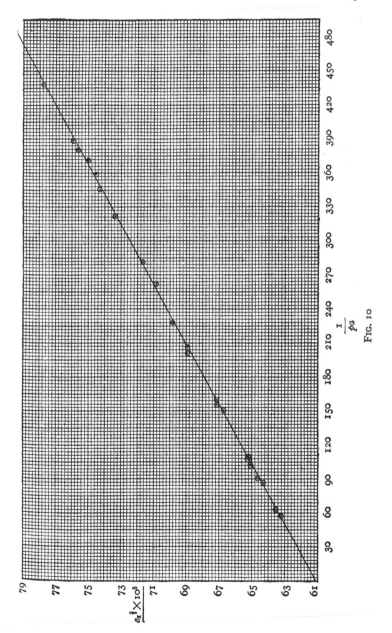

FIG. 10

it would take them just twenty thousand years to finish the task.

Let us now review, with Figs. 5 and 10 before us, the essential elements in the measurement of e. We discover, first, that electricity is atomic, and we measure the electron in terms of a characteristic speed for each droplet. To reduce these speed units to electrical terms, and thus obtain an absolute value of e, it is necessary to know how in a given medium and in a given field the speed due to a given charge on a drop is related to the size of the drop. This we know accurately from Stokes's theory and Arnold's experiments when the holes in the medium, that is, when the values of $\frac{l}{a}$ are negligibly small, but when $\frac{l}{a}$ is large we know nothing about it. *Consequently there is but one possible way to evaluate e, namely, to find experimentally how the apparent value of e, namely, e_1 varies with $\frac{l}{a}$ or $\frac{1}{pa}$, and from the graph of this relation to find what value e_1 approaches as $\frac{l}{a}$ or $\frac{1}{pa}$ approaches zero.* So as to get a linear relation we find by analysis that we must plot $e_1^{\frac{2}{3}}$ instead of e_1 against $\frac{l}{a}$ or $\frac{1}{pa}$. We then get e from the intercept of an experimentally determined straight line on the y-axis of our diagram. This whole procedure amounts simply to reducing our drop-velocities to what they would be if the pressure were so large or $\frac{1}{pa}$ so small that the holes in the medium were all closed up. *For this case and for this case alone we know both from Stokes's and Arnold's work exactly the law of motion of the droplet.*

CHAPTER VI

THE MECHANISM OF IONIZATION OF GASES BY X-RAYS AND RADIUM RAYS

I. EARLY EVIDENCE

Up to the year 1908 the only experiments which threw any light whatever upon the question as to what the act of ionization of a gas consists in were those performed by Townsend[1] in 1900. He had concluded from the theory given on p. 34 and from his measurements on the diffusion coefficients and the mobilities of gaseous ions that both positive and negative ions in gases carry unit charges. This conclusion was drawn from the fact that the value of ne in the equation $ne = \dfrac{v_0 P}{D}$ came out about 1.23×10^{10} electrostatic units, as it does in the electrolysis of hydrogen.

In 1908, however, Townsend[2] devised a method of measuring directly the ratio $\dfrac{v_0}{D}$ and revised his original conclusions. His method consisted essentially in driving ions by means of an electric field from the region between two plates A and B (Fig. 11), where they had been produced by the direct action of X-rays, through the gauze in B, and observing what fraction of these ions was driven by a field established between the plates B and C to the central disk D and what fraction drifted by virtue of diffusion to the guard-ring C.

[1] *Phil. Trans.*, CXCIII (1900), 129.
[2] *Proc. Roy. Soc.*, LXXX (1908), 207.

By this method Townsend found that ne for the negative ions was accurately 1.23×10^{10}, but for the positive ions it was 2.41×10^{10}. From these results the conclusion was drawn that in X-ray ionization *all* of the positive ions are bivalent, i.e., presumably, that the act of ionization by X-rays consists in the detachment from a neutral molecule of two elementary electrical charges.

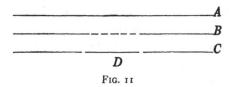

FIG. 11

Townsend accounted for the fact that his early experiments had not shown this high value of ne for the positive ions by the assumption that by the time the doubly charged positive ions in these experiments had reached the tubes in which D was measured, most of them had become singly charged through drawing to themselves the singly charged negative ions with which they were mixed. This hypothesis found some justification in the fact that in the early experiments the mean value of ne for the positive ions had indeed come out some 15 or 20 per cent higher than 1.23×10^{10}—a discrepancy which had at first been regarded as attributable to experimental errors, and which in fact might well be attributed to such errors in view of the discordance between the observations on different gases.

Franck and Westphal,[1] however, in 1909 redetermined ne by a slight modification of Townsend's original method, measuring both v' and D independently, and

[1] *Verh. deutsch. phys. Ges.*, March 5, 1909.

not only found, when the positive and negative ions are separated by means of an electric field so as to render impossible such recombination as Townsend suggested, that D was of exactly the same value as when they were not so separated, but also that ne for the positive ions produced by X-rays was but 1.4×10^{10} instead of 2.41×10^{10}. Since this was in fair agreement with Townsend's original mean, the authors concluded that only a small fraction—about 9 per cent—of the positive ions formed by X-rays are doubles, or other multiples, and the rest singles. In their experiments on the ionization produced by α-rays, β-rays, and γ-rays, they found no evidence for the existence of doubly charged ions.

In summarizing, then, the work of these observers it could only be said that, although both Townsend and Franck and Westphal drew the conclusion that doubly charged ions exist in gases ionized by X-rays, there were such contradictions and uncertainties in their work as to leave the question unsettled. In gases ionized by other agencies than X-rays no one had yet found any evidence for the existence of ions carrying more than a single charge, except in the case of spark discharges from condensers. The spectra of these sparks revealed certain lines called enhanced lines which were thought to be due to doubly ionized atoms. Whether, however, these multiple charges were produced by a single ionizing act or by successive acts was completely unknown.

II.   OIL-DROP EXPERIMENTS ON VALENCY IN GASEOUS IONIZATION

The oil-drop method is capable of furnishing a direct and unmistakable answer to the question as to whether the act of ionization of a gas by X-rays or other agencies consists in the detachment of one, of several, or of many electrons from a single neutral molecule. For it makes it possible to catch the residue of such a molecule practically at the instant at which it is ionized and to count directly the number of charges carried by that residue. The initial evidence obtained from this method seemed to favor the view that the act of ionization may consist in the detachment of quite a number of electrons from a single molecule, for it was not infrequently observed that a balanced oil drop would remain for several seconds unchanged in charge while X-rays were passing between the plates, and would then suddenly assume a speed which corresponded to a change of quite a number of electrons in its charge.

It was of course recognized from the first, however, that it is very difficult to distinguish between the practically simultaneous advent upon a drop of two or three separate ions and the advent of a doubly or trebly charged ion, but a consideration of the frequency with which ions were being caught in the experiments under consideration, a change occurring only once in, say, 10 seconds, seemed at first to render it improbable that the few double, or treble, or quadruple catches observed when the field was on could represent the simultaneous advent of separate ions. It was obvious, however, that the question could be conclusively settled by working with smaller and smaller drops. For the proportion of double or treble to single catches made in a field of strength between 1,000 and 6,000 volts per centimeter should be independent of the size of the drops if the doubles are due to the advent of doubly charged ions, while this proportion should decrease with the square of the radius of the drop if the doubles are due to the simultaneous capture of separate ions.

Accordingly, Mr. Harvey Fletcher and the author,[1] suspended, by the method detailed in the preceding chapter, a very small positively charged drop in the upper part of the field between M and N (Fig. 12), adjusting either the charge upon the drop or the field strength until the drop was nearly balanced. We then produced beneath the drop a sheet of X-ray ionization. With the arrangement shown in the figure, in which M

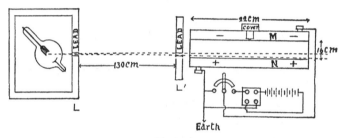

Fig. 12

and N are the plates of the condenser previously described, and L and L' are thick lead screens, the positive ions are thrown, practically at the instant of formation, to the upper plate. When one of them strikes the drop it increases the positive charge upon it, and the amount of the charge added by the ion to the drop can be computed from the observed change in the speed of the drop.

For the sake of convenience in the measurement of successive speeds a scale containing 70 equal divisions was placed in the eyepiece of the observing cathetometer telescope, which in these experiments produced a

magnification of about 15 diameters. The method of procedure was, in general, first, to get the drop nearly
balanced by shaking off its initial charge by holding a
little radium near the observing chamber, then, with a
switch, to throw on the X-rays until a sudden start in
the drop revealed the fact that an ion had been caught,
then to throw off the rays and take the time required
for it to move over 10 divisions, then to throw on the
rays until another sudden quickening in speed indicated
the capture of another ion, then to measure this speed
and to proceed in this way without throwing off the field
at all until the drop got too close to the upper plate, when
the rays were thrown off and the drop allowed to fall
under gravity to the desired distance from the upper
plate. In order to remove the excess of positive charge
which the drop now had because of its recent captures,
some radium was brought near the chamber and the
field thrown off for a small fraction of a second. As
explained in preceding chapters, ions are caught by the
drop many times more rapidly when the field is off than
when it is on. Hence it was in general an easy matter
to bring the positively charged drop back to its balanced
condition, or indeed to any one of the small number of
working speeds which it was capable of having, and then
to repeat the series of catches described above. In this
way we kept the same drop under observation for hours
at a time, and in one instance we recorded 100 successive
captures of ions by a given drop, and determined in each
case whether the ion captured carried a single or a
multiple charge.

The process of making this determination is exceedingly simple and very reliable. For, since electricity is

atomic in structure, there are only, for example, three possible speeds which a drop can have when it carries 1, 2, or 3 elementary charges, and it is a perfectly simple matter to adjust conditions so that these speeds are of such different values that each one can be recognized unfailingly even without a stop-watch measurement. Indeed, the fact that electricity is atomic is in no way more beautifully shown than by the way in which, as reflected in Table XI, these relatively few possible working speeds recur.   After all the possible speeds have been located it is only necessary to see whether one of them is ever skipped in the capture of a new ion in order to know whether or not that ion was a double. Table XI represents the results of experiments made with very hard X-rays produced by means of a powerful 12-inch Scheidel coil, a mercury-jet interrupter, and a Scheidel tube whose equivalent spark-length was about 5 inches. No attempt was made in these experiments to make precise determinations of speed, since a high degree of accuracy of measurement was not necessary for the purpose for which the investigation was undertaken. Table XI is a good illustration of the character of the observations. The time of the fall under gravity recorded in the column headed "t_g" varies slightly, both because of observational errors and because of Brownian movements.   Under the column headed "t_F" are recorded the various observed values of the times of rise through 10 divisions of the scale in the eyepiece. A star (*) after an observation in this column signifies that the drop was moving *with* gravity instead of against it. The procedure was in general to start with the drop either altogether neutral (so that it fell when the field was on with the same speed as when the field was off), or having one single positive charge, and then to throw on positive

TABLE XI

Plate Distance 1.6 cm.   Distance of Fall .0975 cm.   Volts 1,015.
Temperature 23° C.   Radius of Drop .000063 cm.

t_g	t_F	No. of Charges on Drop	No. of Charges on Ion Caught	t_g	t_F	No. of Charges on Drop	No. of Charges on Ion Caught
19.0	100.0	1 P		20.0	10.0*	1 N	
	16.0	2 P	1 P		20.0*	0	1 P
	8.0	3 P	1 P		100.0	1 P	1 P
					20.0*	0	1 N
20.0	16.0	2 P			100.0	1 P	1 P
	8.0	3 P	1 P		16.0	2 P	1 P
					8.0	3 P	1 P
	100.0	1 P	1 P				
	17.0	2 P	1 P		104.0	1 P	
	8.2	3 P	1 P		15.0	2 P	1 P
	6.0	4 P			9.0	3 P	1 P
					6.0	4 P	1 P
	7.0*	2 N	1 P				
	9.8*	1 N	1 N		6.5*	2 N	
	7.0*	2 N			10.0*	1 N	1 P
					20.0*	0	1 P
21.0	20.0*	0	1 P		100.0	1 P	1 P
	95.0	1 P	1 P		15.5	2 P	1 P
	16.5	2 P	1 P		8.0	3 P	1 P
	8.0	3 P	1 P		6.0	4 P	1 P
	6.0	4 P					
					100.0	1 P	1 P
	100.0	1 P	1 P		16.5	2 P	
	16.0	2 P	1 P				
	8.4	3 P			20.0*	0	1 P
					100.0	1 P	1 P
20.0	106.0	1 P	1 P		16.6	2 P	1 P
	16.0	2 P	1 P		8.8	3 P	1 P
	8.4	3 P			5.7	4 P	
	10.0*	1 N	1 P		100.0	1 P	1 N
	20.0*	0	1 P		20.0*	0	1 N
	100.0	1 P	1 P		10.0*	1 N	1 P
	16.0	2 P			20.0*	0	
	100.0	1 P	1 P		44 catches, all singles		
	16.0	2 P	1 P				
	8.0	3 P					

charges until its speed came to the 6.0 second value, then to make it neutral again with the aid of radium, and to begin over again.

It will be seen from Table XI that in 4 cases out of 44 we caught negatives, although it would appear from the arrangement shown in Fig. 12 that we could catch only positives. These negatives are doubtless due to secondary rays which radiate in all directions from the air molecules when these are subjected to the primary X-ray radiation.

Toward the end of Table XI is an interesting series of catches. At the beginning of this series, the drop was charged with 2 negatives which produced a speed in the direction of gravity of 6.5 seconds. It caught in succession 6 single positives before the field was thrown off. The corresponding times were 6.5*, 10*, 20*, 100, 15.5, 8.0, 6.0. The mean time during which the X-rays had to be on in order to produce a "catch" was in these experiments about six seconds, though in some instances it was as much as a minute. The majority of the times recorded in column t_F were actually measured with a stop watch as recorded, but since there could be no possibility of mistaking the 100-second speed, it was observed only four or five times. It will be seen from Table XI that out of 44 catches of ions produced by very hard X-rays there is not a single double. As a result of observing from 500 to 1,000 catches in the manner illustrated in Table XI, we came to the conclusion that, although we had entered upon the investigation with the expectation of proving the existence of valency in gaseous ionization, *we had instead obtained direct, unmistakable evidence that the act of ionization of air* *molecules by both primary and secondary X-rays of widely varying degrees of hardness, as well as by β- and γ-rays, uniformly consists, under all the conditions which we were able to investigate, in the detachment from a neutral molecule of one single elementary electrical charge.*

III. RECENT EVIDENCE AS TO NATURE OF IONIZATION PRODUCED BY ETHER WAVES

Although Townsend and Franck and Westphal dissented from the foregoing conclusion, all the evidence which has appeared since has tended to confirm it. Thus Salles,[1] using a new method due to Langevin of measuring directly the ratio $\left(\dfrac{v_0}{D}\right)$ of the mobility to the diffusion coefficient, concluded that when the ionization is produced by γ-rays there are no ions bearing multiple charges. Again, the very remarkable photographs (see plate opposite p. 190) taken by C. T. R. Wilson in the Cavendish Laboratory of the tracks made by the passage of X-rays through gases show no indication of a larger number of negatively than of positively charged droplets. Such an excess is to be expected if the act of ionization ever consists in these experiments in the detachment of two or more negative electrons from a neutral molecule. Further, if the initial act of ionization by X-rays ever consists in the ejection of two or more corpuscles from a single atom, there should appear in these Wilson photographs a rosette consisting of a group of zigzag lines starting from a common point. A glance at the plate opposite p. 192 shows that this is not the case, each zigzag line having its own individual starting-point.

[1] *Le Radium,* X (1913), 113. 110.

There are two other types of experiments which throw light on this question.

When in the droplet experiments the X-rays are allowed to fall directly upon the droplet, we have seen that they detach negative electrons from it, and if the gas is at so low a pressure that there is very little chance of the capture of ions by the droplet, practically all of its changes in charge have this cause.   Changes produced under these conditions appear, so far as I have yet been able to discover, to be uniformly unit changes.   Also, when the changes are produced by the incidence on the droplet of ultra-violet light, so far as the experiments which have been carried out by myself or my pupils go, they usually, though not always, have appeared to correspond to the loss of one single electron.   The same seems to have been true in the experiments reported by A. Joffé,[1] who has given this subject careful study.

Meyer and Gerlach,[2] it is true, seem very often to observe changes corresponding to the simultaneous loss of several electrons.   It is to be noted, however, that their drops are generally quite heavily charged, carrying from 10 to 30 electrons.   Under such conditions the loss of a single electron makes but a minute change in speed, and is therefore likely not only to be unnoticed, but to be almost impossible to detect *until the change has become more pronounced through the loss of several electrons.   This question, then, can be studied reliably only when the field is powerful enough to hold the droplet balanced with only one or two free electrons upon it.*   Experiments made under such conditions with my apparatus by both Derieux[3] and

[1] *Sitzungsber. d. k. Bayerischen Akad. d. Wiss.* (1913), p. 19.

[2] *Ann. d. Phys.*, XLV, 177; XLVII, 227.   [3] *Phys. Rev.*, X (1918), 283.

Kelly[1] show quite conclusively that the act of photo-emission under the influence of ultra-violet light consists in the ejection of a single electron at each emission.

Table XII contains one series of observations of this sort taken with my apparatus by Mr. P. I. Pierson. The first column gives the volts applied to the plates of the condenser shown in Fig. 7, p. 111. These were made variable so that the drop might always be pulled up with a slow speed even though its positive charge were continually increasing. The second and third columns give the times required to move 1 cm. under gravity and under the field respectively. The fourth column gives the time intervals required for the drop to experience a change in charge under the influence of a constant source of ultra-violet light—a quartz mercury lamp. The fifth column gives the total charge carried by the drop computed from equation (12), p. 91. The sixth column shows the change in charge computed from equation (10), p. 70. This is seen to be as nearly a constant as could be expected in view of Brownian movements and the inexact measurements of volts and times. The mean value of e_1 is seen to be 5.1×10^{-10}, which yields with the aid of equation (16), p. 101, after the value of A found for oil drops has been inserted, $e = 4.77 \times 10^{-10}$, which is in better agreement with the result obtained with oil drops than we had any right to expect. In these experiments the light was weak so that the changes come only after an average interval of 29 seconds and it will be seen that *they are all unit changes*.

So long, then, as we are considering the ionization of neutral atoms through the absorption of an ether wave

[1] *Phys. Rev.*, XVI (1920), 260.

of any kind, the evidence at present available indicates
that the act always consists in the detachment from the

TABLE XII

MERCURY DROPLET OF RADIUS $a = 8 \times 10^{-5}$ CM. DISCHARGING
ELECTRONS UNDER THE INFLUENCE OF ULTRA-VIOLET LIGHT

Volts	Drop No. 1 t_g Sec. per Cm.	F Sec. per Cm.	Time Interval between Discharges in Seconds	$e_n \times 10^{10}$	Change in e	No Electrons Emitted
2,260......	11.0	−1200 }		{49.4}		
3,070......	11.0	+ 32.8}		{50.5}		
			11		4.4	1
1,960......	11.0	− 194.		54.4		
			12.8			
1,960......	....	+ 190		60.8	6.4	1
			23			
1,820......	11.2	+ 220		65.0	4.2	1
			40			
1,690......	....	+ 230		69.8	4.8	1
			15.2			
1,550......	....	+ 332		75.1	5.3	1
3,040......	Drop No. 2 10.4	+ 98		43.5		
			5.6			
2,540......	....	+ 200		49.4	5.9	1
			18.6			
2,230......	....	+ 300		55.2	5.8	1
			35.0			
2,230......	....	+ 76		60.7	5.5	1
			42			
1,930......	....	+ 200		65.0	4.3	1
			54			
1,810.....	....	+ 176		69.6	4.6	1
			70			
1,650......	....	+ 250		75.2	5.6	1
			45			
1,520... ..	....	+ 500		79.4	4.2	1
			9.8			
1,520......	....	119		85.1	5.5	1
		Mean....	29	Mean..	5.1	

atom of one single negative electron, the energy with
which this electron is ejected from the atom depending, as

we shall see in chap. x, in a very definite and simple way upon the frequency of the ether wave which ejects it.

IV. IONIZATION BY β-RAYS

When the ionization is due to the passage of β-rays through matter, the evidence of the oil-drop experiments as well as that of C. T. R. Wilson's experiments (see chap. ix) on the photographing of the tracks of the β-rays is that here, too, the act consists in the detachment of one single electron from a single atom. This experimental result is easy to understand in the case of the β-rays, when it is remembered that Wilson's photographs prove directly the fact, long known from other types of evidence, that a β-ray, in general, ionizes but a very minute fraction of the number of atoms through which it shoots before its energy is expended. If, then, its chance, in shooting through an atom, of coming near enough to one of the electronic constituents of that atom to knock it out is only one in a hundred or one in a thousand, then its chance of getting near enough to two electronic constituents of the same atom to knock them both out is likely to be negligibly small. The argument here rests, however, on the assumption that the electrons within the atom are independent of one another, which is not necessarily the case, so that the matter must be decided after all solely by experiment.

The difference between the act of ionization when produced by a β-ray and when produced by an ether wave seems, then, to consist chiefly in the difference in the energy with which the two agencies hurl the electron from its mother atom. Wilson's photographs show that β-rays do not in general eject electrons with appre-

ciable speeds, while ether waves may eject them with
tremendous energy. Some of Wilson's photographs
showing the effect of passing X-rays through air are
shown in the most interesting plate opposite p. 192.
The original X-rays have ejected electrons with great
speeds from a certain few of the atoms of the gas, and
it is the tracks of these electrons as they shoot through
the atoms of the gas, ionizing here and there as they
go, which constitute the wiggly lines shown in the photograph. Most of the ionization, then, which is produced
by X-rays is a secondary effect due to the negative electrons, i.e., the β-rays which the X-rays eject. If these
β-rays could in turn eject electrons with ionizing speeds,
each of the dots in one of these β-ray tracks would be
the starting-point of a new wiggly line like the original
one. But such is not the case. We may think, then,
of the β-rays as simply shaking loose electronic dust
from some of the atoms through which they pass while we
think of the X-rays as taking hold in some way of the
negative electrons within an atom and hurling them out
with enormous energy.

V. IONIZATION BY α-RAYS

But what happens to the electronic constituents of
an atom when an α-particle, that is, a helium atom,
shoots through it? Some of Bragg's experiments and
Wilson's photographs show that the α-particles shoot
in straight lines through from 3 to 7 cm. of air before
they are brought to rest. We must conclude, then, that
an atom has so loose a structure that another atom, *if
endowed with enough speed*, can shoot straight through it
without doing anything more than, in some instances, to

shake off from that atom an electron or two.   The tracks
shown in Figs. 14 and 15, facing p. 190, are Wilson photographs of the tracks of the α-particles of radium.   They
ionize so many of the atoms through which they pass that
the individual droplets of water which form about the
ions produced along the path of the ray, and which are
the objects really photographed, are not distinguishable
as individuals. The sharp changes in the direction of
the ray toward the end of the path are convincing evidence that the α-particle actually goes through the
atoms instead of pushing them aside as does a bullet.
For if one solar system, for example, endowed with a
stupendous speed, were to shoot straight through another
similar system, but without an actual impact of their
central bodies, the deflection from its straight path which
the first system experienced might be negligibly small *if
its speed were high enough*, and that for the simple reason
that the two systems would not be in each other's
vicinity long enough to produce a deflecting effect. In
technical terms the time integral of the force would be
negligibly small. The slower the speed, however, the
longer this time, and hence the greater the deflection.
Thus it is only when the α-particle shown in Fig. 15 has
lost most of its velocity—i.e., it is only toward the end of
its path—that the nuclei of the atoms through which it
passes are able to deflect it from its straight path. If it
pushed the molecules aside as a bullet does, instead of
going through them, the resistance to its motion would
be greatest when the speed is highest. Now, the facts
are just the opposite of this. The α-particle ionizes
several times more violently toward the end of its path
than toward the beginning, and it therefore loses energy

more rapidly when it is going slowly than when it is going rapidly. Further, it is deflected more readily, then, as the photograph shows. All of this is just as it should be if the α-particle shoots straight through the molecules in its path instead of pushing them aside.

These photographs of Wilson's are then the most convincing evidence that we have that the atom is a sort of miniature stellar system with constituents which are unquestionably just as minute with respect to the total volume occupied by the atom as are the sun and planets and other constituents of the solar system with respect to the whole volume inclosed within the confines of this system. When two molecules of a gas are going as slowly as they are in the ordinary motion of thermal agitation, say a mile a second, when their centers come to within a certain distance—about 0.2 $\mu\mu$ (millionths of a millimeter)—they repel one another and so the two systems do not interpenetrate. This is the case of an ordinary molecular collision. But endow one of these molecules with a large enough energy and it will shoot right through the other, sometimes doubtless without so much as knocking out a single electron. This is the case of an α-particle shooting through air.

But the question to which we are here seeking an answer is, does an individual α-particle ever knock more than one electron from a single atom or molecule through which it passes, so as to leave that atom doubly or trebly charged? The oil-drop method used at low pressures[1] has given a very definite answer to this question. *In no gas or vapor except helium, which we have as yet tried, is there any certain evidence that an individual α-ray in*

[1] *Rays of Positive Electricity* (1913), p. 46.

shooting through an atom is able to remove from that atom more than one single electron at a time.

The foregoing result has been obtained by shooting the α-rays from polonium through a rarefied gas in an oil-drop apparatus of the type sketched in Fig. 12, catching upon a balanced oil drop the positively charged residue of one of the atoms thus ionized, and counting, by the change in speed imparted to the droplet, the number of electrons which were detached from the captured atom by the passage of the α-ray through or near it.[1]

This mode of experimenting extended to helium, however, has yielded the most interesting result[2] that every sixth one on the average of all the passages, or "shots," which detached any electrons at all from the helium atom detached both of the two electrons which the neutral helium atom possesses. Since some of the ionization produced along the path of an α-ray is probably due to slow-speed secondary β-rays produced by the α-ray, it is probable that the fraction of the actual passages through helium atoms of α-rays themselves which detach both electrons is greater than the foregoing one in six. It has been estimated by Fowler at as high as three in four.

The foregoing experimental result of one in six was obtained only at the very end of the range of the α-rays where they have their maximum ionizing power. When these rays were near the beginning of their range, and therefore were moving much more rapidly, the fraction of the number of double catches to total catches was only about half as much, i.e., *the chance of getting both electrons*

[1] Millikan, Gottschalk, and Kelly, *Phys. Rev.*, XI (1920), 157.

[2] Millikan, *Phys. Rev.*, XVIII (1921), 456. Wilkins, *ibid.*, XXIV (1922), 210.

at a single shot is much smaller with a high-speed bullet than with a slow-speed one.   This is to be expected if the two electrons are independent of each other, i.e., if the removal of one does not carry the other out with it.

The foregoing is, I think, the only experiment which has yet been devised in which the act of ionization is isolated and studied as an individual thing.

Since 1913, however, very definite evidence has come in from two different sources that multiply-valent ions are often produced in discharge tubes.   The most unambiguous proof of this result has been furnished by the spectroscope.   Indeed, Mr. Bowen and the author have recently found with great definiteness that high voltage vacuum sparks give rise to spectral lines which are due to singly-, doubly-, trebly-, quadruply-, and quintuply-charged atoms of the elements from lithium to nitrogen, and even to sextuply-charged ones in the case of sulphur.[1] In view of the foregoing studies with X-rays, β-rays, and a-rays, it is probable that these spectroscopically discovered multiply-charged ions are mainly due to successive ionizations such as might be expected to take place in a region carrying a very dense electron current, such as must exist in our "hot-sparks."

Again, J. J. Thomson has brought forward evidence[2] that the positive residues of atoms which shoot through discharge tubes in a direction opposite to that of the cathode rays have suffered multiple ionization.   Indeed, he thinks he has evidence that the act of ionization of atoms of mercury consists either in the detachment of one negative electron or else in the detachment of eight.

[1] *Phys. Rev.*, September or October, 1924.

[2] *Rays of Positive Electricity* (1913), p. 46.

The actual situation, however, is not so simple; for in 1930 Bleakney[1] definitely proved that while in general 80 per cent or more of the ions formed by shooting electrons through mercury vapor are singles, at suitable incident energies 2, 3, 4, or 5 electrons may occasionally be detached in a single encounter. Similar effects occur in neon and argon.

VI. SUMMARY

The results of the studies reviewed in this chapter may be summarized thus:

1. The act of ionization by β-rays usually consists in the shaking off without any appreciable energy of one single electron, sometimes, however, more than one, from an occasional molecule through which the β-ray passes. The faster the β-ray the less frequently does it ionize.

2. The act of ionization by ether waves, i.e., by X-rays or light, seems to consist in the hurling out with an energy which may be very large, but which depends upon the frequency of the incident ether wave, of one single electron from an occasional molecule over which this wave passes.

3. The act of ionization by rapidly moving α-particles consists generally in the shaking loose of one single electron from the atom through which it passes, though in the case of helium, two electrons are certainly sometimes removed at once. It may be, too, that a very slow-moving positive ray, such as J. J. Thomson used, may detach several electrons from a single atom.

[1] Walter Bleakney, *Phys. Rev.*, XXXV (1930), 139, and XXXVI (1930), 1303. See also Overton Luhr, *ibid.*, LXIV (1933), 459.

CHAPTER VII

BROWNIAN MOVEMENTS IN GASES

I. HISTORICAL SUMMARY

In 1827 the English botanist, Robert Brown, first made mention of the fact that minute particles of dead matter in suspension in liquids can be seen in a high-power microscope to be endowed with irregular wiggling motions which strongly suggest "life."[1] Although this phenomenon was studied by numerous observers and became known as the phenomenon of the Brownian movements, it remained wholly unexplained for just fifty years. The first man to suggest that these motions were due to the continual bombardment to which these particles are subjected because of the motion of thermal agitation of the molecules of the surrounding medium was the Belgian Carbonelle, whose work was first published by his collaborator, Thirion, in 1880,[2] although three years earlier Delsaulx[3] had given expression to views of this sort but had credited Carbonelle with priority in adopting them. In 1881 Bodoszewski[4] studied the Brownian movements of smoke particles and other suspensions *in air* and saw in them "an approximate image of the movements of the gas molecules as postulated by the kinetic theory of gases." Others, notably Gouy,[5]

[1] *Phil. Mag.*, IV (1828), 161.

[2] *Revue des questions scientifiques*, Louvain, VII (1880), 5.

[3] *Ibid.*, II (1877), 319.

[4] *Dinglers polyt. Jour.*, CCXXXIX (1881), 325.

[5] *Jour. de Phys.*, VII (1888), 561; *Comptes rendus*, CIX (1889), 102.

urged during the next twenty years the same interpretation, but it was not until 1905 that a way was found to subject the hypothesis to a quantitative test. Such a test became possible through the brilliant theoretical work of Einstein[1] of Bern, Switzerland, who, starting merely with the assumption that the mean kinetic energy of agitation of a particle suspended in a fluid medium must be the same as the mean kinetic energy of agitation of a gas molecule at the same temperature, developed by unimpeachable analysis an expression for the mean distance through which such a particle should drift in a given time through a given medium because of this motion of agitation. This distance could be directly observed and compared with the theoretical value. Thus, suppose one of the wiggling particles is observed in a microscope and its position noted on a scale in the eyepiece at a particular instant, then noted again at the end of τ (for example, 10) seconds, and the displacement Δx in that time along one particular axis recorded. Suppose a large number of such displacements Δx in intervals all of length τ are observed, each one of them squared, and the mean of these squares taken and denoted by $\overline{\Delta x^2}$: Einstein showed that the theoretical value of $\overline{\Delta x^2}$ should be

$$\overline{\Delta x^2} = \frac{2RT}{NK} \tau \ \ldots\ldots\ldots\ldots\ldots (21)$$

in which R is the universal gas constant per gram molecule, namely, $831.4 \times 10^5 \ \frac{\text{ergs}}{\text{degrees}}$, T the temperature on the absolute scale, N the number of molecules in one gram molecule, and K a resistance factor depending

· *Ann. d. Phys.* (4), XVII (1905), 549; XIX (1906), 371; XXII (1907), 569.

upon the viscosity of the medium and the size of the drop, and representing the ratio between the force applied to the particle in any way and the velocity produced by that force. If Stokes's Law, namely, $F = 6\pi\eta a v$, held for the motion of the particle through the medium, then $K = \dfrac{F}{v}$ would have the value $6\pi\eta a$, so that Einstein's formula would become

$$\overline{\Delta x^2} = \frac{RT}{N} \frac{\tau}{3\pi\eta a} \ldots \ldots \ldots \ldots (22)$$

This was the form which Einstein originally gave to his equation, a very simple derivation of which has been given by Langevin.[1] The essential elements of this derivation will be found in Appendix C.

The first careful test of this equation was made on suspensions in liquids by Perrin,[2] who used it for finding N the number of molecules in a gram molecule. He obtained the mean value $N = 68.2 \times 10^{22}$, which, in view of the uncertainties in the measurement of both K and $\overline{\Delta x^2}$, may be considered as proving the correctness of Einstein's equation within the limits of error of Perrin's measurements, which differ among themselves by as much as 30 per cent.

II. QUANTITATIVE MEASUREMENTS IN GASES

Up to 1909 there had been no quantitative work whatever on Brownian movements in gases. Bodoszewski had described them fully and interpreted them correctly

[1] *Comptes rendus*, CXLVI (1908), 530.

[2] *Ibid.*, p. 967; CXLVII (1908), 475, 530, 594; CLII (1911), 1380, 1569; see also Perrin, *Brownian Movements and Molecular Reality*. Engl. tr. by Soddy, 1912.

in 1881. In 1906 Smoluchowski[1] had computed how large the mean displacements in air for particles of radius $a = 10^{-4}$ ought to be, and in 1907 Ehrenhaft[2] had recorded displacements of the computed order with particles the sizes of which he made, however, no attempt to measure, so that he knew nothing at all about the resistance factor K. There was then nothing essentially quantitative about this work.

In March, 1908, De Broglie, in Paris,[3] made the following significant advance. He drew the metallic dust arising from the condensation of the vapors coming from an electric arc or spark between metal electrodes (a phenomenon discovered by Hemsalech and De Watteville[4]) into a glass box and looked down into it through a microscope upon the particles rendered visible by a beam of light passing horizontally through the box and illuminating thus the Brownian particles in the focal plane of the objective. His addition consisted in placing two parallel metal plates in vertical planes, one on either side of the particles, and in noting that upon applying a potential difference to these plates some of the particles moved under the influence of the field toward one plate, some remained at rest, while others moved toward the other plate, thus showing that a part of these particles were positively electrically charged and a part negatively. In this paper he promised a study of the charges on these particles. In May, 1909, in fulfilling this promise[5] he

[1] *Ann. der Phys.*, IV (1906), 21, 756.

[2] *Wiener Berichte*, CXVI (1907), II, 1175.

[3] *Comptes rendus*, CXLVI (1908), 624, 1010.

[4] *Ibid.*, CXLIV (1907), 1338.

[5] *Ibid.*, CXLVIII (1909), 1316.

made the first quantitative study of Brownian movements in gases. The particles used were minute droplets of water condensed upon tobacco smoke. The average rate at which these droplets moved in De Broglie's horizontal electric field was determined. The equation for this motion was

$$Fe = Kv \dots\dots\dots\dots\dots\dots (23)$$

The mean $\overline{\Delta x^2}$ was next measured for a great many particles and introduced into Einstein's equation:

$$\overline{\Delta x^2} = \frac{2RT}{NK}\tau.$$

From these two equations K was eliminated and e obtained in terms of N. Introducing Perrin's value of N, De Broglie obtained from one series of measurements $e = 4.5 \times 10^{-10}$; from another series on larger particles he got a mean value several times larger—a result which he interpreted as indicating multiple charges on the larger particles. Although these results represent merely mean values for many drops which are not necessarily all alike, either in radius or charge, yet they may be considered as the first experimental evidence that Einstein's equation holds approximately, in gases, and they are the more significant because nothing has to be assumed about the size of the particles, if they are all alike in charge and radius, or about the validity of Stokes's Law in gases, the K-factor being eliminated.

The development of the oil-drop method made it possible to subject the Brownian-movement theory to a more accurate and convincing experimental test than had heretofore been attainable, and that for the following reasons:

1. It made it possible to hold, with the aid of the vertical electrical field, one particular particle under observation for hours at a time and to measure as many displacements as desired on it alone instead of assuming the identity of a great number of particles, as had been done in the case of suspensions in liquids and in De Broglie's experiments in gases.

2. Liquids are very much less suited than are gases to convincing tests of any kinetic hypothesis, for the reason that prior to Brownian-movement work we had no satisfactory kinetic theory of liquids at all.

3. The absolute amounts of the displacements of a given particle in air are 8 times greater and in hydrogen 15 times greater than in water.

4. By reducing the pressure to low values the displacements can easily be made from 50 to 200 times greater in gases than in liquids.

5. The measurements can be made independently of the most troublesome and uncertain factor involved in Brownian-movement work in liquids, namely, the factor K, which contains the radius of the particle and the law governing its motion through the liquid.

Accordingly, there was begun in the Ryerson Laboratory, in 1910, a series of very careful experiments in Brownian movements in gases. Svedberg,[1] in reviewing this subject in 1913, considers this "the only exact investigation of quantitative Brownian movements in gases." A brief summary of the method and results was published by the author.[2] A full account was

[1] *Jahrbuch der Radioaktivität und Elektronik*, X (1913), 513.

[2] *Science*, February 17, 1911.

published by Mr. Harvey Fletcher in May, 1911,[1] and further work on the variation of Brownian movements with pressure was presented by the author the year following.[2] The essential contribution of this work as regards method consisted in the two following particulars:

1. By combining the characteristic and fully tested equation of the oil-drop method, namely,

$$e = \frac{mg}{Fv_1}(v_1+v_2)_0 = \frac{K}{F}(v_1+v_2)_0 \ldots \ldots \ldots (24)$$

with the Einstein Brownian-movement equation, namely,

$$\overline{\Delta x^2} = \frac{2RT}{NK}\tau \ldots \ldots \ldots \ldots \ldots (25)$$

it was possible to obtain the product Ne without any reference to the size of the particle or the resistance of the medium to its motion. This quantity could then be compared with the same product obtained with great precision from electrolysis. The experimental procedure consists in balancing a given droplet and measuring, as in any Brownian-movement work, the quantity $\overline{\Delta x^2}$, then unbalancing it and measuring F, v_1 and $(v_1+v_2)_0$; the combination of (24) and (25) then gives

$$\overline{\Delta x^2} = \frac{2RT}{F}\frac{(v_1+v_2)_0}{Ne}\tau \ldots \ldots \ldots \ldots (26)$$

Since it is awkward to square each displacement Δx before averaging, it is preferable to modify by substituting from the Maxwell distribution law, which holds

[1] *Phys. Zeitschr.*, XII (1911), 202–8; see also *Phys. Rev.*, XXXIII (1911), 81.

[2] *Phys. Rev.*, I, N.S. (1913), 218.

for Brownian displacements as well as for molecular velocities, the relation

$$\overline{\Delta x} = \sqrt{\frac{2}{\pi}\overline{\Delta x^2}}\,.$$

We obtain thus

$$\overline{\Delta x} = \sqrt{\frac{4}{\pi}\frac{RT(v_1+v_2)_0}{F(Ne)}\tau}\,\ldots\ldots\ldots\ldots(27)$$

or

$$Ne = \frac{4}{\pi}\frac{RT(v_1+v_2)_0\tau}{F(\overline{\Delta x})^2}\,\ldots\ldots\ldots\ldots(28)$$

The possibility of thus eliminating the size of the particle and with it the resistance of the medium to its motion can be seen at once from the simple consideration that *so long as we are dealing with one and the same particle* the ratio K between the force acting and the velocity produced by it must be the same, whether the acting force is due to gravity or an electrical field, as in the oil-drop experiments, or to molecular impacts as in Brownian-movement work. De Broglie might have made such an elimination and calculation of Ne in his work, had his Brownian displacements and mobilities in electric fields been made on one and the same particle, but when the two sets of measurements are made on different particles, such elimination would involve the very uncertain assumption of identity of the particles in both charge and size. Although De Broglie did actually make this assumption, he did not treat his data in the manner indicated, and the first publication of this method of measuring Ne as well as the first actual determination was made in the papers mentioned above.

Some time later E. Weiss reported similar work to the Vienna Academy.[1]

2. Although it is possible to make the test of Ne in just the method described and although it was so made in the case of one or two drops, Mr. Fletcher worked out a more convenient method, which involves expressing the displacements Δx in terms of the fluctuations in the time required by the particle to fall a given distance and thus dispenses with the necessity of balancing the drop at all. I shall present another derivation which is very simple and yet of unquestionable validity.

In equation (28) let τ be the time required by the particle, if there were no Brownian movements, to fall between a series of equally spaced cross-hairs whose distance apart is d. In view of such movements the particle will have moved up or down a distance Δx in the time τ. Let us suppose this distance to be up. Then the actual time of fall will be $\tau + \Delta t$, in which Δt is now the time it takes the particle to fall the distance Δx. If now Δt is small in comparison with τ, that is, if Δx is small in comparison with d (say $1/10$ or less), then we shall introduce a negligible error (of the order $1/100$ at the most) if we assume that $\Delta x = v_1 \Delta t$ in which v_1 is the mean velocity under gravity. Replacing then in (28) $\overline{(\Delta x)}^2$ by $v_1^2 \overline{(\Delta t)}^2$, in which $\overline{(\Delta t)}^2$ is the square of the average difference between an observed time of fall and the mean time of fall t_g, that is, the square of the average

[1] It was read before the Academy on July 6: *Wiener Berichte*, CXX (1911), II, 1021, but appeared first in print in the August 1st number of the *Phys. Zeitschr.* (1911), p. 63. Fletcher's article is found in brief in an earlier number of the same volume of the *Phys. Zeitschr.*, p. 203, and was printed in full in the July number of *Le Radium*, VIII (1911), 279.

fluctuation in the time of fall through the distance d, we obtain after replacing the ideal time τ by the mean time t_g

$$Ne = \frac{4}{\pi} \frac{RT(v_1 + v_2)_0 t_g}{F v_1^2 (\overline{\Delta t})^2} \quad \ldots \ldots \ldots \ldots (29)$$

In any actual work $\overline{\Delta t}$ will be kept considerably less than $1/10$ the mean time t_g if the irregularities due to the observer's errors are not to mask the irregularities due to the Brownian movements, so that (29) is sufficient for practically all working conditions.[1]

The work of Mr. Fletcher and of the author was done by both of the methods represented in equations (28) and (29). The 9 drops reported upon in Mr. Fletcher's paper in 1911[2] yielded the results shown below in which n is the number of displacements used in each case in determining $\overline{\Delta x}$ or $\overline{\Delta}$

TABLE XIII

$\sqrt{Ne} \times 10^7$	n
1.68	125
1.67	136
1.645	321
1.695	202
1.73	171
1.65	200
1.66	84
1.785	411
1.65	85

When weights are assigned proportional to the number of observations taken, as shown in the last column

---

[1] No error is introduced here if, as assumed, $\overline{\Delta t}$ is small in comparison with t_g. However, for more rigorous equations see Fletcher, *Phys. Rev.*, IV (1914), 442; also Smoluchowski, *Phys. Zeitschr.*, XVI (1915), 321.

[2] *Le Radium*, VIII (1911), 279; *Phys. Rev.*, XXXIII (1911), 107.

of Table XIII, there results for the weighted mean value which represents an average of 1,735 displacements, $\sqrt{Ne} = 1.698 \times 10^7$ or $Ne = 2.88 \times 10^{14}$ *electrostatic units,* as against 2.896×10^{14}, the *value found in electrolysis.* The agreement between theory and experiment is then in this case about as good as one-half of 1 per cent, which is well within the limits of observational error.

This work seemed to demonstrate, with considerably greater precision than had been attained in earlier Brownian-movement work and with a minimum of assumptions, the correctness of the Einstein equation, which is in essence merely the assumption that a particle in a gas, no matter how big or how little it is or out of what it is made, is moving about with a mean translatory kinetic energy which is a universal constant dependent only on temperature. To show how well this conclusion has been established I shall refer briefly to a few later researches.

In 1914 Dr. Fletcher, assuming the value of K which I had published[1] for oil drops moving through air, made new and improved Brownian-movement measurements in this medium and solved for N the original Einstein equation, which, when modified precisely as above by replacing $\overline{\Delta x^2}$ by $\frac{2}{\pi} \left(\overline{\Delta x} \right)^2$ and $\left(\overline{\Delta x} \right)^2 = v_1^2 (\overline{\Delta t})^2$ becomes

$$N = \frac{4}{\pi} \frac{RT t_g}{K v_1^2 (\overline{\Delta t})^2} \quad \ldots \ldots \ldots \ldots \ldots (30)$$

He took, all told, as many as 18,837 Δt's, not less than 5,900 on a single drop, and obtained $N = 60.3 \times 10^{22} \pm 1.2$. This cannot be regarded as an altogether independent determination of N, since it involves my value

[1] *Phys. Rev.*, I (1913), 218.

of K. Agreeing, however, as well as it does with my value of N, it does show with much conclusiveness that both Einstein's equation and my corrected form of Stokes's equation apply accurately to the motion of oil drops of the size here used, namely, those of radius from 2.79×10^{-5} cm. to 4.1×10^{-5} cm. ($280 - 400$ $\mu\mu$).

In 1915 Mr. Carl Eyring tested by equation (29) the value of Ne on oil drops, of about the same size, in hydrogen and came out within .6 per cent of the value found in electrolysis, the probable error being, however, some 2 per cent.

Precisely similar tests on substances other than oils were made by Dr. E. Weiss[1] and Dr. Karl Przibram.[2] The former worked with silver particles only half as large as the oil particles mentioned above, namely, of radii between 1 and 2.3×10^{-5} cm. and obtained $Ne = 10,700$ electromagnetic units instead of 9,650, as in electrolysis. This is indeed 11 per cent too high, but the limits of error in Weiss's experiments were in his judgment quite as large as this. K. Przibram worked on suspensions in air of five or six different substances, the radii varying from 200 $\mu\mu$ to 600 $\mu\mu$, and though his results varied among themselves by as much as 100 per cent, his mean value came within 6 per cent of 9,650. Both of the last two observers took too few displacements on a given drop to obtain a reliable mean displacement, but they used so many drops that their mean Ne still has some significance.

It would seem, therefore, that the validity of Einstein's Brownian-movement equation had been pretty

[1] *Sitzungsber. d. k. Akad. d. Wiss. in Wien*, CXX (1911), II, 1021
[2] *Ibid.*, CXXI (1912), II, 950.

thoroughly established in gases. In liquids too it has recently been subjected to much more precise test than had formerly been attained. Nordlund,[3] in 1914, using minute mercury particles in water and assuming Stokes's Law of fall and Einstein's equations, obtained $N = 59.1 \times 10^{22}$. While in 1915 Westgren at Stockholm[4] by a very large number of measurements on colloidal gold, silver, and selenium particles, of diameter from 65 $\mu\mu$ to 130 $\mu\mu$ (6.5 to 13×10^{-6} cm.), obtained a result which he thinks is correct to one-half of 1 per cent, this value is $N = 60.5 \times 10^{22} \pm .3 \times 10^{22}$, which agrees perfectly with the value which I obtained from the measurements on the isolation and measurement of the electron.

The most recent determination of N by the Brownian movement method has been made at the University of Munich by Kappler[5] who suspends in still air from a very fine quartz fibre an exceedingly light mirror 2 mm. square, reflects a fine pencil of light from it, and records on a moving photographic film the irregular lateral displacements of this beam under the random molecular bombardment of the mirror at the end of this torsional system. From the mean of these displacements he gets $N = 60.59 \pm .036 \times 10^{22}$.

It has been because of such agreements as the foregoing that the last trace of opposition to the kinetic and atomic hypotheses of matter has disappeared from the scientific world, and that even Ostwald has been willing to make such a statement as that quoted on p. 10.

[3] *Ztschr. f. Phys. Chem.*, LXXXVII (1914), 40.

[4] *Die Brown'sche Bewegung besonders als Mittel zur Bestimmung der Avogadroschen Konstante*, inaugural dissertation. Upsala: Almquist & Wiksells Boktryckeri, 1915.

[5] *Ann. der Physik*, XI (1931), 233–56.

CHAPTER VIII

IS THE ELECTRON ITSELF DIVISIBLE?

It would not be in keeping with the method of modern science to make any dogmatic assertion as to the indivisibility of the electron. Such assertions used to be made in high-school classes with respect to the atoms of the elements, but the far-seeing among physicists, like Faraday, were always careful to disclaim any belief in the necessary ultimateness of the atoms of chemistry, and that simply because there existed until recently no basis for asserting anything about the insides of the atom. We knew that there was a smallest thing which took part in chemical reactions and we named that thing the atom, leaving its insides entirely to the future.

Precisely similarly the electron was defined as the smallest quantity of electricity which ever was found to appear in electrolysis, and nothing was then said or is now said about its necessary ultimateness. Our experiments have, however, now shown that this quantity is capable of isolation and exact measurement, and that all the kinds of charges which we have been able to investigate are exact multiples of it. Its value is 4.770×10^{-10} electrostatic units.

I. A SECOND METHOD OF OBTAINING e

I have presented one way of measuring this charge, but there is an indirect method of arriving at it which was worked out independently by Rutherford and Geiger[1]

[1] *Proc. Roy. Soc.*, A LXXXI (1908), 141, 161.

and Regener.[1] The unique feature in this method consists in actually counting the number of α-particles shot off per second by a small speck of radium or polonium through a given solid angle and computing from this the number of these particles emitted per second by one gram of the radium or polonium. Regener made his determination by counting the scintillations produced on a diamond screen in the focal plane of his observing microscope. He then caught in a condenser all the α-particles emitted per second by a known quantity of his polonium and determined the total quantity of electricity delivered to the condenser by them. This quantity of electricity divided by the number of particles emitted per second gave the charge on each particle. Because the α-particles had been definitely proved to be helium atoms[2] and the value of $\frac{e}{m}$ found for them showed that if they were helium they ought to carry double the electronic charge, Regener divided his result by 2 and obtained

$$e = 4.79 \times 10^{-10}.$$

He estimated his error at 3 per cent. Rutherford and Geiger made their count by letting the α-particles from a speck of radium C shoot into a chamber and produce therein sufficient ionization by collision to cause an electrometer needle to jump every time one of them entered. These authors measured the total charge as Regener did and, dividing by 2 the charge on each α-particle, they obtained

$$e = 4.65 \times 10^{-10}.$$

[1] *Sitzungsber. d. k. Preuss. Akad.*, XXXVIII (1909), 948.
[2] Rutherford and Royds, *Phil. Mag.*, XVII (1909), 281.

All determinations of e from radioactive data involve one or the other of these two counts, namely, that of Rutherford and Geiger or that of Regener. Thus, Boltwood and Rutherford[1] measured the total weight of helium produced in a second by a known weight of radium. Dividing this by the number of α-particles (helium atoms) obtained from Rutherford and Geiger's count, they obtain the mass of one atom of helium from which the number in a given weight, or volume since the gas density is known, is at once obtained. They published for the number n of molecules in a gas per cubic centimeter at $0°76$ cm., $n = 2.69 \times 10^{19}$, which corresponds to

$$e = 4.81 \times 10^{-10}.$$

This last method, like that of the Brownian movements, is actually a determination of N, rather than of e, since e is obtained from it only through the relation $Ne = 9,648.9$ electromagnetic units. Indeed, this is true of all methods of estimating e, so far as I am aware, except the oil-drop method and the Rutherford-Geiger-Regener method, and of these two the latter represents the measurement of the *mean* charge on an immense number of α-particles.

Thus a person who wished to contend that the unit charge appearing in electrolysis is only a mean charge which may be made up of individual charges which vary widely among themselves, in much the same way in which the atomic weight assigned to *neon* has recently been shown to be a mean of the weights of at least two different elements inseparable chemically, could not be gainsaid, save on the basis of the evidence contained in

[1] *Phil. Mag.* (6), XXII (1911), 599.

the oil-drop experiments; for these constitute the only method which has been found of measuring directly the charge on each individual ion.   It is of interest and significance for the present discussion, however, that the mean charge on an a-particle has been directly measured and that it comes out, within the limits of error of the measurement, at exactly two electrons—as it should according to the evidence furnished by $\dfrac{e}{m}$ measurements on the a-particles.

II.   THE EVIDENCE FOR THE EXISTENCE OF A SUB-ELECTRON

Now, the foregoing contention has actually been made, and evidence has been presented which purports to show that electric charges exist which are much smaller than the electron.   Since this raises what may properly be called the most fundamental question of modern physics, the evidence needs very careful consideration.   This evidence can best be appreciated through a brief historical review of its origin.

The first measurements on the mobilities in electric fields of swarms of charged particles of microscopically visible sizes were made by H. A. Wilson[1] in 1903, as detailed in chap. iii.   These measurements were repeated with modifications by other observers, including ourselves, during the years immediately following.   De Broglie's modification, published in 1908,[2] consisted in sucking the metallic clouds discovered by Hemsalech and De Watteville,[3] produced by sparks or arcs between

[1] *Phil. Mag.* (6), V (1903), 429.

[2] *Comptes rendus*, CXLVI (1908), 624, 1010.

[3] *Ibid.*, CXLIV (1907), 1338.

metal electrodes, into the focal plane of an ultra-microscope and observing the motions of the individual particles in this cloud *in a horizontal electrical field pro-duced by applying a potential difference to two vertical parallel plates* in front of the objective of his microscope. In this paper De Broglie first commented upon the fact that some of these particles were charged positively, some negatively, and some not at all, and upon the further fact that holding radium near the chamber caused changes in the charges of the particles. He promised quantitative measurements of the charges themselves. One year later he fulfilled the promise,[1] and at practically the same time Dr. Ehrenhaft[2] pub-lished similar measurements made with precisely the arrangement described by De Broglie a year before. Both men, as Dr. Ehrenhaft clearly pointed out,[3] while *observing* individual particles, obtained only a mean charge, since the different measurements entering into the evaluation of *e* were made on different particles. So far as concerns *e*, these measurements, as everyone agrees, were essentially cloud measurements like Wilson's.

In the spring and summer of 1909 I isolated indi-vidual water droplets and determined the charges car-ried by each one,[4] and in April, 1910, I read before the American Physical Society the full report on the oil-drop work in which the multiple relations between charges were established, Stokes's Law corrected, and *e* accurately

[1] *Ibid.*, CXLVIII (1909), 1316.

[2] *Phys. Zeitschr.*, X (1909), 308.

[3] *Ibid.*, XI (1910), 619.

[4] This paper was published in abstract in *Phys. Rev.*, XXX (1909), 560, and *Phil. Mag.*, XIX (1910), 209.

determined.[1] In the following month (May, 1910) Dr. Ehrenhaft,[2] having seen that a vertical condenser arrangement made possible, as shown theoretically and experimentally in the 1909 papers mentioned above, the independent determination of the charge on each individual particle, read the first paper in which he had used this arrangement in place of the De Broglie arrangement which he had used theretofore. He reported results identical in all essential particulars with those which I had published on water drops the year before, save that where I obtained consistent and simple multiple relations between charges carried by different particles he found no such consistency in these relations. The absolute values of these charges obtained on the assumption of Stokes's Law fluctuated about values considerably lower than 4.6×10^{-10}. Instead, however, of throwing the burden upon Stokes's Law or upon wrong assumptions as to the density of his particles, he remarked in a footnote that Cunningham's theoretical correction to Stokes's Law,[3] which he (Ehrenhaft) had just seen, would make his values come still lower, and hence that no failure of Stokes's Law could be responsible for his low values. He considered his results therefore as opposed to the atomic theory of electricity altogether, and in any case as proving the existence of charges much smaller than that of the electron.[4]

[1] This paper was published in abstract in *Phys. Rev.*, XXXI (1910), 92; *Science*, XXXII (1910), 436; *Phys. Zeitschr.*, XI (1910), 1097.

[2] *Wien. Ber.*, CXIX (1910), II, 809. This publication was apparently not issued before December, 1910, for it is not noted in *Naturae Novitates* before this date.

[3] *Proc. Roy. Soc.*, LXXXIII (1910), 360.

[4] These results were presented and discussed at great length in the fall of 1910; see *Phys. Zeitschr.*, XI (1910), 619, 940.

The apparent contradiction between these results and mine was explained when Mr. Fletcher and myself showed[1] experimentally that Brownian movements produced just such apparent fluctuations as Ehrenhaft observed when the e is computed, as had been done in his work, from one single observation of a speed under gravity and a corresponding one in an electric field. We further showed that the fact that his values fluctuated about too low an average value meant simply that his particles of gold, silver, and mercury were less dense because of surface impurities, oxides or the like, than he had assumed. The correctness of this explanation would be well-nigh demonstrated if the values of Ne computed by equations (28) or (29) in chap. vii from a large number of observations on Brownian movements always came out as in electrolysis, for in these equations no assumption has to be made as to the density of the particles. As a matter of fact, all of the nine particles studied by us and computed by Mr. Fletcher[2] showed the correct value of Ne, while only six of them as computed by me fell on, or close to, the line which pictures the law of fall of an oil drop through air (Fig. 5, p. 106). This last fact was not published in 1911 because it took me until 1913 to determine with sufficient certainty a second approximation to the complete law of fall of a droplet through air; in other words, to extend curves of the sort given in Fig. 5 to as large values of $\frac{l}{a}$ as correspond to particles small enough to show large Brownian movements. As soon as I had done this I computed all the nine drops which gave correct values of Ne and

[1] *Phys. Zeitschr.*, XII (1911), 161; *Phys. Rev.*, XXXII (1911), 394.

[2] *Le Radium*, VIII (1911), 279; *Phys. Rev.*, XXXIII (1911), 107.

found that two of them fell far below the line, one more fell somewhat below, while one fell considerably above it. This meant obviously that these four particles were not spheres of oil alone, two of them falling much too slowly to be so constituted and one considerably too rapidly. There was nothing at all surprising about this result, since I had explained fully in my first paper on oil drops[1] that until I had taken great precaution to obtain dust-free air "the values of e_1 came out differently, even for drops showing the same velocity under gravity." In the Brownian-movement work no such precautions to obtain dust-free air had been taken because we wished to test the general validity of equations (28) and (29). That we actually used in this test two particles which had a mean density very much smaller than that of oil and one which was considerably too heavy, was fortunate since it indicated that our result was indeed independent of the material used.

It is worthy of remark that in general, even with oil drops, almost all of those behaving abnormally fall too slowly, that is, they fall below the line of Fig. 5 and only rarely does one fall above it. This is because the dust particles which one is likely to observe, that is, those which remain long in suspension in the air, are either in general lighter than oil or else expose more surface and hence act as though they were lighter. When one works with particles made of dense metals this behavior will be still more marked, since all surface impurities of whatever sort will diminish the density. The possibility, however, of freeing oil-drop experiments from all such sources of error is shown by the fact that although during

[1] *Phys. Rev.*, XXXIII (1911), 366, 367.

the year 1915–16 I studied altogether as many as three hundred drops, there was not one which did not fall within less than 1 per cent of the line of Fig. 5.   It will be shown, too, in this chapter, that in spite of the failure of the Vienna experimenters, it is possible under suitable conditions to obtain mercury drops which behave, even as to law of fall, in practically all cases with perfect consistency and normality.

When E. Weiss in Prag and K. Przibram in the Vienna laboratory itself, as explained in chap. vii, had found that Ne for all the substances which they worked with, including silver particles like those used by Ehrenhaft, gave about the right value of Ne, although yielding much too low values of e when the latter was computed from the law of fall of *silver* particles, the scientific world practically universally accepted our explanation of Ehrenhaft's results and ceased to concern itself with the idea of a sub-electron.[1]

In 1914 and 1915, however, Professor Ehrenhaft[2] and two of his pupils, F. Zerner[3] and D. Konstantinowsky,[4] published new evidence for the existence of such a subelectron and the first of these authors has kept up some discussion of the matter up to the present.   These experimenters make three contentions.   The first is essentially that they have now determined Ne for their particles by equation (29); and although in many instances it comes out as in electrolysis, in some instances it comes out from

[1] See R. Pohl, *Jahrbuch der Radioactivität und Elektronik*, VII (1912), 431.

[2] *Wien. Sitzungsber.*, CXXIII (1914), 53–155; *Ann. d. Phys.*, XLIV (1914), 657.

[3] *Phys. Zeitschr.*, XVI (1915), 10.

[4] *Ann. d. Phys.*, XLVI (1915), 261.

20 per cent to 50 per cent too low, while in a few cases it is as low as one-fourth or one-fifth of the electrolytic value. Their procedure is in general to publish, not the value of Ne, but, instead, the value of e obtained from Ne by inserting Perrin's value of N (70×10^{22}) in (29) and then solving for e. This is their method of determining e "from the Brownian movements."

Their second contention is the same as that originally advanced, namely, that, in some instances, when e is determined with the aid of Stokes's Law of fall (equation 12, p. 91), even when Cunningham's correction or my own (equation 15, p. 101) is employed, the result comes out very much lower than. 4.77×10^{-10}. Their third claim is that the value of e, determined as just explained from the Brownian movements, is in general higher than the value computed from the law of fall, and that the departures become greater and greater the smaller the particle. These observers conclude therefore that we oil-drop observers failed to detect sub-electrons because our droplets were too big to be able to reveal their existence. The minuter particles which they study, however, seem to them to bring these sub-electrons to light. *In other words, they think the value of the smallest charge which can be caught from the air actually is a function of the radius of the drop on which it is caught, being smaller for small drops than for large ones.*

Ehrenhaft and Zerner even analyze our report on oil droplets and find that these also show in certain instances indications of sub-electrons, for they yield in these observers' hands too low values of e, whether computed from the Brownian movements or from the law of fall.

When the computations are made in the latter way e is found, according to them, to decrease with decreasing radius, as is the case in their experiments on particles of mercury and gold.

III. CAUSES OF THE DISCREPANCIES

Now, the single low value of Ne which these authors find in the oil-drop work is obtained by computing Ne from some twenty-five observations on the times of fall, and an equal number on the times of rise, of a particle which, before we had made any Ne computations at all, we reported upon[1] for the sake of showing that the Brownian movements would produce just such fluctuations as Ehrenhaft had observed when the conditions were those under which he worked. When I compute Ne by equation (29), using merely the twenty-five times of fall, I find the value of Ne comes out 26 per cent low, just as Zerner finds it to do. If, however, I omit the first reading it comes out but 11 per cent low. In other words, the omission of one single reading changes the result by 15 per cent. Furthermore, Fletcher[2] has shown that these same data, though treated entirely legitimately, but with a slightly different grouping than that used by Zerner, can be made to yield exactly the right value of Ne. This brings out clearly the futility of attempting to test a statistical theorem by so few observations as twenty-five, which is nevertheless more than Ehrenhaft usually uses on his drops. Furthermore, I shall presently show that unless one observes under carefully chosen conditions, his own errors of observation

[1] *Phys. Zeitschr.*, XII (1911), 162.
[2] *Ibid.*, XVI (1915), 316.

and the slow evaporation of the drop tend to make Ne obtained from equation (29) come out too low, and these errors may easily be enough to vitiate the result entirely. *There is, then, not the slightest indication in any work which we have thus far done on oil drops that Ne comes out too small.*

Next consider the apparent *variation in e* when it is computed from the law of fall. Zerner computes e from my law of fall in the case of the nine drops published by Fletcher, in which Ne came out as in electrolysis, and finds that one of them yields $e = 6.66 \times 10^{-10}$, one $e = 3.97 \times 10^{-10}$, one $e = 1.32 \times 10^{-10}$, one $e = 1.7 \times 10^{-10}$, while the other five yield about the right value, namely, 4.8×10^{-10}. In other words (as stated on p. 165 above), five of these drops fall exactly on my curve (Fig. 5), one falls somewhat above it, one somewhat below, while two are entirely off and very much too low. These two, therefore, I concluded were not oil at all, but dust particles. Since Zerner computes the radius from the rate of fall, these two dust particles which fall much too slowly, and therefore yield too low values of e, must, of course, yield correspondingly low values of a. Since they are found to do so, Zerner concludes that our oil drops, as well as Ehrenhaft's mercury particles, yield decreasing values of e with decreasing radius. His own tabulation does not show this. It merely shows three erratic values of e, two of which are very low and one rather high. But a glance at all the other data which I have published on oil drops shows the complete falsity of this position,[1] *for these data show that after I had eliminated dust all of my particles yielded exactly the same value of "e" whatever their*

[1] *Phys. Rev.*, II (1913), 138.

size[1]. The only possible interpretation then which could
be put on these two particles which yielded correct values
of *Ne*, but too slow rates of fall, was that which I put
upon them, namely, that they were not spheres of oil.

As to the Vienna data on mercury and gold, Dr.
Ehrenhaft publishes, all told, data on just sixteen particles and takes for his Brownian-movement calculations
on the average *fifteen times of fall and fifteen of rise on
each, the smallest number being 6 and the largest 27*. He
then computes his statistical average $(\overline{\Delta t})^2$ from observations of this sort. Next he assumes Perrin's value of N,
namely, 70×10^{22}, which corresponds to $e = 4.1$, and
obtains instead by the Brownian-movement method, i.e.,
the *Ne* method, the following values of e, the exponential
term being omitted for the sake of brevity: 1.43, 2.13,
1.38, 3.04, 3.5, 6.92, 4.42, 3.28, .84. Barring the first
three and the last of these, the mean value of e is just
about what it should be, namely, 4.22 instead of 4.1.
Further, the first three particles are the heaviest ones,
the first one falling between his cross-hairs in 3.6 seconds,
and its fluctuations in time of fall are from 3.2 to 3.85
seconds, that is, three-tenths of a second on either side
of the mean value. *Now, these fluctuations are only
slightly greater than those which the average observer will
make in timing the passage of a uniformly moving body
across equally spaced cross-hairs*. This means that in
these observations two nearly equally potent causes were
operating to produce fluctuations. The observed Δt's
were, of course, then, larger than those due to Brownian
movements alone, and might easily, with but a few
observations, be two or three times as large. Since

[1] See *Phys. Rev.*, II (1913), 134–35.

$(\overline{\Delta t})^2$ appears in the denominator of equation (29), it will be seen at once that because of the observer's timing errors a series of observed Δt's will always tend to be larger than the Δt due to Brownian movements alone, and hence that the Brownian-movement method always tends to yield too low a value of Ne, and accordingly too low a value of e. *It is only when the observer's mean error is wholly negligible in comparison with the Brownian-movement fluctuations that this method will not yield too low a value of e.* The overlooking of this fact is, in my judgment, one of the causes of the low values of e recorded by Dr. Ehrenhaft.

Again, in the original work on mercury droplets which I produced both by atomizing liquid mercury and by condensing the vapor from boiling mercury,[1] I noticed that such droplets evaporated for a time even more rapidly than oil, and other observers who have since worked with mercury have reported the same behavior.[2] The amount of this effect may be judged from the fact that one particular droplet of mercury recently under observation in this laboratory had at first a speed of 1 cm. in 20 seconds, which changed in half an hour to 1 cm. in 56 seconds. The slow cessation, however, of this evaporation indicates that the drop slowly becomes coated with some sort of protecting film. Now, if any evaporation whatever is going on while successive times of fall are being observed—and as a matter of fact changes due to evaporation or condensation are always taking place to some extent—the apparent $(\overline{\Delta t})^2$ will be larger than that due to Brownian movements, even

[1] *Phys. Rev.*, XXXII (1911), 389.
[2] See Schidlof et Karpowicz, *Comptes rendus*, CLVIII (1914), 1912.

though these movements are large enough to prevent the
observer from noticing, in taking twenty or thirty readings, that the drop is continually changing. These
changes combined with the fluctuations in t due to the
observer's error are sufficient, I think, to explain all of the
low values of e obtained by Dr. Ehrenhaft by the
Brownian-movement method. Indeed, I have myself
repeatedly found Ne coming out less than half of its
proper value *until I corrected for the evaporation of the
drop*, and this was true when the evaporation was so
slow that its rate of fall changed but 1 or 2 per cent in a
half-hour. But it is not merely evaporation which introduces an error of this sort. The running down of the
batteries, the drifting of the drop out of focus, or anything which causes changes in the times of passage
across the equally spaced cross-hairs tends to decrease
the apparent value of Ne. *There is, then, so far as I can
see, no evidence at all in any of the data published to date
that the Brownian-movement method actually does yield too
low a value of "e," and very much positive evidence that it
does not was given in the preceding chapter.*

Indeed, the same type of Brownian-movement work
which Fletcher and I did upon oil-drops in 1910 and 1911
(see preceding chapter) was done in 1920 and 1921 in
Vienna with the use of particles of selenium, and with results which are in complete harmony with our own. The
observer, E. Schmid,[1] takes as many as 1,500 "times of
fall" upon a given particle, the radius of which is in one
case as low as 5×10^{-6} cm.—quite as minute as any used
by Dr. Ehrenhaft—and obtains in all cases values of e

[1] E. Schmid, *Wien. Akad. Ber.*, CXXIX (1920), 813, and *ZfP*, V
(1921), 27.

by "the Brownian-movement method" which are in as good agreement with our own as could be expected in view of the necessary observational error. This complete check of our work in Vienna itself should close the argument so far as the Brownian movements are concerned.

That e and a computed from the law of fall become farther and farther removed from the values of e and a computed from the Brownian movements, the smaller these particles appear to be, is just what would be expected if the particles under consideration have surface impurities or non-spherical shapes or else are not mercury at all.[1] If, further, *exact multiple relations hold for them*, as at least a dozen of us, including Dr. Ehrenhaft himself, now find that they invariably do, there is scarcely any other interpretation possible except that of incorrect assumptions as to density.[1] Again, the fact that these data are all taken when the observers are working with the exceedingly dense substances, mercury and gold, *volatilized in an electric arc*, and when, therefore, anything not mercury or gold, but assumed to be, would yield very low values of e and a, is in itself a very significant circumstance. The further fact that Dr. Ehrenhaft implies that normal values of e very frequently appear in his work,[2] while these low erratic drops represent only a part of the data taken, is suggestive. When

[1] R. Bär, in a series of articles recently summarized in *Die Naturwissenschaften*, Vols. XIV and XV, 1922, has emphasized this point. His data serve merely as a new check upon the work found in our preceding tables.

[2] "Die bei grösseren Partikeln unter gewissen Umständen bei gleicher Art der Erzeugung häufig wiederkehrenden höheren Quanten waren dann etwa als stabilere räumliche Gleichgewichtsverteilungen

one considers, too, that in place of the beautiful consistency and duplicability shown in the oil-drop work,
Dr. Ehrenhaft and his pupils never publish data on any
two particles which yield the same value of e, but instead
find only irregularities and erratic behavior,[1] just as
they would expect to do with non-uniform particles, or
with particles having dust specks attached to them,
one wonders why any explanation other than the foreign-
material one, which explains all the difficulties, has ever
been thought of.   As a matter of fact, in our work with
mercury droplets, we have found that the initial rapid
evaporation gradually ceases, just as though the droplets
had become coated with some foreign film which prevents further loss.   Dr. Ehrenhaft himself, in speaking
of the Brownian movements of his metal particles, com-
ments on the fact that they seem at first to show large
movements which grow smaller with time.[2]   This is
just what would happen if the radius were increased by
the growth of a foreign film.

Now what does Dr. Ehrenhaft say to these very
obvious suggestions as to the cause of his troubles?
Merely that he has avoided all oxygen, and hence that

---

dieser Sub-electron anzusehen, die sich unter gewissen Umständen
ergeben."—*Wien. Ber.*, CXXIII, 59.

[1] Their whole case is summarized in the tables in *Ann. d. Phys.*,
XLIV (1914), 693, and XLVI (1915), 292, and it is recommended that
all interested in this discussion take the time to glance at the data on
these pages, for the data themselves are so erratic as to render discussion needless.

[2] "Wie ich in meinen früheren Publikationen erwähnt habe, zeigen
die ultramikroskopischen Metallpartikel, unmittelbar nach der Erzeugung beobachtet, eine viel lebhaftere Brownsche Bewegung als nach
einer halben Stunde."—*Phys. Zeitschr.*, XII, 98.

an oxide film is impossible. *Yet he makes his metal particle by striking an electric arc between metal electrodes.* This, as everyone knows, brings out all sorts of occluded gases. Besides, chemical activity in the electric arc is tremendously intense, so that there is opportunity for the formation of all sorts of higher nitrides, the existence of which in the gases coming from electric arcs has many times actually been proved. Dr. Ehrenhaft says further that he photographs big mercury droplets and finds them spherical and free from oxides. But the fact that some drops are pure mercury is no reason for assuming that all of them are, and it is only the data on those which are not which he publishes. Further, because big drops which he can see and measure are of mercury is no justification at all for assuming that sub-microscopic particles are necessarily also spheres of pure mercury. In a word, Dr. Ehrenhaft's tests as to sphericity and purity are all absolutely worthless as applied to the particles in question, which according to him have radii of the order 10^{-6} cm.—a figure a hundred times below the limit of sharp resolution.

IV. THE BEARING OF THE VIENNA WORK ON THE QUESTION OF THE EXISTENCE OF A SUB-ELECTRON

But let us suppose that these observers do actually work with particles of pure mercury and gold, as they think they do, and that the observational and evaporational errors do not account for the low values of Ne. Then what conclusion could legitimately be drawn from their data? Merely this and nothing more, that (1) Einstein's Brownian-movement equation is not universally applicable, and (2) that the law of motion of their very minute charged particles through air is not yet fully known.[1] So long as they find exact multiple relationships, as Dr. Ehrenhaft now does, between the charges carried by a given particle when its charge is changed by the capture of ions or the direct loss of electrons, the charges on these ions must be the same as the ionic charges which I have accurately and consistently measured and found equal to 4.77×10^{-10} electrostatic units; for they, in their experiments, capture exactly the same sort of ions, produced in exactly the same way as those which I captured and measured in my experiments. That these same ions have one sort of a charge when captured by a big drop and another sort when captured by a little drop is obviously absurd. *If they are not the same ions which are caught, then in order to reconcile the results with the existence of the exact multiple relationship found by Dr. Ehrenhaft as well as ourselves, it would be necessary to assume that there exist in the air an infinite number of different kinds of ionic charges corresponding to the infinite number of possible radii of drops, and that when a powerful electric field drives all of these ions toward a given drop this drop selects in each instance just the charge which corresponds to its particular radius.* Such an assumption is not only too grotesque for serious consideration, but it is directly contradicted by my experiments, for I have repeatedly pointed out that with a given value of $\dfrac{l}{a}$ I obtain exactly the same value of e_1, whether I work with big drops or with little ones.

---

[1] In my own opinion this is a conclusion contrary to fact, since in a recent paper (see *Phys. Rev.*, July, 1923) I have fully established the "Complete Law of Fall."

V. NEW PROOF OF THE CONSTANCY OF *e*

For the sake of subjecting the constancy of *e* to the most searching test, I have made new measurements of the same kind as those heretofore reported, but using now a range of sizes which overlaps that in which Dr. Ehrenhaft works. I have also varied through wide limits the nature and density of both the gas and the drops. Fig. 13 (I) contains new oil-drop data taken in air; Fig. 13 (II) similar data taken in hydrogen. The radii of these drops, computed by the very exact method given in the *Physical Review*,[1] vary tenfold, namely, from .000025 cm. to .00023 cm. Dr. Ehrenhaft's range is from .000008 cm. to .000025 cm. It will be seen that these drops fall in every instance on the lines of Fig. 13, I and II, and hence that they all yield exactly the same value of e^z, namely, 61.1×10^{-8}. The details of the measurements, which are just like those previously given, will be entirely omitted. There is here *not a trace of an indication that the value of "e" becomes smaller as "a" decreases*. The points on these two curves represent consecutive series of observations, not a single drop being omitted in the case of either the air or the hydrogen. This shows the complete uniformity and consistency which we have succeeded in obtaining in the work with oil drops.

That mercury drops show a similar behavior was somewhat imperfectly shown in the original observations which I published on mercury.[2] I have since fully confirmed the conclusions there reached. That mercury drops can with suitable precautions be made to behave

[1] II (1913), 117.
[2] *Ibid.*, CCC (1911), 389–90.

practically as consistently as oil is shown in Fig. 13 (III), which represents data obtained by blowing into the observing chamber above the pinhole in the upper plate a cloud of mercury droplets formed by the condensation of the vapor arising from boiling mercury. These results have been obtained in the Ryerson Laboratory with my apparatus by Mr. John B. Derieux. Since the pressure was here always atmospheric, the drops progress in the order of size from left to right, the largest having a diameter about three times that of the smallest, the radius of which is .00003244 cm. The original data may be found in the *Physical Review*, December, 1916. In Fig. 13 (IV) is found precisely similar data taken with my apparatus by Dr. J. Y. Lee on solid spheres of shellac falling in air.[1] Further, very beautiful work, of this same sort, also done with my apparatus, has recently been published by Dr. Yoshio Ishida (*Phys. Rev.*, May, 1923), who, using many different gases, obtains a group of lines like those shown in Fig. 13, *all of which though of different slopes, converge upon one and the same value of "$e^{\frac{2}{3}}$," namely, 61.08×10^{-8}.*

These results establish with absolute conclusiveness the correctness of the assertion that the apparent value of the electron is not in general a function of the gas in which the particle falls, of the materials used, or of the radius of the drop on which it is caught, even when that drop is of mercury, and even when it is as small as some of those

[1] The results shown in Fig. 13 do not lay claim to the precision reached in those recorded in Table X and Fig. 10. No elaborate precautions were here taken in the calibration of the Hipp chronoscope and the voltmeter, and it is due to slight errors discovered later in these calibrations that the slope of line I in Fig. 13 is not quite in agreement with the slope in Fig. 10.

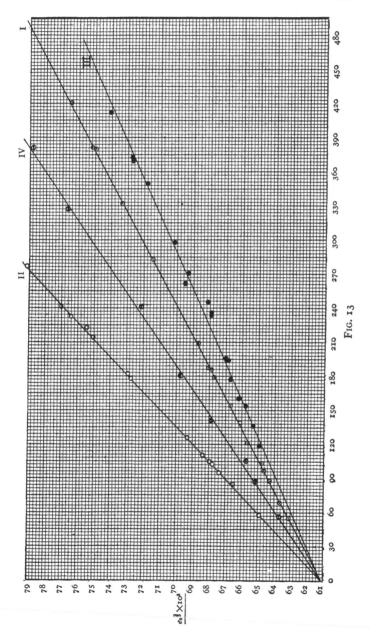

FIG. 13

with which Dr. Ehrenhaft obtained his erratic results. If it appears to be so with his drops, the cause cannot possibly be found in actual fluctuations in the charge of the electron without denying completely the validity of my results. But these results have now been checked, in their essential aspects, by scores of observers, including Dr. Ehrenhaft himself. Furthermore, it is not my results alone with which Dr. Ehrenhaft's contention clashes. The latter is at variance also with all experiments like those of Rutherford and Geiger and Regener on the measurement of the charges carred by α- and β-particles, for these are infinitely smaller than any particles used by Dr. Ehrenhaft; and if, as he contends, the value of the unit out of which a charge is built up is smaller and smaller the smaller the capacity of the body on which it is found, then these α-particle charges ought to be extraordinarily minute in comparison with the charges on our oil drops. Instead of this, the charge on the α-particle comes out exactly twice the charge which I measure in my oil-drop experiments.

While then it would not be in keeping with the spirit or with the method of modern science to make any dogmatic assertion about the existence or non-existence of a sub-electron, it can be asserted with entire confidence that there has not appeared up to the present a scrap of evidence for the existence of charges smaller than the electron. If all of Dr. Ehrenhaft's assumptions as to the nature of his particles were correct, then his experiments would mean simply that Einstein's Brownian-movement equation is not of universal validity and that the law of motion of minute charged particles is quite different from that which he has assumed. It is exceedingly unlikely that either of these results can be drawn from his experiments, for Nordlund[1] and Westgren[2] have apparently verified the Einstein equation in liquids with very much smaller particles than Dr. Ehrenhaft uses; and, on the other hand, while I have worked with particles as small as 2×10^{-5} cm. and with values of $\frac{l}{a}$ as large as 135, which is very much larger than any which appear in the work of Dr. Ehrenhaft and his pupils, I have thus far found no evidence of a law of motion essentially different from that which I published in 1913, and further elaborated and refined in 1923.

There has then appeared up to the present time no evidence whatever for the existence of a sub-electron. The chapter having to do with its discussion is now considered for the present at least to have been closed,[3] but it constitutes an interesting historical document worthy of study as an illustration on the one hand of the solidity of the foundations upon which the atomic theory of electricity now rests, and on the other hand of the severity of the gauntlet of criticism which new results must run before they gain admission to the body of established truth in physics.

[1] *Zeit. für Phys. Chem.*, LXXXVII (1914), 40.

[2] Inaugural Dissertation von Arne Westgren, *Untersuchungen über Brownsche Bewegung*, Stockholm, 1915.

[3] R. Bär, "Der Streit um das Elektron," *Die Naturwissenschaften*, 1922.

CHAPTER IX

THE STRUCTURE OF THE ATOM

We have shown in the preceding chapters how within the last three decades there has been discovered beneath the nineteenth-century world of molecules and atoms a wholly new world of electrons, the very existence of which was undreamed of forty years ago. We have seen that these electrons, since they can be detached by X-rays from all kinds of neutral atoms, must be constitutents of all atoms. Whether or not they are the sole constituents we have thus far made no attempt to determine. We have concerned ourselves with studying the properties of these electrons themselves and have found that they are of two kinds, negative and positive, which are, however, exactly alike in strength of charge but wholly different in inertia or mass, the negative being commonly associated with a mass which is but $1/1,835$ of that of the lightest known atom, that of hydrogen while the positive appears never to be associated with a mass smaller than that of the hydrogen atom. We have found how to isolate and measure accurately the electronic charge and have found that this was the key which unlocked the door to many another otherwise inaccessible physical magnitude. It is the purpose of this chapter to consider certain other fields of exact knowledge which have been opened up through the measurement of the electron, and in particular to discuss what the physicist, as he has peered with his newly discovered agencies, X-rays, radioactivity, ultra-violet light, etc., into the insides of atoms, has been able to discover regarding the numbers and sizes and relative positions and motions of these electronic constituents, and to show how far he has gone in answering the question as to whether the electrons are the sole building-stones of the atoms.

I. THE SIZES OF ATOMS

One of the results of the measurement of the electronic charge was to make it possible to find the quantity which is called the diameter of an atom with a definiteness and precision theretofore altogether unattained.

It was shown in chap. v that the determination of e gave us at once a knowledge of the exact number of molecules in a cubic centimeter of a gas. Before this was known we had fairly satisfactory information as to the relative diameters of different molecules, for we have known for a hundred years that different gases when at the same temperature and pressure possess the same number of molecules per cubic centimeter (Avogadro's rule). From this it is at once evident that, as the molecules of gases eternally dart hither and thither and ricochet against one another and the walls of the containing vessel, the average distance through which one of them will go between collisions with its neighbors will depend upon how big it is. The larger the diameter the less will be the mean distance between collisions—a quantity which is technically called "the mean free path." Indeed, it is not difficult to see that in different gases the mean free path l is an inverse measure of the molecular cross-section. The exact relation is easily deduced (see Appendix E). It is

$$l = \frac{1}{\pi n d^2 \sqrt{2}} \dots\dots\dots\dots\dots\dots (31)$$

in which d is the molecular diameter and n is the number of molecules per cubic centimeter of the gas. Now, we have long had methods of measuring l, for it is upon this that the coefficient of viscosity of the gas largely depends. When, therefore, we have measured the viscosities of different gases we can compute the corresponding l's, and then from equation (31) the relative diameters d, since n is the same for all gases at the same temperature and pressure. But the absolute value of d can be found only after the absolute value of n is known. If we insert in equation (31) the value of n found from e by the method presented in chap. v, it is found that the average diameter of the atom of the monatomic gas helium is 2×10^{-8} cm., that of the diatomic hydrogen molecule is a trifle more, while the diameters of the molecules of the diatomic gases, oxygen and nitrogen, are 50 per cent larger.[1] This would make the diameter of a single atom of hydrogen a trifle smaller, and that of a single atom of oxygen or nitrogen a trifle larger than that of helium. By the average molecular diameter we mean the average distance to which the centers of two molecules approach one another in such impacts as are continually occurring in connection with the motions of thermal agitation of gas molecules—this and nothing more.

As will presently appear, the reason that two molecules thus rebound from one another when in their motion of thermal agitation their centers of gravity approach to a distance of about 2×10^{-8} cm. is presumably that the atom is a system with negative electrons in its outer regions. When these negative electrons in two

[1] R. A. Millikan, *Phys. Rev.*, XXXII (1911), 397.

different systems which are coming into collision approach to about this distance, the repulsions between these similarly charged bodies begin to be felt, although at a distance the atoms are forceless. With decreasing distance this repulsion increases very rapidly until it becomes so great as to overcome the inertias of the systems and drive them asunder.

II. THE RADIUS OF THE ELECTRON FROM THE ELECTRO-MAGNETIC THEORY OF THE ORIGIN OF MASS

The first estimates of the volume occupied by a single one of the electronic constituents of an atom were obtained from the electromagnetic theory of the origin of mass, and were therefore to a pretty large degree speculative, but since these estimates are strikingly in accord with results which follow from direct experiments and are independent of any theory, and since, further, they are of extraordinary philosophic as well as historic interest, they will briefly be presented here.

Since Rowland proved that an electrically charged body in motion is an electrical current the magnitude of which is proportional to the speed of motion of the charge, and since an electric current, by virtue of the property called its self-induction, opposes any attempt to increase or diminish its magnitude, it is clear that an electrical charge, as such, possesses the property of inertia. But inertia is the only invariable property of matter. It is the quantitative measure of matter, and matter quantitatively considered is called *mass*. It is clear, then, theoretically, that an electrically charged pith ball must possess more mass than the same pith ball when uncharged. But when we compute how much the mass of a pith ball is increased by any charge which we can actually get it to hold, we find that the increase is so extraordinarily minute as to be hopelessly beyond the possibility of experimental detection. However, the method of making this computation, which was first pointed out by Sir J. J. Thomson in 1881,[1] is of unquestioned validity, so that we may feel quite sure of the correctness of the result. Further, when we combine the discovery that an electric charge possesses the distinguishing property of matter, namely, inertia, with the discovery that all electric charges are built up out of electrical specks all alike in charge, we have made it entirely legitimate to consider an electric current as *the passage of a definite, material, granular substance along the conductor. In other words, the two entities, electricity and matter, which the nineteenth century tried to keep distinct, begin to look like different aspects of one and the same thing.*

But, though we have thus justified the statement that electricity is material, have we any evidence as yet that all matter is electrical—that is, that all inertia is of the same origin as that of an electrical charge? The answer is that we have *evidence*, but as yet no *proof*. The theory that this is the case is still a speculation, but one which rests upon certain very significant facts. These facts are as follows:

If a pith ball is spherical and of radius a, then the mass m due to a charge E spread uniformly over its surface is given, as is shown in Appendix D by,

$$m = \frac{2}{3}\frac{E^2}{a} \quad \dots\dots\dots\dots\dots\dots (32)$$

[1] J. J. Thomson, *Phil. Mag.*, XI (1881), 229.

The point of especial interest in this result is that the mass is inversely proportional to the radius, so that the smaller the sphere upon which we can condense a given charge E the larger the mass of that charge. If, then, we had any means of measuring the minute increase in mass of a pith ball when we charge it electrically with a known quantity of electricity E, we could compute from equation (32) the size of this pith ball, even if we could not see it or measure it in any other way. This is much the sort of a position in which we find ourselves with respect to the negative electron. We can measure its mass, and it is found to be accurately $1/1,835$ of that of the hydrogen atom. We have measured accurately its charge and hence can compute the radius a of *the equivalent sphere*, that is, the sphere over which e would have to be uniformly distributed to have the observed mass, provided we assume that the observed mass of the electron is all due to its charge.

The justification for such an assumption is of two kinds. First, since we have found that electrons are constituents of all atoms and that mass is a property of an electrical charge, it is of course in the interests of simplicity to assume that all the mass of an atom is due to its contained electrical charges, rather than that there are two wholly different kinds of mass, one of electrical origin and the other of some other sort of an origin. Secondly, if the mass of a negative electron is all of electrical origin, then we can show from electromagnetic theory that this mass ought to be independent of the speed with which the electron may chance to be moving unless that speed approaches close to the speed of light. But from one-tenth the speed of light up to that speed the mass ought to vary with speed in a definitely predictable way.

Now, it is a piece of rare good fortune for the testing of this theory that radium actually does eject negative electrons with speeds which can be accurately measured and which do vary from three-tenths up to ninety-eight hundredths of that of light. *It is further one of the capital discoveries of the twentieth century[1] that within these limits the observed rate of variation of the mass of the negative electron with speed agrees accurately with the rate of variation computed on the assumption that this mass is all of electrical origin.* Such is the experimental argument for the electrical origin of mass.[2]

Solving then equation (32) for a, we find that the radius of the sphere over which the charge e of the negative electron would have to be distributed to have the observed mass is but 2×10^{-13} cm., or but one fifty-thousandth of the radius of the atom (10^{-8} cm.). From this point of view, then, the negative electron represents a charge of electricity which is condensed into an exceedingly minute volume. In fact, its radius cannot be larger in comparison with the radius of the atom than is the radius of the earth in comparison with the radius of her orbit about the sun.

In the case of the proton (the positive electron associated with the mass of the nucleus of the atom of hydrogen) there is no direct experimental justification for the assumption that the mass is also wholly of electrical origin, for we cannot by any means whatever ob-

[1] Bucherer, *Annalen der Physik*, XXVIII (1909), 513.

[2] The inadequacy in this argument arises from the fact that Einstein's Theory of Relativity requires that all mass, whether of electromagnetic origin or not, varies in just this way with speed.

tain as yet any protons which are endowed with speeds greater than a few tenths of that of light.   But in view of the experimental results obtained with the negative electron, the carrying over of the same assumption to the proton has been at least natural.   Further, if this step be taken, it is clear from equation (32), since m for this positive is nearly two thousand times larger than m for the negative, that a for the positive can be only $1/2,000$ of what it is for the negative.   In other words, the size of the proton would be to the size of the negative electron as a sphere having a two-mile radius would be to the size of the earth.   From the standpoint, then, of the electromagnetic theory of the origin of mass, the dimensions of the negative and positive constituents of atoms in comparison with the dimensions of the atoms themselves are like the dimensions of the planets and asteroids in comparison with the size of the solar system.   All of these computations, whatever their value, are rendered possible by the fact that e is now known.

Now we know from methods which have nothing to do with the electromagnetic theory of the origin of mass, that the excessive minuteness predicted by that theory for both the positive and the negative constituents of atoms is in fact correct, though we have no evidence as to whether the foregoing ratio is right.

III.   DIRECT EXPERIMENTAL PROOF OF THE EXCESSIVE MINUTENESS OF THE ELECTRONIC CONSTITUENTS OF ATOMS

For at least thirty years we have had direct experimental proof[1] that the fastest of the α-particles, or

[1] Bragg, *Phil. Mag.*, VIII (1904), 719, 726; X (1905), 318; XI (1906), 617.

helium atoms, which are ejected by radium, shoot in practically straight lines through as much as 7 cm. of air at atmospheric pressure before being brought to rest. This distance is then called the "range" of these α-rays. Figs. 14 and 15 show actual photographs of the tracks of such particles. We know too, for the reasons given on p. 139, that these α-particles do not penetrate the air after the manner of a bullet, namely, by pushing the molecules of air aside, but rather that they actually shoot through all the molecules of air which they encounter. The number of such passages through molecules which an α-particle would have to make in traversing seven centimeters of air would be about a hundred and thirty thousand.

Further, the very rapid β-particles, or negative electrons, which are shot out by radium have been known for a still longer time to shoot in straight lines through much greater distances in air than 7 cm., and even to pass practically undeflected through appreciable thicknesses of glass or metal.

We saw in chap. vi that the tracks of both the α- and the β-particles through air could be photographed because they ionize some of the molecules through which they pass. These ions then have the property of condensing water vapor about themselves, so that water droplets are formed which can be photographed by virtue of the light which they reflect. Fig. 17 shows the track of a very high-speed β-ray. A little to the right of the middle of the photograph a straight line can be drawn from bottom to top which will pass through a great many pairs of specks. These specks are the

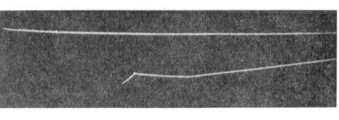

FIG. 15.

FIG. 14.—Photograph taken by Blackett and Lees (Cambridge, Eng.) of the tracks of α rays of mean ranges 82.7 mm. and 44.3 mm. The crosswise track is that of a proton ejected from a nitrogen nucleus by the impinging α particle.

FIG. 15.—Photograph of α ray tracks taken by C. T. R. Wilson. The heel at the point of impact is due to the velocity imparted by the blow to the nitrogen nucleus which in this case was not disrupted.

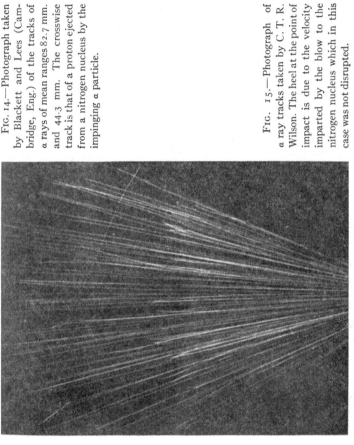

FIG. 14.

PHOTOGRAPHS OF THE TRACKS OF α-PARTICLES SHOOTING THROUGH AIR

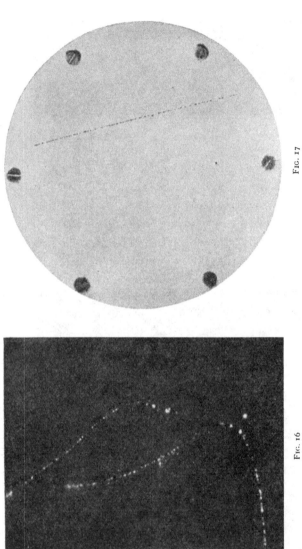

FIG. 16.

FIG. 17.

FIG. 16.—Photograph taken by C. T. R. Wilson of the track of a β ray from radium.

FIG. 17.—Photograph of the track of a very high-speed negative electron or β particle. It was actually released by a cosmic ray and has an energy of 1.3 billion electron-volts. Such a particle shoots in practically a straight line through 20 cm. of lead. The photograph was taken in the Norman Bridge Laboratory at Pasadena.

PHOTOGRAPHS OF THE TRACKS OF β-PARTICLES SHOOTING THROUGH AIR

water droplets formed about the ions which were produced at these points.   Since we know the size of a
molecule and the number of molecules per cubic centimeter, we can compute, as in the case of the α-particle,
the number of molecules through which a β-particle
must pass in going a given distance.   The extraordinary
situation revealed by this photograph is that this particular particle shot through on an average as many as
300 atoms before it came near enough to an electronic constituent of any one of these atoms to detach
it from its system and form an ion.   *This shows conclusively that the electronic or other constituents of atoms
can occupy but an exceedingly small fraction of the space
inclosed within the atomic system.   Practically the whole
of this space must be empty to an electron going with
this speed.*

The left panel in the lower half of the plate (**Fig.** 16)
shows the track of a negative electron of much slower
speed, and it will be seen, first, that it ionizes much
more frequently, and, secondly, that instead of continu-
ing in a straight line it is deflected at certain points
from its original direction.   The reason for both of these
facts can readily be seen from the considerations on
p. 139, which it may be worth while to extend to the case
in hand as follows.

If a new planet or other relatively small body were
to shoot with stupendous speed through our solar system, the time which it spent within our system might
be so small that the force between it and the earth or
any other member of the solar system would not have
time either to deflect the stranger from its path or to
pull the earth out of its orbit.   If the speed of the strange

body were smaller, however, the effect would be more disastrous both to the constituents of our solar system and to the path of the strange body, for the latter would then have a much better chance of pulling one of the planets out of our solar system and also a much better chance of being deflected from a straight path itself. The slower a negative electron moves, then, the more is it liable to deflection and the more frequently does it ionize the molecules through which it passes.

This conclusion finds beautiful experimental confirmation in the three panels of the plate opposite this page, for the speed with which X-rays hurl out negative electrons from atoms has long been known to be much less than the speed of β-rays from radium, and the zigzag tracks in these photographs are the paths of these corpuscles. It will be seen that they bend much more often and ionize much more frequently than do the rays shown in Figs. 16 and 17.

But the study of the tracks of the α-particles (Figs. 14 and 15, opposite p. 190) is even more illuminating as to the structure of the atom. For the α-particle, being an atom of helium eight thousand times more massive than a negative electron, could no more be deflected by one of the latter in an atom through which it passes than a cannon ball could be deflected by a pea. Yet Figs. 14 and 15 show that toward the end of its path the α-particle does in general suffer several sudden deflections. Such deflections could be produced only by a very powerful center of force within the atom whose mass is at least comparable with the mass of the helium atom.

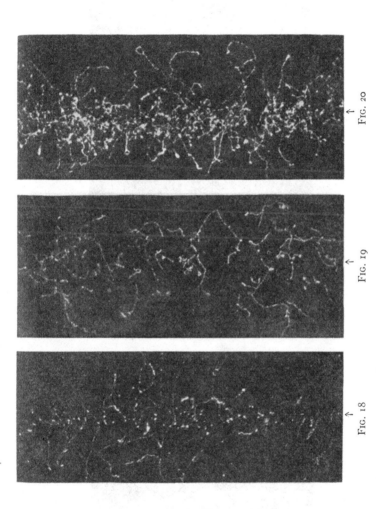

FIG. 18 ←   FIG. 19 ←   FIG. 20 ←

PHOTOGRAPHS OF THE TRACKS OF β-PARTICLES EJECTED BY X-RAYS FROM MOLECULES OF AIR

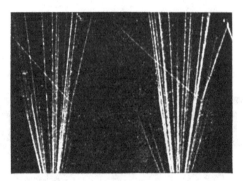

Collision of α particle with hydrogen atom

Collision of α-particle with helium atom

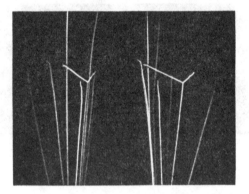

Collision of α-particle with oxygen atom
BLACKETT'S PHOTOGRAPHS OF α-PARTICLE COLLISIONS

These sharp deflections, which occasionally amount to as much as 150° to 180°, lend the strongest of support to the view that the atom consists of a heavy positively charged nucleus about which are grouped enough electrons to render the whole atom neutral. But the fact that in these experiments the a-particle goes through 130,000 atoms without approaching near enough to this central nucleus to suffer appreciable deflection more than two or three times constitutes the most convincing evidence that this central nucleus which holds the negative electrons within the atomic system occupies an excessively minute volume, just as we computed from the electromagnetic theory of the origin of mass that the proton, for example, should do. Indeed, knowing as he did by direct measurement the speed of the a-particle, Rutherford, who is largely responsible for the nucleus-atom theory, first computed,[1] with the aid of the inverse square law, which we know to hold between charged bodies of dimensions which are small compared with their distances apart, how close the a-particle would approach to the nucleus of a given atom like that of gold before it would be turned back upon its course (see Appendix F). The result was in the case of gold, one of the heaviest atoms, about 10^{-12} cm., and in the case of hydrogen, the lightest atom, about 10^{-13} cm.   These are merely upper limits for the dimensions of the nuclei.

However uncertain, then, we may feel about the sizes of positive and negative electrons computed from the electromagnetic theory of the origin of the mass, we may regard it as fairly well established by such direct experiments as these that the electronic constituents

[1] *Phil. Mag.*, XXI (1911). 669.

of atoms are as small, in comparison with the dimensions of the atomic systems, as are the sun and planets in comparison with the dimensions of the solar system. Indeed, when we reflect that we can shoot helium atoms by the billion through a thin-walled highly evacuated glass tube without leaving any holes behind, i.e., without impairing in the slightest degree the vacuum or perceptibly weakening the glass, we see from this alone that the atom itself must consist mostly of "hole"; in other words, that an atom, like a solar system, must be an exceedingly loose structure whose impenetrable portions must be extraordinarily minute in comparison with the penetrable portions. The notion that an atom can appropriate to itself all the space within its boundaries to the exclusion of all others is then altogether exploded by these experiments. A particular atom can certainly occupy the same space at the same time as any other atom if it is only endowed with sufficient kinetic energy. Such energies as correspond to the motions of thermal agitation of molecules are not, however, sufficient to enable one atom to penetrate the boundaries of another, hence the seeming impenetrability of atoms in ordinary experiments in mechanics. That there is, however, a portion of the atom which is wholly impenetrable to the alpha particles is definitely proved by experiments of the sort we have been considering; for it occasionally happens that an alpha particle hits this nucleus "head on," and, when it does so, it is turned straight back upon its course. As indicated above, the size of this impenetrable portion, which may be defined as the size of the nucleus, is in no case larger than 1/10,000 the diameter of the atom, *and yet there may be contained within it, as will presently be shown, several hundred positive and negative electrons, so that the excessive minuteness of these bodies is established, altogether without reference to any theory as to what they are.*

IV. THE NUMBER OF ELECTRONS IN AN ATOM

If it be considered as fairly conclusively established by the experiments just described that an atom consists of a heavy but very minute positively charged nucleus which holds light negative electrons in some sort of a configuration about it, then the number of negative electrons outside the nucleus must be such as to have a total charge equal to the free positive charge of the nucleus, since otherwise the atom could not be neutral.

But the positive charge on the nucleus has been approximately determined as follows: With the aid of the knowledge, already obtained through the determination of e, of the exact number of atoms in a given weight of a given substance, Sir Ernest Rutherford[1] first computed the chance that a single helium atom in being shot with a known speed through a sheet of gold foil containing a known number of atoms per unit of area of the sheet would suffer a deflection through a given angle. This computation can easily be made in terms of the known kinetic energy and charge of the α-particle, the known number of atoms in the gold foil, and the unknown charge on the nucleus of the gold atom (see Appendix F). Geiger and Marsden[2] then actually counted in Rutherford's laboratory, by means of the scintillations produced on a zinc-sulphide screen, what

[1] *Phil. Mag.*, XXI (1911), 669–88.

[2] *Ibid.*, XXV (1913), 604.

fraction of, say, a thousand α-particles, which were shot normally into the gold foil, were deflected through a given angle, and from this observed number and Rutherford's theory they obtained the number of free positive charges on the nucleus of the gold atom.

Repeating the experiment and the computations with foils made from a considerable number of other metals, they found that in every case *the number of free positive charges on the atoms of different substances was approximately equal to half its atomic weight.* This means that the aluminum atom, for example, has a nucleus containing about thirteen free positive charges and that the nucleus of the atom of gold contains in the neighborhood of a hundred. This result was in excellent agreement with the conclusion reached independently by Barkla[1] from experiments of a wholly different kind, namely, experiments on the scattering of X-rays. These indicated that the number of scattering centers in an atom—that is, its number of free negative electrons—was equal to about half the atomic weight. But this number must, of course, equal the number of free positive electrons in the nucleus.

V. MOSELEY'S REMARKABLE DISCOVERY

The foregoing result was only approximate. Indeed, there was internal evidence in Geiger and Marsden's work itself that a half was somewhat too high. The answer was made very definite and very precise in 1913 through the extraordinary work of a brilliant young Englishman, Moseley, who, at the age of twenty-seven, had accomplished as notable a piece of research in

[1] *Phil. Mag.*, XXI (1911), 648.

physics as has appeared during the last fifty years. Such a mind was one of the early victims of the world-war. He was shot and killed instantly in the trenches in Gallipoli in the summer of 1915.

Laue in Munich had suggested in 1912 the use of the regular spacing of the molecules of a crystal for the analysis, according to the principle of the grating, of ether waves of very short wave-length, such as X-rays were supposed to be, and the Braggs[1] had not only perfected an X-ray spectrometer which utilized this principle, but had determined accurately the wave-lengths of the X-rays which are characteristic of certain metals. The accuracy with which this can be done is limited simply by the accuracy in the determination of e, so that the whole new field of exact X-ray spectrometry is made available through our exact knowledge of e. Moseley's discovery,[2] made as a result of an elaborate and difficult study of the wave-lengths of the character-istic X-rays which were excited when cathode rays were made to impinge in succession upon anticathodes em-bracing most of the known elements, was that these characteristic wave-lengths of the different elements, or, better, their characteristic frequencies, are related in a very simple but a very significant way. *These frequencies were found to constitute the same sort of an arithmetical progression as do the charges which we found to exist on our oil drops.* It was the square root of the frequencies rather than the frequencies themselves which showed this beauti-fully simple relationship, but this is an unimportant detail. The significant fact is that, *arranged in the order of increas-*

[1] Bragg, *X-Rays and Crystal Structure*, 1915.
[2] *Phil. Mag.*, XXVI (1912), 1024; XXVII (1914), 703.

*ing frequency of their characteristic X-ray spectra, all the
known elements which have been examined constitute a
simple arithmetical series each member of which is obtained
from its predecessor by adding always the same quantity.*

The plate opposite this page shows photographs of
the X-ray spectra of a number of elements whose atomic
numbers—that is, the numbers assigned them in Moseley's arrangement of the elements on the basis of increasing X-ray frequency—are given on the left. These
photographs were taken by Siegbahn.[1] The distance
from the "central image"—in this case the black line
on the left—to a given line of the line spectrum on the
right is approximately proportional to the wave-length
of the rays producing this line. The photographs show
beautifully, first, how the atoms of all the elements
produce spectra of just the same type, and, secondly, how
the wave-lengths of corresponding lines decrease, or
the frequencies increase, with increasing atomic number.
The photograph on the left shows this progression for
the highest frequency rays which the atoms produce, the
so-called K series, while the one on the right shows the
same sort of a progression for the rays of next lower frequency, namely, those of the so-called L series, which have
uniformly from seven to eight times the wave-length
of the K series. The plate opposite p. 199 shows some
very beautiful photographs taken by De Broglie in Paris[2]
in October, 1916. The upper one is the X-ray emission
spectrum of tungsten. It consists of general radiations, corresponding to white light, scattered throughout the whole length of the spectrum as a background

[1] *Jahrbuch der Radioaktivität u. Elektronik*, XIII (1916), 326.

[2] *Comptes rendus*, CLXV (1916), 87, 352.

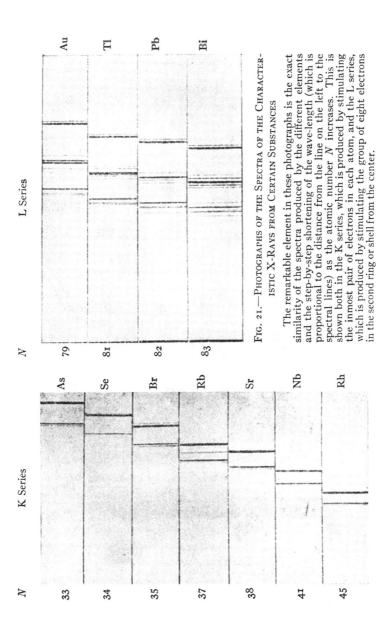

FIG. 21.—PHOTOGRAPHS OF THE SPECTRA OF THE CHARACTER-
ISTIC X-RAYS FROM CERTAIN SUBSTANCES

The remarkable element in these photographs is the exact
similarity of the spectra produced by the different elements
and the step-by-step shortening of the wave-length (which is
proportional to the distance from the line on the left to the
spectral lines) as the atomic number N increases. This is
shown both in the K series, which is produced by stimulating
the inmost pair of electrons in each atom, and the L series,
which is produced by stimulating the group of eight electrons
in the second ring or shell from the center.

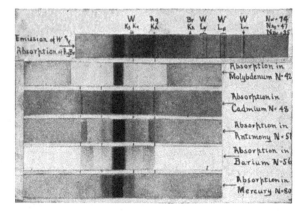

FIG. 22.—X-RAY ABSORPTION SPECTRA, K SERIES

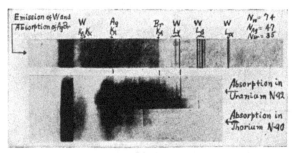

FIG. 23.—X-RAY ABSORPTION SPECTRA, L SERIES

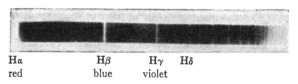

Hα   Hβ   Hγ   Hδ
red   blue   violet

FIG. 24.—HYDROGEN SPECTRUM FROM THE STAR VEGA

and superposed upon these two groups of lines.   The two K lines are here close to the central image, for the K wave-lengths are here very short, since tungsten has a high atomic number (74).   Farther to the right is the L series of tungsten lines which will be recognized because of its similarity to the L series in the plate opposite p. 198.   Between the K and the L lines are two absorption edges marked $\frac{Ag}{K_\wedge}$ and $\frac{Br}{K_\wedge}$.   The former represents the frequency above which the silver absorbs all the general radiation of tungsten but below which it lets it all through.   The latter is the corresponding line for bromine.   In a print from a photograph absorption in the plate itself obviously appears as a darkening, transmission as a lightening.   Just below is the spectrum obtained by inserting a sheet of molybdenum in the path of the beam, i.e., before the slit of the spectrometer. Absorption in the molybdenum will obviously appear as a lightening, transmission as a darkening.   It will be seen that the molybdenum absorbs all the frequencies in the X-ray emission of tungsten higher than a partic- ular frequency and lets through all frequencies lower than this value.   This remarkable characteristic of the absorption of X-rays was discovered by Barkla in 1909.[1] The absorption edge at which, with increasing frequency, absorption suddenly begins is very sharply marked. This edge coincides with the highest emission frequency of which molybdenum is theoretically capable, and is a trifle higher than the highest observed emission fre- quency.   De Broglie has measured accurately these critical absorption frequencies for all the heavy elements

[1] Barkla and Sadler, *Phil. Mag.*, XVII (May, 1909), 749.

up to thorium, thus extending the K series from atomic number $N = 60$ where he found it, to $N = 90$, a notable advance. The two absorption edges characteristic of the silver and the bromine in the photographic plate appear in the same place on all the photographs in which they could appear. The other absorption edges vary from element to element and are characteristic each of its particular element. The way in which this critical absorption edge moves toward the central image as the atomic number increases in the steps Br 35, Mo 42, Ag 47, Cd 48, Sb 51, Ba 56, W 74, Hg 80, is very beautifully shown in De Broglie's photographs all the way up to mercury, where the absorption edge is somewhat inside the shortest of the characteristic K radiations of tungsten. There must be twelve more of these edges between mercury ($N = 80$) and uranium ($N = 92$) and De Broglie has measured them up to thorium ($N = 90$). They become, however, very difficult to locate in this K region of frequencies on account of their extreme closeness to the central image. But the L radiations, which are of seven times longer wave-length, may then be used, and Fig. 23 of the plate opposite page 199 shows the L-ray absorption edges, of which there are three, as obtained by De Broglie in both uranium and thorium, so that the position in the Moseley table of each element all the way to the heaviest one, uranium, is fixed in this way by direct experiment. Fig. 25 shows the progression of square-root frequencies as it appears from measurements made on the successive absorption edges of De Broglie's photographs and on a particular one of Siegbahn's emission lines. It will be noticed that, in going from bromine (35) to uranium (92), the length of

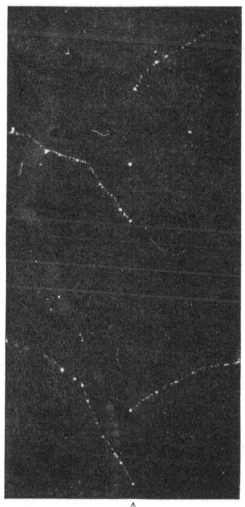

↑

These beta ray tracks were taken by C. T. R. Wilson. The electrons are photoelectrically released and hence receive the full energy of the incident photons, here over 25,000 volts. When these photons are of sufficiently low energy the electrons are thrown out at right angles to the direction of the incident photon (see Figs. 18, 19, 20), but as the energy increases the electron tends to take on a stronger and stronger forward component as here shown. The X-ray beam was here a very narrow one passing upward in a line just to the right of the middle of the figure (see arrow).

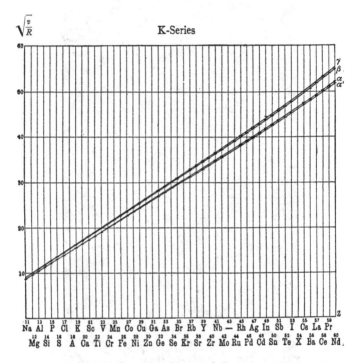

FIG. 25a. MOSELEY'S LAW

This figure shows in graphical form the linear step-by-step progression of square-root X-ray frequencies with atomic number. It is this relationship, now extending in 92 steps from hydrogen up to uranium, that tells us that the physical world as we know it is built up out of just 92 elements, an element being defined as a body possessing a given nuclear charge, and therefore in its neutral state having a corresponding number of extra-nuclear electrons. Fermi, in Rome, announced in 1934 that there is some evidence for the existence of element of atomic number 93. Each element may have a number of isotopes because the foregoing charge relations can be built up by combinations of protons and negative electrons that give different weights to the nucleus.

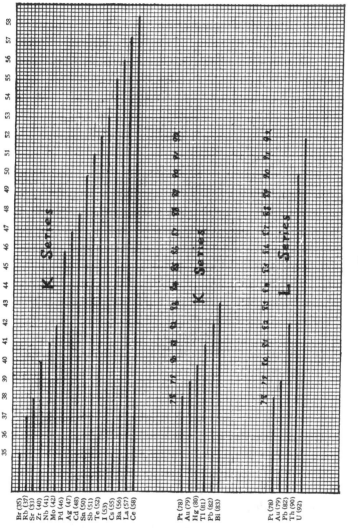

FIG. 25

the step does change by a few per cent. The probable cause of this will be considered later.

According to modern theory an absorption edge appears where the incident energy—which is proportional to the incident frequency—has become just large enough to lift the particular electron which absorbs it entirely out of the atom. If this removed electron should then fall back to its old place in the atom, it would emit in so doing precisely the frequency which was absorbed in the process of removal.

Since these enormously high X-ray frequencies must arise from electrons which fall into extraordinarily powerful fields of force, such as might be expected to exist in the inner regions of the atom close to the nucleus, Moseley's discovery strongly suggests that the charge on this nucleus is produced in the case of each atom by adding some particular invariable charge to the nucleus of the atom next below it in Moseley's table. This suggestion gains added weight when it is found that with one or two trifling exceptions, to be considered later, *Moseley's series of increasing X-ray frequencies is exactly the series of increasing atomic weights*. It also receives powerful support from the following discovery.

Mendeleéff's periodic table shows that the progression of chemical properties among the elements coincides in general with the progression of atomic weights. Now it was pointed out twenty years ago that whenever a radioactive substance loses a doubly charged α-particle it moves two places to the left in the periodic table, while whenever it loses a singly charged β-particle it moves one place to the right,[1] thus showing that the chemical

[1] Soddy, *The Chemistry of the Radioelements*, Part II, 1914.

character of a substance depends upon the number of free positive charges in its nucleus.

One of the most interesting and striking characteristics of Moseley's table is that all the known elements between sodium (atomic number 11, atomic weight 23) and lead (atomic number 82, atomic weight, 207.2) have been fitted into it and there are now left no vacancies within this range. Below sodium there are just 10 known elements, and very recent study[1] of their spectra in the extreme ultra-violet has fixed the place of each in the Moseley progression, though in this region the progression of atomic weights and of chemical properties is also altogether definite and unambiguous. It seems highly probable, then, from Moseley's work that we have already found every one of the complete series of different types of atoms from hydrogen to lead, i.e., from 1 to 82, of which the physical world is built. From 82 to 92 comes the group of radioactive elements which are continually transmuting themselves into one another, and above 92 (uranium) it is not likely that any elements exist.

That hydrogen is indeed the base of the Moseley series can be seen independently of all theory by the following simple computation. If we write Moseley's discovery that the square roots of the highest frequencies, n_1, n_2, etc., emitted by different atoms are proportional to the nuclear charges, E_1, E_2, etc., in the following form:

$$\sqrt{\frac{n_1}{n_2}} = \frac{E_1}{E_2} \text{ or } \frac{\lambda_2}{\lambda_1} = \frac{E_1^2}{E_2^2} \dots \dots \dots \dots \dots (33)$$

[1] Millikan and Bowen, "Extreme Ultra-Violet Spectra," *Phys. Rev.*, January, 1924.

and substitute for λ_2 the observed wave-length of the highest frequency line emitted by tungsten—a wave-length which has been accurately measured and found to be 0.179×10^{-8} cm.; and, further, if we substitute for E_2, 74, the atomic number of tungsten, and for E_1, 1, if the Moseley law were exact we should obtain, by solving for λ_1, the wave-length of the highest frequency line which can be emitted by the element whose nucleus contains but one single positive electron. The result of this substitution is $\lambda_1 = 98.0$ $\mu\mu$ (millionths millimeters). Now the wave-length corresponding to the highest observed frequency in the ultra-violet series of hydrogen lines recently discovered by Lyman is 97.4 $\mu\mu$ and there is every reason to believe from the form of this series that its convergence wave-length—this corresponds to the highest frequency of which the hydrogen atom is theoretically capable—is 91.2 $\mu\mu$. The agreement is only approximate, but it is as close as could be expected in view of the lack of exact equality in the Moseley steps. *It is well-nigh certain, then, that this Lyman ultra-violet series of hydrogen lines is nothing but the K X-ray series of hydrogen.* Similarly, it is equally certain that the L X-rays series of hydrogen is the ordinary Balmer series in the visible region, the head of which is at $\lambda = 365$ $\mu\mu$. In other words, hydrogen's ordinary radiations are its X-rays and nothing more.

There is also an M series for hydrogen discovered by Paschen in the ultra-red, which in itself would make it probable that there are series for all the elements of longer wave-length than the L series, and that the complicated optical series observed with metallic arcs are parts of these longer wave-length series. As a

matter of fact, an M series has been found for a considerable group of the elements of high atomic number.

Thus the Moseley experiments have gone a long way
toward solving the mystery of spectral lines. They
reveal to us clearly and certainly the whole series of
elements from hydrogen to uranium, all producing
spectra of remarkable similarity, at least so far as the K
and L radiations are concerned, but scattered regularly
through the whole frequency region, from the ultra-
violet, where the K lines for hydrogen are found, all
the way up to frequencies $(92)^2$ or 8,464 times as high.
There is scarcely a portion of this whole field which is not
already open to exploration. How brilliantly, then,
have these recent studies justified the predictions of the
spectroscopists that the key to atomic structure lay
in the study of spectral lines!

Moseley's work is, in brief, evidence from a wholly
new quarter that all these elements constitute a family,
each member of which is related to every other member
in a perfectly definite and simple way. It looks as if
the dream of Thales of Miletus had actually come true
and that we have found a primordial element out of
which all substances are made, or better two of them.
For the succession of steps from one to ninety-two, each
corresponding to the addition of an extra free positive
charge upon the nucleus, suggests at once that the unit
positive charge is itself a primordial element, and this
conclusion is strengthened by recently discovered atomic-
weight relations. It is well known that Prout thought
a hundred years ago that the atomic weights of all elements were exact multiples of the weight of hydrogen,
and hence tried to make hydrogen itself the primordial

element. But fractional atomic weights like that of chlorine (35.5) were found, and were responsible for the later abandonment of the theory. Within the past ten years, however, it has been shown that, within the limits of observational error, practically all of those elements which had fractional atomic-weights are mixtures of substances, so called *isotopes*, each of which has an atomic weight that is very nearly an exact multiple of the unit of the atomic-weight table, so that Prout's hypothesis is now very much alive again.

Indeed, all results so far obtained are consistent with the view that within every atomic nucleus each positive electron is at least associated with the mass characteristic of the nucleus of the hydrogen atom, so that to within one part in eighteen hundred thirty-five the mass of every atom may be thought of as simply the mass of the whole number of hydrogen nuclei, or protons (each charged with one positive electron) which are contained within its nucleus. Now the atomic weight of helium is four, while its atomic number, the *free* positive charge upon its nucleus, is only two. The helium atom must therefore contain *inside its nucleus* two negative electrons which neutralize two of these positives and serve to hold together the four positives which would otherwise fly apart under their mutual repulsions. Into that tiny nucleus of helium, then, that infinitesimal speck not as big as a pin point, even when we multiply all dimensions ten billion fold so that the diameter of the helium atom, the orbit of its two outer negatives, has become a yard, into that still almost invisible nucleus there must be packed four positive and two negative electrons.

By the same method it becomes possible to count the exact number of both positive and negative electrons which are packed into the nucleus of every other atom. In uranium, for example, since its atomic weight is 238, we know that there must be 238 positive electrons in its nucleus. But since its atomic number, or the measured number of free unit charges upon its nucleus, is but 92, it is obvious that $(238-92=)$ 146 of the 238 positive electrons in the nucleus must be neutralized by 146 negative electrons *which are also within that nucleus;* and so, in general, *the atomic weight minus the atomic number gives at once the number of negative electrons which are contained within the nucleus of any atom.* That these negative electrons are actually there within the nucleus is independently demonstrated by the facts of radioactivity, for in the radioactive process we find negative electrons, so called β-rays, actually being ejected from the nucleus. They can come from nowhere else, for the chemical properties of the radioactive atom are found to change with every such ejection of a β-ray, and change in chemical character always means change in the free charge contained in the nucleus.

We have thus been able to look with the eyes of the mind, not only inside an atom, a body which becomes but a meter in diameter when looked at through an instrument of ten billion fold magnification, but also inside its nucleus, which, even with that magnification, is still a mere pin point, and to count within it just how many positive and how many negative electrons are there imprisoned, numbers reaching 238 and 146, respectively, in the case of the uranium atom. And let it be remembered, the dimensions of these atomic nuclei are about one-billionth of those of the smallest object which has ever been seen or can ever be seen and measured in a microscope. From these figures it will be obvious that, for practical purposes, we may neglect the dimensions of electrons altogether and consider them as mere point charges.

But what a fascinating picture of the ultimate structure of matter has been presented by this voyage to the land of the infinitely small! Only two ultimate entities have we been able to see there, namely, positive and negative electrons; alike in the magnitude of their charge but apparently differing much in mass; the positive being eighteen hundred and thirty-five times heavier than the negative; both being so vanishingly small that hundreds of them can somehow get inside a volume which is still a pin point after all dimensions have been swelled ten billion times: the ninety-two different elements of the world determined simply by the difference between the number of positives and negatives which have been somehow packed into the nucleus; all these elements transmutable, ideally at least, into one another by a simple change in this difference. Has nature a way of making these transmutations in her laboratories? She is doing it under our eyes in the radioactive process—a process, however, which is confined to the two heaviest elements, uranium and thorium, and their disintegration products, save that it is possessed in very slight degree by potassium and rhubidium. Does the process go on in both directions, heavier atoms being continually formed as well as continually disintegrating into lighter ones? Not on the earth so far as we can see. Perhaps in the depths of space or in the stars. Some day we may find out.

Can we on the earth artificially control the process? To a certain degree we know already how to disintegrate artificially and also how to build up. For, ever since 1919 Rutherford and his co-workers at the Cavendish Laboratory have been disintegrating all the elements from boron to potassium, save only carbon and oxygen, by bombarding them with a rays (see opp. p. 190)—a process which knocks hydrogen nuclei out of these elements. Also beginning in 1932, Cockroft and Walton in England, Lawrence in Berkeley, and Lauritsen and Crane in Pasadena have built up certain new atoms by throwing hydrogen nuclei with great energy, imparted by powerful electric fields, into other nuclei. How far we can go in this artificial transmutation of the elements is not yet certain. (See chap. xv.)

VI. THE BOHR ATOM

Thus far nothing has been said as to whether the electrons within the atom are at rest or in motion, or, if they are in motion, as to the character of these motions. In the hydrogen atom, however, which contains, according to the foregoing evidence, but one positive and one negative electron, there is no known way of preventing the latter from falling into the positive nucleus unless centrifugal forces are called upon to balance attractions, as they do in the case of the earth and moon. Accordingly it seems to be necessary to assume that the negative electron is rotating in an orbit about the positive. But such a motion would normally be accompanied by a continuous radiation of energy of continuously increasing frequency as the electron, by virtue of its loss of energy, approached closer and closer to the nucleus. Yet experiment reveals no such behavior, for, so far as we know, hydrogen does not radiate at all unless it is ionized, or has its negative electron knocked, or lifted, from its normal orbit to one of higher potential energy, and, when it does radiate, it gives rise, not to a continuous spectrum, as the foregoing picture would demand, but rather to a line spectrum in which the frequencies corresponding to the various lines are related to one another in the very significant way shown in the photograph of Fig. 24 and represented by the so-called Balmer-Ritz equation,[1] which has the form

$$\nu = N\left(\frac{1}{n_1^2} - \frac{1}{n_2^2}\right) \dots\dots\dots\dots (34)$$

In this formula ν represents frequency, N a constant, and n_1, for all the lines in the visible region, has the value 2, while n_2 takes for the successive lines the values 3, 4, 5, 6, etc. In the hydrogen series in the infra-red discovered by Paschen[2] $n_1 = 3$ and n_2 takes the successive values 4, 5, 6, etc. It is since the development of the Bohr theory that Lyman[3] discovered his hydrogen series in the ultra-violet in which $n_1 = 1$ and $n_2 = 2, 3, 4,$ etc. Since 1 is the smallest whole number, this series should correspond, as indicated heretofore, to the highest frequencies of which hydrogen is capable, the upper limit toward which these frequencies tend being reached when $n_1 = 1$ and $n_2 = \infty$, that is, when $\nu = N$.

---

[1] Balmer (1885) expressed the formula in wave-lengths. Ritz (1908) first replaced wave-lengths by wave-numbers, or frequencies, and thereby saw his "combination-principle," while Rydberg discovered the general significance of what is now known as the Rydberg constant N.

[2] Paschen, *Ann. d. Phys.*, XXVII (1908), 565.

[3] *Spectroscopy of the Extreme Ultraviolet*, p. 78.

Guided by all of these facts except the last, Niels Bohr, a young mathematical physicist of Copenhagen, in 1913 devised[1] an atomic model which has had some very remarkable successes. This model was originally designed to cover only the simplest possible case of one single electron revolving around a positive nucleus. In order to account for the large number of lines which the spectrum of such a system reveals (see Fig. 24), Bohr's first assumption was that the electron may rotate about the nucleus in a whole series of different orbits, as shown in Fig. 26, and

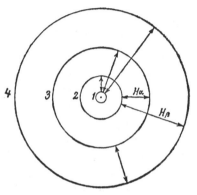

FIG. 26.—The original Bohr model of the hydrogen atom.

that each of these orbits is governed by the well-known Newtonian law, which when mathematically stated takes the form:

$$\frac{eE}{a^2} = (2\pi n)^2 ma \dots \dots \dots (35)$$

in which e is the change of the electron, E that of the nucleus, a the radius of the orbit, n the orbital frequency, and m the mass of the electron. This is merely the assumption that the electron rotates in a circular orbit which is governed by the laws which are known, from

[1] N. Bohr, *Phil. Mag.*, XXVI (1913), 1 and 476 and 857; XXIX (1915), 332; XXX (1915), 394; Sommerfeld, *Atomic Structure and Spectral Lines.* New York: Dutton, 1923.

the work on the scattering of the alpha particles, to hold inside as well as outside the atom. The radical element in it is that it permits the negative electron to maintain this orbit or to persist in this so-called "*stationary state*" without radiating energy even though this appears to conflict with ordinary electromagnetic theory. But, on the other hand, the facts of magnetism[1] and of optics, in addition to the successes of the Bohr theory which are to be detailed, appear at present to lend experimental justification to such an assumption.

Bohr's second assumption is that radiation takes place only when an electron jumps from one to another of these orbits. If A_2 represents the energy of the electron in one orbit and A_1 that in any other orbit, then it is clear from considerations of energy alone that when the electron passes from the one orbit to the other the amount of energy radiated must be A_2-A_1; further, this radiated energy obviously must have some frequency ν, and, in view of the experimental work presented in the next chapter, Bohr placed it proportional to ν, and wrote:

$$h\nu = A_2 - A_1 \dots\dots\dots\dots\dots\dots (36)$$

h being the so-called Planck constant to be discussed later. It is to be emphasized that this assumption gives no physical picture of the way in which the radiation takes place. It merely states the energy relations which must be satisfied when it occurs. The red hydrogen line H_a is, according to Bohr, due to a jump from orbit 3 to orbit 2 (Fig. 26), the blue line H_β to a jump from 4 to 2,

[1] Einstein and De Haas, *Verh. der deutsch. phys. Ges.*, XVII (1915), 152; also Barnett, *Phys. Rev.*, VI (1915), 239; also Epstein, *Science*, LVII (1923), 532.

H_γ to a jump from 5 to 2, etc.; while the Lyman ultra-violet lines correspond to a series of similar jumps into the inmost orbit 1 (see Fig. 26).

Bohr's third assumption is that the various possible circular orbits are determined by assigning to each orbit a kinetic energy T such that

$$T = \tfrac{1}{2}\tau hn \dots\dots\dots\dots\dots\dots (37)$$

in which τ is a whole number, n the orbital frequency, and h is again Planck's constant. This value of T is assigned so as to make the series of frequencies agree with that actually observed, namely, that represented by the Balmer series of hydrogen.

It is to be noticed that, if circular electronic orbits exist at all, no one of these assumptions is arbitrary. Each of them is merely the statement of the existing *experimental* situation. It is not surprising, therefore, that they predict the sequence of frequencies found in the hydrogen series. They have been purposely made to do so. But they have not been made with any reference whatever to the exact numerical values of these frequencies.

The evidence for the soundness of the conception of non-radiating electronic orbits is to be looked for, then, first, in the success of the constants involved, and, second, in the physical significance, if any, which attaches to the third assumption. If these constants come out right within the limits of experimental error, then the theory of non-radiating electronic orbits has been given the most crucial imaginable of tests, especially if these constants are accurately determinable.

What are the facts? The constant of the Balmer series in hydrogen, that is, the value of N in equation (34), is known with the great precision attained in all wave-length determinations and is equal to 3.28988×10^{15}. From the Bohr theory it is given by the simplest algebra (Appendix G) as

$$N = \frac{2\pi^2 e^4 m}{h^3} = \frac{2\pi^2 e^5}{h^3 \dfrac{e}{m}} \quad \dots\dots\dots\dots\dots (38)$$

As already indicated, in 1917 I redetermined[1] e with an estimated accuracy of one part in 1,000 and obtained for it the value, $4,770 \times 10^{-10}$. As will be shown in the next chapter, I have also determined h photo-electrically[2] with an error, in the case of sodium, of no more than one-half of 1 per cent, the value for sodium, upon which I got the most reliable data, being 6.56×10^{-27}. The value found by Duane's X-ray method,[3] which is thought to yield a result correct to one part in 700, is exceedingly close to mine, namely, 6.555×10^{-27}. Substituting this in (38), we get with the aid of Houston's value of $\dfrac{e}{m}$ (1.7570×10^{7}), which is probably correct to 0.1 per cent, $N = 3.285 \times 10^{15}$, *which agrees within a tenth of 1 per cent with the observed value.* This agreement constitutes most extraordinary justification of the theory of non-radiating electronic orbits. It demonstrates that the behavior of the negative electron in the hydrogen atom is at least correctly described by the *equation* of a circular non-radiating orbit. If this equation can be

[1] R. A. Millikan, *Phil. Mag.*, XXXIV (1917), 1.

[2] R. A. Millikan, *Phys. Rev.*, VII (1916), 362.

[3] Blake and Duane, *ibid.* (1917), 624.

obtained from some other physical condition than that of an actual orbit, it is obviously incumbent upon those who so hold to show what that condition is. The so-called wave mechanics, which is rather a modification than an abandonment of orbits, does equally well.

Again, the radii of the stable orbits for hydrogen are easily found from Bohr's assumptions to take the mathematical form (Appendix G)

$$a = \frac{\tau^2 h^2}{4\pi^2 m e^2}. \quad\ldots\ldots\ldots\ldots\ldots (39)$$

In other words, since τ is a whole number, the radii of these orbits bear the ratios 1, 4, 9, 16, 25. If normal hydrogen is assumed to be that in which the electron is in the inmost possible orbit, namely, that for which $\tau = 1$, $2a$, the diameter of the normal hydrogen *atom*, comes out 1.1×10^{-8}. The best determination for the diameter of the hydrogen *molecule* yields 2.2×10^{-8} in extraordinarily close agreement with the prediction from Bohr's theory.

Further, the fact that normal hydrogen does not absorb at all the Balmer series lines which it emits is beautifully explained by the foregoing theory, since, according to it, normal hydrogen has no electrons in the orbits corresponding to the lines of the Balmer series. Again, the fact that *hydrogen emits its characteristic radiations only when it is ionized or excited* favors the theory that the process of emission is a process of settling down to a normal condition through a series of possible intermediate states, and is therefore in line with the view that a change in orbit is necessary to the act of radiation.

Another triumph of the theory is that the third assumption, devised to fit a purely empirical situation, viz., the observed relations between the frequencies of the Balmer series, is found to have a very simple and illuminating physical meaning and one which has to do with *orbital* motion. It is that all the possible values of the *angular momentum* of the electron rotating about the positive nucleus are exact multiples of a particular value of this angular momentum. Angular momentum then has the property of *atomicity*. Such relationships do not in general drop out of *empirical* formulae. When they do, we see in them at least general interpretations of the formulae—not merely coincidences.

Again, the success of a theory is often tested as much by its adaptability to the explanation of deviations from the behavior predicted by its most elementary form as by the exactness of the fit between calculated and observed results. The theory of electronic orbits has had remarkable successes of this sort. Thus it predicts the Moseley law (33). But this law, discovered afterward, was found inexact, and it should be inexact when there is more than one electron in the atom, as is the case save for H atoms and for such He atoms as have lost one negative charge, and that because of the way in which the electrons influence one another's fields. By taking account of these influences, the inexactnesses in Moseley's law have been very satisfactorily explained.

Another very beautiful quantitative argument for the correctness of Bohr's orbital conception comes from the prediction of a slight difference between the positions in the spectrum of two sets of lines, one due to ionized helium and the other to hydrogen. These two sets of lines, since they are both due to a single electron rotating about a simple nucleus, ought to be exactly coincident, i.e., they ought to be one and the same set of lines, *if it were not for the fact that the helium nucleus is four times as heavy as the hydrogen nucleus.*

To see the difference that this causes it is only necessary to reflect that, when an electron revolves about a hydrogen nucleus, the real thing that happens is that the two bodies revolve about their common center of gravity. But since the nucleus is two thousand times heavier than the electron, this center is exceedingly close to the hydrogen nucleus.

When, now, the hydrogen nucleus is replaced by that of helium, which is four times as heavy, the common center of gravity is still closer to the nucleus, so that the helium-nucleus describes a much smaller circle than did that of hydrogen. This situation is responsible for a slight but accurately predictable difference in the energies of the two orbits, which should cause the spectral lines produced by electron-jumps to these two different orbits to be slightly displaced from one another.

This predicted slight displacement between the hydrogen and helium lines is not only found experimentally, but the most refined and exact of recent measurements has shown that *the observed displacement agrees with the predicted value to within a small fraction of 1 per cent.*

This not only constitutes general evidence for the orbit theory, but it seems to be irreconcilable with a ring-electron theory once favored by some authors, since it seems to require that the mass of the electron be concentrated at a point.

The next amazing success of the orbit theory came when Sommerfeld[1] showed that the "quantum" principle underlying the Bohr theory ought to demand two different hydrogen orbits corresponding to the second quantum state—second orbit from the nucleus—one a circle and one an ellipse. And by applying the general relativity theory (which involves the change in mass of the electron with its change in speed as it moves through the different portions of its orbit), he showed that the circular and elliptical orbits should have slightly different energies, and consequently that both the hydrogen and the helium lines corresponding to the second quantum state should be close doublets.

Now *not only is this found to be the fact, but the measured separation of these two doublet lines agrees closely with the predicted value*, so that this again constitutes extraordinary evidence for the validity of the orbit-conceptions underlying the computation.

In Fig. 27 the two orbits which are here in question are those which are labeled 2_2 and 2_1, the large numeral denoting the total quantum number, and the subscript the auxiliary, or *azimuthal*, quantum number which determines the ellipticity of the orbit. The figure is introduced to show the types of stationary orbits which the extended Bohr theory permits. For total quantum number 1 there was but one possible orbit, a circle. For total quantum numbers 2, 3, 4, etc., there are 2, 3, 4, etc., possible orbits, respectively. The ratio of the auxiliary to the total quantum number gives the ratio of the minor and major axes of the ellipse. The fourth quantum

[1] A. Sommerfeld, *Ann. d. Phys.*, III (1916), 1. Also Paschen, *ibid.*, p. 901.

state, for example, has four orbits, 4_1, 4_2, 4_3, 4_4, all of which have the same major axis, but minor axes which increase in the ratios 1, 2, 3, 4 up to equality, in the circle (4_4), with the major axis. It is this multiplicity of orbits which predicts with beautiful accuracy the "fine-structure" of all of the lines due to atomic hydrogen and to helium.

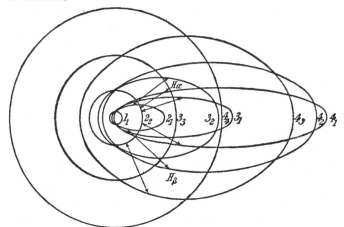

FIG. 27.—Bohr-Sommerfeld model of the hydrogen atom with stationary orbits corresponding to principal quantum numbers and auxiliary or azimuthal quantum numbers.

The next quantitative success of the Bohr theory came when Epstein,[1] of the California Institute, applied his unusual grasp of orbit theory to the exceedingly difficult problem of computing the perturbations in electron orbits, and hence the change in energy of each, due to exciting hydrogen and helium atoms to radiate in an electrostatic field. He thus predicted the whole complex character of what we call the "Stark effect,"

[1] P. Epstein, *ibid.*, L (1916), 489.

showing just how many new lines were to be expected and where each one should fall, and then *the spectroscope yielded, in practically every detail, precisely the result which the Epstein theory demanded.*

Another quantitative success of the orbit theory is one which Mr. I. S. Bowen and the author,[1] at the California Institute, have brought to light. Through creating what we call "hot sparks" in extreme vacuum we have succeeded in stripping in succession, 1, 2, 3, 4, 5, and 6 of the valence, or outer, electrons from the atoms studied. In going from lithium, through beryllium, boron and carbon to nitrogen, we have thus been able to work with stripped atoms of all these substances.

Now these stripped atoms constitute structures which are all exactly alike save that the fields in which the single electron is radiating as it returns toward the nucleus increase in the ratios 1, 2, 3, 4, 5, as we go from stripped lithium to stripped nitrogen. *We have applied the relativity-doublet formula, which, as indicated above, Sommerfeld had developed for the simple nucleus-electron system found in hydrogen and ionized helium, and have found that it not only predicts everywhere the observed doublet-separation of the doublet-lines produced by all these stripped atoms, but that it enables us to compute how many electrons are in the inmost, or K shell, screening the nucleus from the radiating electron. This number comes out just 2, as we know from radioactive and other data that it should.* However, it will be shown in chapter xii that these facts require notable modification of the orbit theory.

Further, when we examine the spectra due to the stripped atoms of the group of elements from sodium

[1] See *Phys. Rev.*, July, 1924.

to sulphur, one electron having been knocked off from sodium, two from magnesium, three from aluminum, four from silicon, five from phosphorus, and six from sulphur, we ought to find that the number of screening electrons in the two inmost shells combined is 2+8 = 10, *and it does come out 10, quite as predicted*, and all this through the simple application of the relativity principle in quite the same way in which it is applied to the quantitative computation of the orbit of Mercury.

The physicist has thus piled Ossa upon Pelion in his quantitative test of the correctness of the orbital equations of Bohr. About the *shapes* of these orbits he has some little information (Fig. 27) but about their *orientations* he is as yet pretty largely in the dark. The diagrams[1] on the accompanying pages, Figs. 28, 29, and 31, represent hypothetical conceptions, due to Bohr in 1922, of the electronic orbits in a group of atoms. Since, however, these orbits are some sort of space configurations, the accompanying plane diagrams are merely schematic. They may be studied in connection with Fig 27, Table XIV, and Bohr's diagram[2] of the periodic system of the elements shown in Fig. 30. These contain the most essential additions which Bohr made in 1922 and 1923 to the simple theory developed in 1913.

The most characteristic feature of these additions is the conception of the penetration, in the case of the less simple atoms, of electrons in highly elliptical orbits into the region inside the shells of lower quantum number.

[1] These appeared in an article by Kramers in *Naturwissenschaften*, 1923.

[2] Bohr and Coster, *Zeit. f. Physik*, XII (1923), 344.

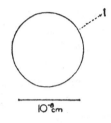

HYDROGEN (1)

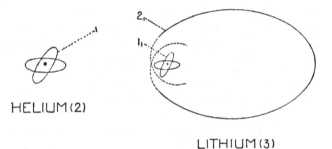

HELIUM (2)

LITHIUM (3)

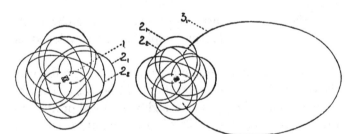

NEON (10)　　SODIUM (11)

Fig. 28.—Hypothetical atomic structures of the year 1923, modified a few years later in the manner shown in Table XIV, then rendered less definite by the development of wave-mechanics.

This gives, so Bohr believes, these penetrating electron-orbits in some cases a smaller mean potential energy, and therefore a higher stability, than some of the orbits corresponding to the smaller quantum numbers.

A glance at the group of elements beginning with argon, the last element in shell 3, in both Table XIV and Fig. 30, will make clear the meaning of this statement. The fourth column of Table XIV shows that we may assign to argon two very elliptical orbits of shape 3_1 and six of shape 3_2. Glancing down the same column to copper, or lower, one sees that there are eighteen possible third-shell orbits, namely, two of shape 3_1, six of shape 3_2, and ten of shape 3_3, i.e., there are in the third shell in argon ten unfilled orbits. But when a new electron is added, as we pass from argon to potassium, it goes, according to Bohr, into the 4_1 orbit, thus giving potassium univalent properties like lithium and sodium (see Fig. 28). Similarly, calcium is shown in Table XIV as taking its two extra electrons into its 4_1 orbits. But as now the nuclear charge gets stronger and stronger with increasing atomic number, the empty third-shell orbits gain in stability over the fourth-shell ones, and a stage of reconstruction sets in with scandium (Fig. 30) and continues down to copper, all the added electrons now going *inside* to fill the ten empty orbits in the third shell, with the result that the chemical properties, which depend on the outer or valence electrons, do not change much while this is going on. With copper (see Table XIV) the eighteen third-shell orbits are completely filled and one electron is in the 4_1 orbit (see also Fig. 29), and from there down to krypton the chemical properties progress normally much as they do from Mg to Ar.

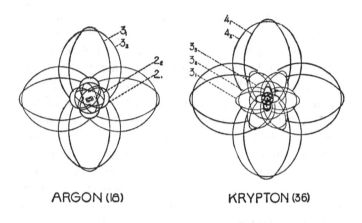

ARGON (18)   KRYPTON (36)

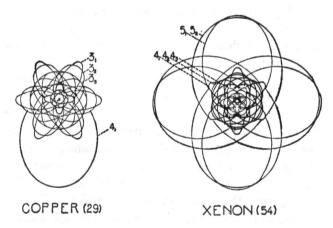

COPPER (29)   XENON (54)

FIG. 29.—Hypothetical atomic structures of the year 1923, modified a few years later in the manner shown in Table XIV, then rendered less definite by the development of wave-mechanics.

Precisely the same procedure is repeated in the fifth period of eighteen elements between krypton and xenon, the rare-earth group which intervenes between strontium (Sr) and silver (Ag) corresponding to the elements in which, with increasing atomic number, the added electrons are filling up the empty orbits in the fourth shell instead of going into what is now the outer or fifth shell (see Table XIV).

Now in considering the sixth period of thirty-two elements from xenon (Xe) to niton (Nt), a glance at Table XIV shows that the fourth shell in xenon contained only eighteen electrons, whereas in niton there are thirty-two, i.e., there are fourteen unfilled orbits in xenon in the fourth shell; and a similar glance at the fifth shell shows $18-8=10$ vacant orbits there. The first two elements in this group, viz., caesium (Cs) and barium (Ba), take the added electrons in 6_1 orbits, then the electrons begin to go inside until gold is reached, when the fourth and fifth shells become full and from gold (Au) to niton (Nt), as the added electrons go to the outer shell, the chemical properties again progress as from sodium to argon, or from copper to krypton.

It will be noticed that in Fig. 30 element 72 is hafnium, the element discovered in 1923 by Coster and Hevesy[1] by means of X-ray analysis. It is because its chemical properties resemble so closely those of zirconium that it had not been found earlier by chemical means. Hevesy estimates that it represents one one hundred-thousandth of the earth's crust, which makes it more plentiful than lead or tin.

[1] Coster and Hevesy, *Nature*, III (1923), 79; *Ber. d. chem. Ges.*, LVI (1923), 1503.

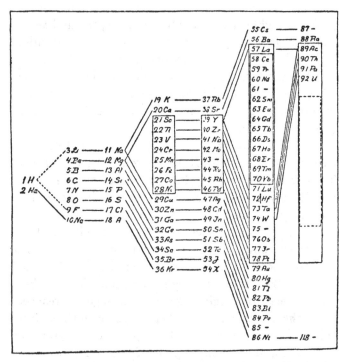

FIG. 30.—Bohr's form of the periodic table, the most illuminating
thus far devised. The elements which are in process of orbital reconstruction, because of the passage of electrons into thus far unfilled
inner quantum orbits, are inclosed in frames. Lines connect elements
which have similar properties.

TABLE XIV

NUMBER OF ELECTRONS IN DIFFERENT n_k ORBITS

(n_k is now generally replaced by n_l where l is defined as $(k-1)$ [see chap. xii])

Period	Z	1_1	$2_1\,2_2$	$3_1\,3_2\,3_3$	$4_1\,4_2\,4_3\,4_4$	$5_1\,5_2\,5_3\,5_4\,5_5$	$6_1\,6_2\,6_3\,6_4\,6_5\,6$	$7_1\,7_2$
1...	1 H	1						
	2 He							
2...	3 Li	2	1					
	4 Be	2	2					
	5 B	2	2 1					
	10 Ne	2	2 6					
3...	11 Na	2	2 6	1				
	12 Mg	2	2 6	2				
	13 Al	2	2 6	2 1				
	18 A	2	2 6	2 6				
4...	19 K	2	2 6	2 6	1			
	20 Ca	2	2 6	2 6	2			
	21 Sc	2	2 6	2 6 1	2			
	22 Ti	2	2 6	2 6 2	2			
	29 Cu	2	2 6	2 6 10	1			
	30 Zn	2	2 6	2 6 10	2			
	31 Ga	2	2 6	2 6 10	2 1			
	36 Kr	2	2 6	2 6 10	2 6			
5..	37 Rb	2	2 6	2 6 10	2 6	1		
	38 Sr	2	2 6	2 6 10	2 6	2		
	39 Y	2	2 6	2 6 10	2 6 1	(2)		
	40 Zr	2	2 6	2 6 10	2 6 2	(2)		
	47 Ag	2	2 6	2 6 10	2 6 10	1		
	48 Cd	2	2 6	2 6 10	2 6 10	2		
	49 In	2	2 6	2 6 10	2 6 10	2 1		
	54 Xe	2	2 6	2 6 10	2 6 10	2 6		
6...	55 Cs	2	2 6	2 6 10	2 6 10	2 6	1	
	56 Ba	2	2 6	2 6 10	2 6 10	2 6	2	
	57 La	2	2 6	2 6 10	2 6 10	2 6 1	(2)	
	58 Ce	2	2 6	2 6 10	2 6 10 1	2 6 1	(2)	
	59 Pr	2	2 6	2 6 10	2 6 10 2	2 6 1	(2)	
	71 Lu	2	2 6	2 6 10	2 6 10 14	2 6 1	(2)	
	72 Hf	2	2 6	2 6 10	2 6 10 14	2 6 2	(2)	
	79 Au	2	2 6	2 6 10	2 6 10 14	2 6 10	1	
	80 Hg	2	2 6	2 6 10	2 6 10 14	2 6 10	2	
	81 Ti	2	2 6	2 6 10	2 6 10 14	2 6 10	2 1	
	86 Nt	2	2 6	2 6 10	2 6 10 14	2 6 10	2 6	
7...	87 —	2	2 6	2 6 10	2 6 10 14	2 6 10	2 6	1
	88 Ra	2	2 6	2 6 10	2 6 10 14	2 6 10	2 6	(2)
	89 Ac	2	2 6	2 6 10	2 6 10 14	2 6 10	2 6 1	(2)
	90 Th	2	2 6	2 6 10	2 6 10 14	2 6 10	2 6 2	(2)
	118 (?)	2	2 6	2 6 10	2 6 10 14	2 6 10 14	2 6 10	2 6

The seventh period begins (Fig. 30) with an unknown element of atomic number 87, which, with its single 7_1 orbit, should have a valency of 1, then passes to radium with its two 7_1 orbits (see Fig. 31) and valency 2, and breaks off suddenly with uranium because the *nucleus* has here become unstable.

It should be clearly understood that the detailed theory as here presented, and above all the models of complicated atoms, are to a very considerable degree hypothetical and speculative. But it is highly probable that they give a more or less correct *general* picture of the way electrons behave in atoms. So far as the general conception of orbits which behave in the main, especially in the simpler atoms, in accordance with the Bohr assumptions, is concerned, if the test of truth in a physical theory is large success both in the prediction of new relationships and in correctly and exactly accounting for old ones, the theory of non-radiating orbits is one of the well-established truths of modern physics. However, all mechanical pictures like the foregoing, while useful as mnemonic devises, have definite limitations (see chapter xii) and must not be thought of as corresponding in any accurate way to reality.

I am well aware that the facts of organic chemistry seem to demand that the valence electrons be grouped in certain definite equilibrium positions about the periphery of the atom, and that at first sight this demand appears difficult to reconcile with the theory of electronic orbits. But a little reflection shows that there is here no necessary clash. With a suitable *orientation* of orbits, these localized valencies of chemistry are about as easy to reconcile with an orbit theory as with a fixed electron

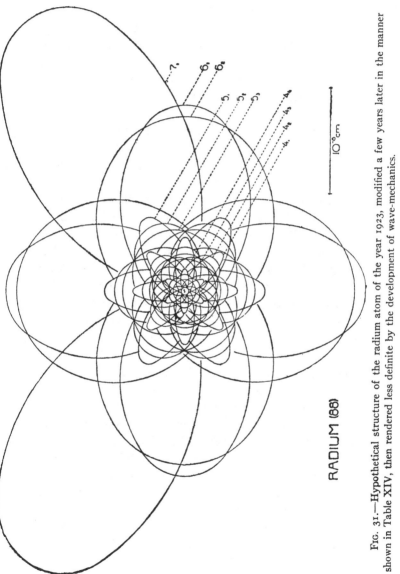

RADIUM (88)

FIG. 31.—Hypothetical structure of the radium atom of the year 1923, modified a few years later in the manner shown in Table XIV, then rendered less definite by the development of wave-mechanics.

theory. It is only for *free atoms* that spectroscopic evidence has led us to built up orbit pictures of the foregoing sort. When atoms unite into molecules, or into solid bodies, these orbits will undoubtedly be very largely readjusted under the mutual influence of the two or more nuclei which are now acting simultaneously upon them.

It has been objected, too, that the Bohr theory is not a radiation theory because it gives us no picture of the mechanism of the production of the frequency ν. This is true, and therein lies its strength, just as the strength of the first and second laws of thermodynamics lies in the fact that they are true irrespective of a mechanism. The Bohr theory is a theory of atomic structure; it is not a theory of radiation, for it merely states what energy relations must exist when radiation, whatever its mechanism, takes place. It is the first attempt to determine in the light of well-established experimental facts what the electrons inside the atom are doing, and as such a first attempt it must be regarded as, thus far, a success, though it has by no means got beyond the hypothetical stage. Its chief difficulty arises from the apparent contradiction involved in a non-radiating electronic orbit, and there appears to be no solution to this difficulty save in the denial of the *universal* applicability of the classical electromagnetic laws. But why assume the universal applicability of these laws, even in the hearts of atoms, when this is the first opportunity which we have had to test them out in the region of the infinitely small?

There is one other very important relation predicted by the Bohr theory and beautifully verified by experiment, but not involving at all its orbital feature. The frequency value of the inmost, or K level, can be exactly determined by measuring the K absorption edge so beautifully shown on the De Broglie photographs opposite p. 199. Let us call this frequency ν_{KA}. Similarly, to each orbit in the second or L quantum state, there corresponds a definite absorption edge ν_{LA}. Two of these are shown clearly in Fig. 23. The difference between the K absorption frequency and each L absorption frequency should obviously, according to Bohr, correspond exactly to the frequency ν_{K_a} of an *emission* line in the K X-ray spectrum, i.e.,

$$\nu_{KA} - \nu_{LA} = \nu_{K_a} \quad \dots \dots \dots \dots \dots (40)$$

This so-called Kossel relation is of course applicable to all X-ray and optical spectra. Indeed, in the latter field it appeared before the Bohr theory under the name of the "Ritz combination principle." It has been one of the most important keys to the unlocking of the meaning of spectra and the revealing of *atomic structure.*

CHAPTER X

THE NATURE OF RADIANT ENERGY

The problems thus far discussed have all been in the domain of molecular physics, but the discovery and measurement of the electron have also exerted a powerful influence upon recent developments in the domain of ether physics. These developments are of extraordinary interest and suggestiveness, but they lead into regions in which the physicist sees as yet but dimly—perhaps more dimly than he thought he saw forty years ago.

But while the beauty of a problem solved excites the admiration and yields a certain sort of satisfaction, it is after all the unsolved problem, the quest of the unknown, the struggle for the unattained, which is of most universal and most thrilling interest. I make no apologies, therefore, for introducing one of the incompletely solved problems of modern physics, nor for leaving it with somewhat tentative suggestions toward a solution.

I. THE CORPUSCULAR AND THE ETHER THEORIES OF RADIATION

The newest of the problems of physics is at the same time the oldest. For nothing is earlier in the experiences either of the child or of the race than the sensation of receiving light and heat from the sun. But how does light get to us from the sun and the stars through the empty interstellar spaces? The Greeks answered this query very simply and very satisfactorily from the standpoint of people who were content with plausible explanations but had not yet learned perpetually to question **nature** experimentally as to the validity **or** invalidity of a conclusion. They said that the sun and all radiators of light and heat must shoot off minute corpuscles whose impact upon the eye or skin produces the sensations of light and warmth.

This corpuscular theory was the generally accepted one up to 1800 A.D. It was challenged, it is true, about 1680 by the Dutch physicist Huygens, who, starting with the observed phenomena of the transmission of water waves over the surface of a pond or of sound waves through the air, argued that light might be some vibratory disturbance transmitted by some medium which fills all interstellar space. He postulated the existence of such a medium, which was called the luminiferous or light-bearing ether.

Partly no doubt because of Newton's espousal of the corpuscular theory, the ether or wave theory gained few adherents until some facts of interference began to appear about 1800 which baffled explanation from the standpoint of the corpuscular theory, but which were easily handled by its rival. During the nineteenth century the evidence became stronger and stronger, until by its close the corpuscular theory had been completely eliminated for four different reasons: (1) The facts of interference were not only found inexplicable in terms of it, but they were completely predicted by the wave theory. (2) The fact that the speed of propagation of light was experimentally found to be greater in air than in water was in accord with the demands of the ether theory, but directly contrary to the demands of the corpuscular theory. (3) Wireless waves had appeared and had been shown to be just like light waves save for wave-length, and they had been found to pass over continuously, with increasing wave-length, into static electrical fields such as could not apparently be explained from a corpuscular point of view. (4) The speed of light had been shown to be independent of the speed of the source as demanded by the ether theory and denied by the corpuscular theory.

By 1900, then, the ether theory had become apparently impregnably intrenched. A couple of years later it met with some opposition of a rather ill-considered sort, as it seems to me, from a group of extreme advocates of the relativity theory, but this theory is now commonly regarded, I think, as having no bearing whatever upon the question of the existence or non-existence of an ether as I use the term. For such an ether was called into being solely for the sake of furnishing a carrier for electro-magnetic waves, and so defined it stands or falls with the existence of such waves *in vacuo*, and this has never been questioned by anyone so far as I am aware.

II. DIFFICULTIES CONFRONTING THE WAVE THEORY

Up to 1903, then, the theory which looked upon an electromagnetic wave as a disturbance which originated at some point in the ether at which an electric charge was undergoing a change in speed, and was propagated from that point outward as a spherical wave or pulse, the total energy of the disturbance being always spread uniformly over the wave front, had met with no serious question from any source. Indeed, it had been extraordinarily successful, not only in accounting for all the known facts, but in more than one instance in predicting new ones. The first difficulty appeared after the discovery of the electron and in connection with the relations of the electron to the absorption or emission of such electromagnetic waves. It was first pointed out in 1903 by Sir J. J. Thomson in his Silliman lectures at Yale. It may be stated thus:

X-rays unquestionably pass over all but an exceedingly minute fraction, say one in a thousand billion, of the atoms contained in the space traversed without spending any energy upon them or influencing them in any observable way. But here and there they find an atom from which, as is shown in the photographs opposite p. 192, they hurl a negative electron with enormous speed. This is the most interesting and most significant characteristic of X-rays, and one which distinguishes them from the α- and β-rays just as sharply as does the property of non-deviability in a magnetic field; for Figs. 14 and 15 and the plate opposite p. 190 show that neither α- nor β-rays in general eject electrons from the atoms through which they pass, with speeds comparable with those produced by X-rays, else there would be new long zigzag lines branching out from points all along the paths of the α- and β-particles shown in these photographs.

But this property of X-rays introduces a serious difficulty into a wave theory. For if the electric intensity in the wave front of the X-ray is sufficient thus to hurl a corpuscle with huge energy from one particular atom, why does it not at least detach corpuscles from all of the atoms over which it passes?

Again when ultra-violet light falls on a metal it, too, like X-rays, is found to eject negative electrons. This phenomenon of the emission of electrons under the influence of light is called the photo-electric effect. Lenard[1] first made the astonishing discovery that the energy of ejection of the electron **is** altogether independent of the intensity of the light which causes the ejection, no matter whether this intensity is varied by varying the distance of the light or by introducing absorbing screens. I have myself[2] subjected this relation to a very precise test and found it to hold accurately. Furthermore, this sort of independence has also been established for the negative electrons emitted by both X- and γ-rays.

Facts of this sort are evidently difficult to account for on any sort of a spreading-wave theory. But it will be seen that they lend themselves to easy interpretation in terms of a corpuscular theory, for if the energy of an escaping electron comes from the absorption of a light-corpuscle, then the energy of emission of the ejected electron ought to be independent of the distance of the source, as it is found to be, and furthermore corpuscular rays would hit but a very minute fraction of the atoms contained in the space traversed by them. This would explain, then, both the independence of the energy of emission upon intensity and the smallness of the number of atoms ionized.

In view, however, of the four sets of facts mentioned above, Thomson found it altogether impossible to go back to the old and exploded form of corpuscular theory for an explanation of the new facts as to the emission of electrons under the influence of ether waves. He accordingly attempted to reconcile these troublesome new facts with the wave theory by assuming a fibrous structure in the ether and picturing all electromagnetic

---

[1] *Ann. d. Phys.* (4), VIII (1902), 149.   [2] *Phys. Rev.*, I (1913), 73.

energy as traveling along Faraday lines of force conceived of as actual strings extending through all space. Although this concept, which we shall call the ether-string theory, is like the corpuscular theory in that the energy, after it leaves the emitting body, remains localized in space, and, when absorbed, is absorbed as a whole, yet it is after all essentially an ether theory. For in it the speed of propagation is determined by the properties of the medium—or of space, if one prefers a mere change in name—and has nothing to do with the nature or condition of the source. Thus the last three of the fatal objections to a corpuscular theory are not here encountered. As to the first one, no one has yet shown that Thomson's suggestion is reconcilable with the facts of interference, though so far as I know neither has its irreconcilability been as yet absolutely demonstrated.

But interference aside, all is not simple and easy for Thomson's theory. For one encounters serious difficulties when he attempts to visualize the universe as an infinite cobweb whose threads never become tangled or broken however swiftly the electrical charges to which they are attached may be flying about.

III. EINSTEIN'S QUANTUM THEORY OF RADIATION

Yet the boldness and the difficulties of Thomson's "ether-string" theory did not deter Einstein[1] in 1905 from making it even more radical. In order to connect it up with some results to which Planck of Berlin had been led in studying the facts of black-body radiation, Einstein assumed that the energy emitted by any radiator not only kept together in bunches or quanta as it traveled through space, as Thomson had assumed it to do, but that a given source could emit and absorb radiant energy only in units which are all exactly equal to $h\nu$, ν being the natural frequency of the emitter and h a constant which is the same for all emitters.

I shall not attempt to present the basis for such an assumption, for, as a matter of fact, it had almost none at the time. But whatever its basis, it enabled Einstein to predict at once that the energy of emission of electrons under the influence of light would be governed by the equation

$$\tfrac{1}{2}mv^2 = Ve = h\nu - p \ldots\ldots\ldots\ldots\ldots(41)$$

in which $h\nu$ is the energy absorbed by the electron from the light wave or light quantum, for, according to the assumption it was the whole energy contained in that quantum, p is the work necessary to get the electron out of the metal, and $\tfrac{1}{2}mv^2$ is the energy with which it leaves the surface—an energy evidently measured by the product of its charge e by the potential difference V against which it is just able to drive itself before being brought to rest.

At the time at which it was made this prediction was as bold as the hypothesis which suggested it, for at that time there were available no experiments whatever for determining anything about how the positive potential V necessary to apply to the illuminated electrode to stop the discharge of negative electrons from it under the influence of monochromatic light varied with the frequency ν of the light, or whether the quantity h to which Planck had already assigned a numerical value appeared at all in connection with photo-electric discharge. We are confronted, however, by the astonishing situation that after ten years of work at the Ryerson Laboratory (1904-15) and elsewhere upon the discharge of electrons by light this equation of Einstein's was found to predict accurately all of the facts which had been observed.

IV. THE TESTING OF EINSTEIN'S EQUATION

The method which was adopted in the Ryerson Laboratory for testing the correctness of Einstein's equation involved the performance of so many operations upon the highly inflammable alkali metals in a vessel which was freed from the presence of all gases that it is not inappropriate to describe the experimental arrangement as a machine-shop *in vacuo*. Fig. 32 shows a photograph of the apparatus, and Fig. 33 is a drawing of a section which should make the necessary operations intelligible.

One of the most vital assertions made in Einstein's theory is that the kinetic energy with which mono-chromatic light ejects electrons from any metal is proportional to the frequency of the light, i.e., if violet light is of half the wave-length of red light, then the violet light should throw out the electron with twice the energy imparted to it by the red light. In order to test whether any such linear relation exists between the energy of the escaping electron and the light which throws it out it was necessary to use as wide a range of frequencies as possible. This made it necessary to use the alkali metals, sodium, potassium, and lithium, for electrons are thrown from the ordinary metals only by ultra-violet light, while the alkali metals respond in this way to any waves shorter than those of the red, that is, they respond throughout practically the whole visible spectrum as well as the ultra-violet spectrum. Cast cylinders of these metals were therefore placed on the wheel W (Fig. 33) and fresh clean surfaces were obtained by cutting shavings from each metal in an excellent vacuum with the aid of the knife K, which was operated

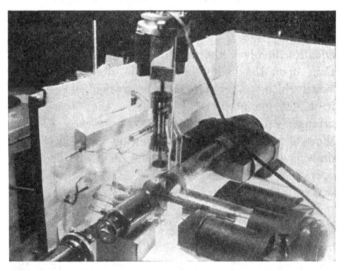

F𝗂𝗀. 32

by an electromagnet F outside the tube. After this the freshly cut surface was turned around by another electromagnet until it was opposite the point O of Fig. 33 and a beam of monochromatic light from a spectrometer was let in through O and allowed to fall on the new surface. The energy of the electrons ejected by it was measured by applying to the surface a positive potential just strong enough to prevent any of the discharged electrons from reaching the gauze cylinder opposite (shown in dotted lines) and thus communicating an observable negative

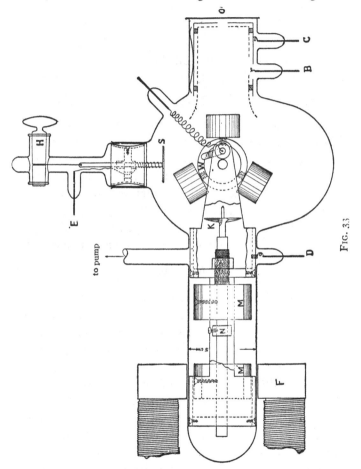

FIG. 33

charge to the quadrant electrometer which was attached to this gauze cylinder. For a complete test of the equation it was necessary also to measure the contact-electromotive force between the new surface and a test plate S. This was done by another electromagnetic device shown in Fig. 32, but for further details the original paper may be consulted.[1] Suffice it here to say that Einstein's equation demands a linear relation between the applied positive volts and the frequency of the light, and it also demands that the slope of this line should be exactly equal to $\left(\dfrac{h}{e}\right)$. Hence from this slope, since e is known, it should be possible to obtain h. How perfect a linear relation is found may be seen from Fig. 34, which also shows that from the slope of this line h is found to be 6.26×10^{-27}, which is as close to the value obtained by Planck from the radiation laws as is to be expected from the accuracy with which the experiments in radiation can be made. The most reliable value of h obtained from a consideration of the whole of this work is

$$h = 6.56 \times 10^{-27}.$$

In the original paper will be found other tests of the Einstein equation, but the net result of all this work is to confirm in a very complete way the equation which Einstein first set up on the basis of his semi-corpuscular theory of radiant energy. And if this equation is of general validity it must certainly be regarded as one of the most fundamental and far-reaching of the equations of physics, and one which is destined to play in the future a scarcely less important rôle than Maxwell's equations have played in the past, for it must govern the transformation of all short-wave-length electromagnetic energy into heat energy.

[1] *Phys. Rev.*, VII (1916), 362.

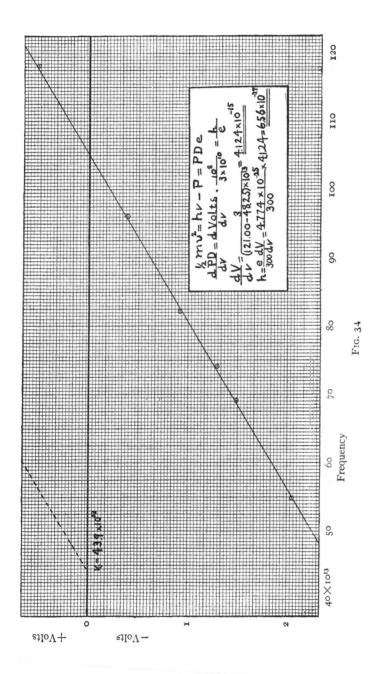

$$\tfrac{1}{2}mv^2 = h\nu - P = PDe$$

$$\frac{d\,PD}{d\nu} = d\,\text{Volts} \cdot \frac{10^8}{3\times10^{10}} = \frac{h}{e}$$

$$\frac{dV}{d\nu} = \frac{3}{(21.00-48.25)\times10^{13}} = \frac{4.124\times10^{-15}}{}$$

$$h = \frac{e}{300}\frac{dV}{d\nu} = \frac{4.774\times10^{-25}}{300}\times 4.124 = 6.56\times10^{-27}$$

$V_0 = 4.39\times10^{13}$

Frequency

40×10^{13} 50   60   70   80   90   100   110   120

+Volts   0
−Volts   1
−Volts   2

Fig. 34

V. HISTORY OF EINSTEIN'S EQUATION

The whole of this chapter up to this point has been left practically as it was written for the first edition of *The Electron* (1917). Now the altogether overwhelming proof that Einstein's equation is an exact equation of very general validity was perhaps the most conspicuous achievement of experimental physics during the next decade. Its history is briefly as follows.

As early as 1900 Planck[1] had been led from theoretical considerations to the conclusion that atoms radiated energy discontinuously in units which were equal to, or multiples of, $h\nu$, in which ν is the natural frequency of the radiator, and h a universal constant which is now called Planck's h. He adopted the view that the seat of the discontinuity was *in the radiator*, not in the radiation after it had left the radiator, and in the second edition of his book modified the formulation of his theory so as to make this appear without any ambiguity.

It was in 1905, as stated above, that Einstein definitely put the discontinuity into the radiation itself, assuming that light itself consisted of darts of localized energy, "light-quants," of amount $h\nu$. He further assumed that one of these light-quants could transfer its energy undiminished to an electron, so that, in the photo-electric effect, the electron shot out from the metal with the energy $h\nu - p$, where p represents the work necessary to get it out of the metal.

In 1913 Bohr, in the development of his theory of spectra, without accepting Einstein's view as to the seat of the discontinuity, assumed an *equation* which was

[1] *Warme Strahlung*, 1st. ed.

precisely the inverse of Einstein's, i.e., he assumed that the energy lost when an electron jumps from one stationary state to another is wholly transformed into monochromatic radiation whose frequency is determined by equating the loss in energy $E_1 - E_2$ to $h\nu$. In other words, *Einstein and Bohr together have set up a reciprocal and reversible relation between electronic and radiant energy.*

Up to 1914 no direct experimental proof had appeared for the correctness of this relation. In the photo-electric field discussion was active as to whether any definite maximum velocity of emission of electrons under the influence of monochromatic light existed, and although linear relations between energy and frequency had been reported by Ladenburg, Richardson and Compton, and Hughes, the range of frequencies available had been so small as to leave uncertainties in the minds of reviewers[1] who showed that $\nu \propto \sqrt{E}$ fitted existing observations quite as well as $\nu \propto E$ (E denoting stopping potential).

The unambiguous experimental proofs of the correctness of the foregoing theoretical relation came with the publication of the accompanying photoelectric results[2] reported briefly in 1914, and submitted *in extenso* in September, 1915. These were in a form to prove the correctness of the Einstein equation; for monochromatic light of widely differing frequencies fell upon a metal and the maximum energy of electronic ejection was found to be exactly determined by $h\nu = \frac{1}{2}mv^2 - p$ as Einstein's equation required.

[1] Cf. R. Pohl u. P. Pringsheim, *Verh. der deutsch. phys. Ges.*, XV (1913), 637; Sommerfeld, *Atombau*, etc. (3d ed. 1922), p. 47; also *Phys. Rev.*, VII (1916), 18, 362.

[2] *Phys. Rev.*, IV (1914), 73; VI (1915), 55; and VII (1916), 362.

A year or two later Duane[1] and his associates had found unambiguous proof of the inverse effect. A target had been bombarded by electrons of known and constant energy $(Ve = \frac{1}{2}mv^2)$ and the maximum frequency of the emitted ether waves (general x radiation) was found to be precisely given by $\frac{1}{2}mv^2 = h\nu$.

D. L. Webster then proved that the characteristic X-ray frequencies of atoms begin to be excited at exactly the potential at which the energy of the stream of electrons which is bombarding the atoms has reached the value given by $h\nu = \frac{1}{2}mv^2$ in which ν is now the frequency of an absorption edge.[2] This checks Bohr's formulation of frequency-energy relations, since it shows that when an electron within an atom receives just enough energy by bombardment to be entirely removed from the atom, the total energy values of the frequencies emitted during its return are equal to the electronic energy of the original bombardment.

De Broglie,[3] Ellis,[4] and Watson[5] on the other hand, have measured with great accuracy, through deviabilities in a magnetic field, the velocities of electrons ejected from different sorts of atoms by monochromatic X-rays, and have completely confirmed by such photo-electric work in the X-ray field my previous results obtained with ultra-violet light. They here verify in great detail and with much elaboration the Einstein formulation

[1] *Phys. Rev.*, VI (1915), 166; *Proc. Nat. Acad.*, II (1916), 90; *Phys. Rev.*, VII (1916), 599; IX, 568; X (1917), 93 and 624.

[2] D. L. Webster and H. Clark, *Proc. Nat. Acad.*, III (1917), 18. Also Webster, *ibid.*, VI (1920), 26 and 639.

[3] Paper read before the Third Solvay Congress, 1921.

[4] *Proc. Roy. Soc.*, XCIX (1921), 261. See *ibid.*, January, 1924.

[5] Watson and Van den Akker, *Proc. Roy. Soc. A.*, CXXVI (1929), 138.

$\frac{1}{2}mv^2 = h\nu - p$ where p now represents the work necessary to lift the electron out of any particular level in the atom.

Parallel to this very complete establishment of the validity in the X-ray field of the Einstein photoelectric equation, and of its inverse the Bohr equation, has come the rapid working out in the domain of optics of the very large field of ionizing and radiating potentials which has also involved the utilization and verification of the same reciprocal relation. This will be seen at once from the definition of the ionizing potential of an atom as the electronic energy which must be thrown into it by bombardment to just remove from it one of its outer electrons. Through the return of such removed electrons there is in general a whole spectral series emitted. Similarly the radiating potential of an atom is defined as the bombarding energy which must be supplied to it to just lift one of its outer electrons from its normal orbit to the first virtual orbit outside that normal orbit. When this electron drops back there is in general the emission of a single-line spectrum. All this work took its origin in the fundamental experiments of Franck and Hertz[1] on mercury vapor in 1914. From 1916–22 the field was worked out in great detail, especially in America by Foote and Mohler, Wood, McLennan, Davis and Goucher, and others.

Suffice it to say that whether the energy comes in the form of ether waves which through absorption in an atom lift an electron out of a normal orbit, so that the atom passes over to an excited or to an ionized state, or whether the energy enters in the form of a bombarding

[1] *Verh. der deutsch. phys. Ges.*, XV and XVI, 1914.

electron and reappears as a radiated frequency, *the reciprocal relation represented in the Einstein-Bohr equation $E_1 - E_2 = h\nu$ has been found fulfilled in the most complete manner.*

In view of all these methods and experiments the general validity of the Einstein equation, first proved photo-electrically between 1912 and 1916, is now universally conceded. *The incident energy is called a "photon."*

VI. OBJECTIONS TO AN ETHER-STRING THEORY

In spite of the credentials which have just been presented for Einstein's equation, the essentially corpuscular theory out of which he got it has only recently met with general acceptance even by physicists of Bohr's type; for there seemed at first to be no possibility of bringing it into harmony with a whole group of well-established facts of physics, and even now the difficulties are great.

The recent practically complete bridging of the gap between X-rays and light,[1] as well as that between heat waves and wireless waves,[2] with the perfectly continuous passage of the latter over into static electrical fields, appears to demand that, if we attempt to interpret high-frequency electromagnetic waves—X-rays and light—in terms of undulatory "darts of light," we also interpret wireless waves in the same way, and this in turn requires us to use a similar mechanism in the interpretation of static electrical fields. This brings us back to something like an ether-string theory, which seems to be a necessary part of Einstein's conception, if it is to have any physical basis whatever.

[1] Millikan and Bowen, *Phys. Rev.*, January, 1924.
[2] Nichols and Tear, *ibid.*, 1923.

Two very potent objections, however, may be urged against all forms of ether-string theory. The first is that no one has ever yet been able to show that such a theory can predict any one of the facts of interference. The second is that there is direct positive evidence against the view that the ether possesses a fibrous structure. For if a static electrical field has a fibrous structure, as postulated by any form of ether-string theory, "each unit of positive electricity being the origin and each unit of negative electricity the termination of a Faraday tube,"[1] then the force acting on one single electron between the plates of an air condenser cannot possibly vary *continuously* with the potential difference between the plates. Now in the oil-drop experiments[2] we actually study the behavior in such an electric field of one single, isolated electron and we find, over the widest limits, exact proportionality between the field strength and the force acting on the electron as measured by the velocity with which the oil drop to which it is attached is dragged through the air.

When we maintain the field constant and vary the charge on the drop, the granular structure of electricity is proved by the discontinuous changes in the velocity, but when we maintain the charge constant and vary the field the lack of discontinuous change in the velocity disproves the contention of a fibrous structure in the field, unless the assumption be made that there are an enormous number of ether strings ending in one electron. Such an assumption takes most of the virtue out of an ether-string theory.

[1] J. J. Thomson, *Electricity and Matter*, p. 9.

[2] *Phys. Rev.*, II (1913), 109.

Despite, then, the apparently complete success of the Einstein equation, the physical theory of which it was designed to be the symbolic expression is thus far so irreconcilable with a whole group of well-established facts that most modern physicists have abandoned the attempt to visualize it, and we are somewhat in the position of having built a very perfect structure and then knocked out entirely the underpinning without causing the building to fall. It stands complete and apparently well tested, but without any visible means of support. These supports must obviously exist, and the most fascinating problem of modern physics is to find them. Experiment has outrun theory, or, better, guided by thus far non-visualizable theory, it has discovered relationships which seem to be of the greatest interest and importance, but the physical reasons for them are as yet not at all understood.

VII. ATTEMPTS TOWARD A SOLUTION

It is possible, however, to go a certain distance toward a solution and to indicate some conditions which must be satisfied by the solution when it is found. For the energy $h\nu$, with which the electron is found by experiment to escape from the atom, must have come either from the energy stored up inside of the stom or else from the light. There is no third possibility. Now the fact that the energy of emission is the same, whether the body from which it is emitted is held within an inch of the source, where the light is very intense, or a mile away, where it is very weak, would seem to indicate that the light simply pulls a trigger in the atom which itself furnishes all the energy with which the electron escapes, as was originally suggested by Lenard in 1902,[1] or else, if the light furnishes the energy, that light itself must consist of bundles of energy which keep together as they travel through space, as suggested in the Thomson-Einstein theory.

Yet the fact that the energy of emission is directly proportional to the frequency ν of the incident light spoils Lenard's form of trigger theory, since, if the atom furnishes the energy, it ought to make no difference what kind of a wave-length pulls the trigger, while it ought to make a difference what kind of a gun, that is, what kind of an atom, is shot off.   But both of these expectations are the exact opposite of the observed facts.   *The energy of the escaping electron must come, then, in some way or other, from the incident light, or from other light of its frequency, since it is characteristic of that frequency alone.*

When, however, we attempt to compute on the basis of a spreading-wave theory how much energy an electron can receive from a given source of light, we find it difficult to find anything more than a very minute fraction of the amount which it actually acquires.

Thus, the total luminous energy falling per second from a standard candle on a square centimeter at a distance of 3 m. is 1 erg.[2]   Hence the amount falling per second on a body of the size of an atom, i.e., of cross-section 10^{-15} cm., is 10^{-15} ergs, but the energy $h\nu$ with which an electron is ejected by light of wave-length 500 $\mu\mu$ (millionths millimeter) is 4×10^{-12} ergs, or four thousand times as much.   Since not a third of the incident energy is in wave-lengths shorter than 500 $\mu\mu$, a

[1] *Ann. d. Phys.* (4), VIII (1902), 149.

[2] Drude, *Lehrbuch der Optik* (1906), p. 472.

surface of sodium or lithium which is sensitive up to
500 $\mu\mu$ should require, even if all this energy were in one
wave-length, which it is not, at least 12,000 seconds
or 4 hours of illumination by a candle 3 m. away before
any of its atoms could have received, all told, enough
energy to discharge an electron. Yet the electron is
observed to shoot out the instant the light is turned on.
It is true that Lord Rayleigh has shown[1] that an atom
may conceivably absorb wave-energy from a region of
the order of magnitude of the square of a wave-length
of the incident light rather than of the order of its own
cross-section. This in no way weakens, however, the
cogency of the type of argument just presented, for it is
only necessary to apply the same sort of analysis to the
case of γ-rays, the wave-length of which is sometimes as
low as a hundredth of an atomic diameter (10^{-8} cm.), and
the difficulty is found still more pronounced. Thus
Rutherford[2] estimates that the total γ-ray energy radiated per second by one gram of radium cannot possibly
be more than 4.7×10^4 ergs. Hence at a distance of 100
meters, where the γ-rays from a gram of radium would
be easily detectable, the total γ-ray energy falling per
second on a square millimeter of surface, the area of
which is ten-thousand billion times greater than that
of an atom, would be $4.7 \times 10^4 \div 4\pi \times 10^{10} = 4 \times 10^{-7}$ ergs.
This is very close to the energy with which β-rays are
actually observed to be ejected by these γ-rays, the
velocity of ejection being about nine-tenths that of light.
Although, then, it should take ten thousand billion
seconds for the atom to gather in this much energy from

[1] *Phil. Mag.*, XXXII (1916), 188.

[2] *Radioactive Substances and Their Radiations*, p. 288.

the γ-rays, on the basis of classical theory, the β-ray is observed to be ejected with this energy as soon as the radium is put in place. This shows that if we are going to abandon the Thomson-Einstein hypothesis of localized energy, which is of course competent to satisfy these energy relations, there is no alternative but to assume that at some previous time the electron had absorbed and stored up from light of this wave-length enough energy so that it needed but a minute addition at the time of the experiment to be able to be ejected from the atom with the energy $h\nu$. What sort of an absorbing and energy-storing mechanism an atom might have which would give it the weird property of storing up energy to the value $h\nu$, where ν is the frequency of the *incident* light, and then shooting it all out at once, is terribly difficult to conceive. Or, if the absorption is thought of as due to resonance it is equally difficult to see how there can be, in the atoms of a solid body, electrons having all kinds of natural frequencies so that some are always found to absorb and ultimately be ejected by impressed light of any particular frequency.

However, then, we may interpret the phenomenon of the emission of electrons under the influence of ether waves, whether upon the basis of the Thomson-Einstein assumption of bundles of localized energy traveling through the ether, or upon the basis of a peculiar property of the inside of an atom which enables it to absorb continuously incident energy and emit only explosively, *the observed characteristics of the effect seem to furnish proof that the emission of energy by an atom is a discontinuous or explosive process.* This was the fundamental assumption of Planck's so-called quantum theory of radiation. The Thomson-Einstein theory makes both the absorption and the emission sudden or discontinuous, while the loading theory first suggested by Planck makes the absorption continuous and only the emission explosive.

The new facts in the field of radiation which have been discovered through the study of the properties of the electron seem, then, to require in any case a very fundamental revision or extension of classical theories of absorption and emission of radiant energy. The Thomson-Einstein theory throws the whole burden of accounting for the new facts upon the unknown nature of the ether, or, if one dislikes the word, "of the field," and makes radical assumptions about its structure. The loading theory leaves the ether alone and puts the burden of an explanation upon the unknown conditions and laws which exist inside the atom.

In the first edition of *The Electron*, of date 1917, I expressed the view that the chances were in favor of the ultimate triumph of the second alternative. In 1921, however, I presented at the Third Solvay Congress some new photo-electric experiments[1] which seemed at the time to point strongly the other way.

These experiments consisted in showing with greater certainty than had been possible in earlier years[2] that the stopping potentials of different metals *A*, *B*, *C*, when brought in succession before the same Faraday cylinder *F* (see Fig. 35) and illuminated with a given frequency, were strictly identical. The significance of these results

[1] Millikan, *Phys. Rev.*, XVIII (1921), 236.

[2] Page, *Amer. Jour. Sci.*, XXXVI (1913), 501; Hennings and Kadesch, *Phys. Rev.*, VIII (1916), 217.

for the theory of quanta lay in the fact that I deduced
from them the conclusion that in the photo-electric
effect, contrary to preceding views including my own,
the *energy "hv" is transferred without loss from the ether-
waves to the free, i.e., the conduction electrons of the metal,
and not merely to those bound in atoms.* This seemed to
take the absorbing mechanism out of the atom entirely,
and to make the property of imparting the energy hv to
an electron, whether free or bound, an intrinsic property
of light itself.

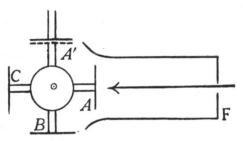

FIG. 35.—Showing how photo-electric stopping potentials of different
metals are compared by rotating B and C *in vacuo* into the position of A.

But a beautiful discovery by Klein and Rosseland[1] a
little later, in Bohr's Institute, made this conclusion
unnecessary. For it showed that there was an intermediate process, namely, a so-called *collision of the second
kind*, by means of which the energy hv might be transferred without loss, *indirectly* from the light-wave to the
conduction electron, thus obviating the necessity of a
direct transfer. In other words, the Klein and Rosseland
discovery proved that the energy hv could be transferred
from the light-wave to the conduction electron by being

[1] *Zeitschrift für Physik*, 4 (1921), 46.

absorbed first by an atom, which would thus be changed from the normal to the excited state, i.e., the state in which one of its electrons has been lifted from a normal to an outer orbit. This excited atom could then return to its normal state *without radiation* by a collision "of the second kind," which consists in transferring its whole absorbed energy $h\nu$ to a free or conduction electron. The reality of this phenomenon has been experimentally checked by Franck and Cario.[1] This important discovery then left the evidence for localized light-quanta precisely where it was before.[2]

In the years 1923–25, however, the American physicist, Dr. A. H. Compton, of the University of Chicago, discovered another new phenomenon which constitutes quite as good evidence as the photoelectric effect in favor of Einstein's hypothesis of localized light-quanta.

Compton's procedure is as follows. Assuming, for the sake of obtaining quantitative relations, the correctness of Einstein's hypothesis, he argues that when such a "light-quant" collides with a *free* electron the impact should be governed by the laws which hold for the collision between any material bodies. These are two in number, namely: (1) the principle of the conservation of energy; (2) the principle of the conservation of momentum (Newton's Third Law).

Now the energy of a light-quant, as heretofore shown, is $h\nu$. It moves with the speed of light, c, and if its momentum is taken as mc, it follows at once from the Einstein relativity relation between energy and mass,

[1] *Zeitschrift für Physik*, 10 (1922), 185.

[2] This was first called to my attention by Dr. Epstein, of the California Institute.

namely, energy$/c^2 = m$, that its momentum is $\dfrac{h\nu}{c}$. This is seen by substituting in the foregoing Einstein relation $h\nu$ for energy. Or, if preferred, the same expression for momentum may be deduced easily from the established laws of light-pressure.

The qualitative results of the preceding assumptions are immediately seen to be as follows. The light-quant, by colliding with the free electron necessarily transfers some of its energy to it, and therefore, if it arrives with the energy $h\nu_0$, it must recoil from the impact at some angle θ with a smaller energy $h\nu_\theta$, and therefore a lower frequency ν_θ, than that with which it impinged. In other words, *light waves should be changed from a higher frequency to a lower—from blue toward red—by impact with a free electron.*

A second qualitative result is that, since the mass of the light-quant, as defined above, is even for the hardest X-rays ($\lambda = 0.1$ Angstrom), of the order of a tenth of the mass of the electron, it is impossible from the laws of elastic impact that it transfer more than a small part of its energy to it. In other words, *if Compton's assumptions are correct, the photo-electric effect, in which there certainly is such a complete transfer, cannot possibly represent the interaction between a light-wave and a free electron.* When the electron is *bound* in the atom there is no difficulty of this sort, for the huge mass of the atom then permits the momentum equation to be satisfied without forbidding the practically complete transfer of the energy to one of its electrons. From this point of view, then, the photo-electric effect represents the interaction between ether-waves and *bound* electrons—the Compton effect the interaction between ether-waves and *free* electrons.

The quantitative results which can be deduced from Compton's assumptions are definite and simple. Combining the energy and momentum equations in the manner shown in Appendix H he obtains easily the result

$$\Delta\lambda = .0484 \sin^2 \tfrac{1}{2}\theta,$$

in which $\Delta\lambda$ represents the increase in wave-length due to the "scattering" of the incident beam by free electrons, and θ is the angle between the original direction of the beam and the direction at which the scattered waves come to the measuring apparatus.

Compton then tested this relation experimentally,[1] using as his incident waves the characteristic X-rays from a molybdenum target, and as his scattering substance the free (or substantially free) electrons found in graphite. *He found indeed that the a-line of molybdenum was shifted toward longer wave-lengths just as predicted, and in approximately the correct amount.* There was also an unshifted line presumably due to scattering by bound electrons.

Compton had used an ionization-chamber spectrometer for locating his lines. Ross[2] repeated these experiments at Stanford University, California, using the more accurate photographic plate for locating his lines, but still using graphite as the scattering substance. His published photograph shows a line shifted the correct amount and also an unshifted one, but he commented on the fact that the shifted line shows no sign of a separation of the a_1 and a_2 components while they are clearly separate in the direct picture.

[1] A. H. Compton, *Phys. Rev.*, XXI (1923), 483, 715; XXII (1923), 409.
[2] P. A. Ross, *Proc. Nat. Acad.*, VII (1923), 246.

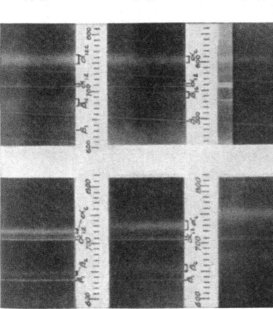

Molybdenum K primary
radiation scattered at
63°30′ ± 30′

Molybdenum K primary
radiation scattered at
90° ± 30′

Molybdenum K primary
radiation scattered at
156°27′ ± 15′

Dependence of broadening
on primary wave-length.

Molybdenum K primary
radiation scattered at
156°27′ ± 15′

Silver K primary
radiation scattered at
155°21′ ± 15′

Tungsten K primary
radiation scattered at
154°15′ ± 1°30′

Dependence of broadening
on scattering angle.

Fig. 36

FIG. 36.—The DuMond Kirkpatrick Experiment. The six spectrograms were made with X-rays which had been scattered at well-defined scattering angles by a graphite scattering body and analyzed by means of a multi-crystal X-ray spectrograph[1] containing fifty crystals, each an X-ray spectrograph in itself, and all focussing their spectra in exact register on a single photographic film. The accuracy of register is evident from the sharpness of the fine lines in the unmodified scattered positions.

The modified lines designated by the subscript c are seen to be greatly broadened. The broadening is (to a classical approximation) simply a Doppler effect of the rapidly moving electrons inside the atoms of the scattering body. It is more accurate to refer to the *momenta* rather than to the velocities associated with these electrons, however, since the kinematic terminology is meaningless when, as in this case, the electron cannot be sharply localized in space. The dynamic quantity momentum rather than the kinematic quantity velocity is what is actually measured in this experiment. The Compton effect which is the *shift* of the modified line away from the unmodified position may be thought of as an effect of the recoil of the electrons under the impact imparted by the X-rays. The DuMond-Kirkpatrick effect which is the *broadening* of the modified line is on the contrary an effect of the *initial momenta* possessed by the electrons in their quantum dynamical states.

The six spectrograms shown establish the validity of this interpretation of the breadth of the modified line. By a simple analysis based only on the laws of conservation of energy and momentum DuMond predicted[2] that this breadth of the modified line in wave-length units should to a first approximation vary directly as the primary wave-length and as the sine of half the scattering angle. This variation can be seen in the above spectra and the predictions mentioned have been completely verified by microphotometer analysis. The broadening is thus shown to behave in all observable respects like a Doppler broadening.[2,3]

Curves of the statistical distribution of momentum among the electrons derived from the microphotometer analysis of these spectra turn out to be in good agreement with the Bohr theory of the atom and in even better agreement with the quantum dynamical improvements upon that theory.[5]

The scattering bodies employed were solids. The electrons involved in producing this effect are therefore not only those associated with individual atoms but also the structure electrons shared between atoms and in the case of electrical conductors the "free electrons" responsible for conductivity. Experiments of this type by DuMond[4] with a beryllium scatterer have furnished the first experimental evidence for the correctness of the Fermi statistics in contradistinction to the Maxwell-Boltzmann statistics in the case of such conduction electrons.

[1] *Rev. Scient. Inst.*, I (1930), 88–105.

[2] *Phys Rev.*, XXXVII (1931), 136–59.

[3] *Ibid.*, XXXVIII (1931), 1094–1108.

[4] *Ibid.*, XXXIII (1929), 643–58.

[5] *Rev. Mod. Phys.*, V (1933), 1–33.

VIII. DUMOND'S DISCOVERY

The reason these components are not separated in any experiments like those of Ross was brought to light in very beautiful work at the California Institute of Technology by Dr. Jesse DuMond in 1931. DuMond designed and built a multicrystal X-ray spectrograph with the aid of which he and Harry A. Kirkpatrick were able for the first time to analyze the structure of the Compton shifted line. They found it not a line at all (see Fig. 36) but rather a broad band the width of which DuMond most skilfully and convincingly interpreted as due to the actual distribution of velocities among the electrons, within the metals, which suffered Compton-encounters with the incident X-ray photons. He showed that the displacement of each element of this band from the central image represented the superposition of the simple Compton effect arising from the impact of a photon with a stationary electron and the Doppler effect arising from the relative velocities of the incident photon and the moving electron actually encountered. This DuMond effect thus gives us a means of determining not only the distribution of velocities among the free electrons of a metal but also *the orbital velocities of the bound electrons.* It presents direct evidence of the existence of both of these velocities, or momenta, since it is the electronic momentum that actually appears in DuMond's equations. It will of course be seen that these interpretations are all made in terms of the photon, or light-dart, theory of radiation, and thus lend the strongest of support both to it and to the conceptions of Bohr as to the dynamic as opposed to static nature of the atom, since these are also implied in the interpretations.

CHAPTER XI

WAVES AND PARTICLES

The foregoing history and discussion show how inevitably the physicist has been forced whenever he is dealing with the interaction between so-called electromagnetic waves and individual electrons to use and to believe in the Einstein hypothesis of localized light quanta, or light-darts, more generally termed "photons."

But what, then, has become of the wave theory which seemed to be established "beyond the shadow of a doubt" by all the vast amount of work that has been done during the past hundred years in the field of *interference?* Whether we are dealing with long wireless waves or with short ones, with heat waves, with light waves, with X-rays, or with gamma rays from radium, the classical electromagnetic wave theory always predicts correctly the observed interference pattern. How can we reconcile these old facts with the more recently discovered facts detailed above which seem to demonstrate so conclusively that radiant energy travels in the form of localized bunches of energy or photons, each of energy $h\nu$?

This was the problem upon which Louis de Broglie was working when he published a paper in 1925 in which he tried to bring about some reconciliation between wave ideas and particle ideas.[1] To do this he at first said something like this, let us conceive the electronic orbits of Bohr as due to the travel of waves of some kind about the

[1] L. de Broglie, *Ann. de Physique*, X, 1925.

nucleus, each of the Bohr non-radiating orbits being of such radius that its circumference contains a whole number of wave-lengths of these hypothetical waves, and thus constitutes a system of "standing electronic waves." This hypothesis required, in accordance with the Bohr equation, that the wave-lengths of these assumed electron waves be given by $\lambda = \dfrac{h}{mv}$ in which m is the mass of the electron and v is the velocity with which it is moving. These wave ideas of de Broglie were taken up and modified by Schrödinger, Dirac, and others, and developed into what is now known as the theory of "wave mechanics," but with all these changes in mode of approach the wave-length is always given by $\lambda = \dfrac{h}{mv}$.

Then the experimentalist came upon the scene, and the wave mechanics began to be more than merely a theory. One of its earliest successes in predicting results that had seemed quite inexplicable from the standpoint of preceding modes of thought came through its application by Oppenheimer at the Norman Bridge Laboratory to the interpretation of the laws of "field currents," or cold emission which had been worked out experimentally by Eyring, Lauritsen, and the author.[1] We had first found that steady streams of negative electrons could be pulled out of cold metal surfaces by the use of sufficiently powerful electrical fields, a result which we felt no difficulty in interpreting from the standpoint of the usual electron theory of metals. But when we found that the law governing the relation of this current, i, to the applied field,

[1] Millikan and Eyring, *Phys. Rev.*, XXVII (1926), 51, and Lauritsen, *Proc. Nat. Acad. Sci.*, XIII (1928), 45–49.

F, was of the precise form $i = Ce^{-\frac{h}{F}}$, a form analogous to that governing thermionic emission save that the field F here plays the same rôle that the temperature T plays in thermionic effects, we were completely unable to give the law any theoretical significance until Oppenheimer[1] deduced it from the fundamental postulate of the wave mechanics, viz., that there is a certain *probability* at all temperatures that an electron will pass *through*, rather than *over*, a "potential barrier like that existing at the surface of a metal."

But even before this there had begun to appear much more spectacular successes of the new "wave equations," for a bit earlier in 1927 Davisson and Germer in the Bell Laboratories in New York had found in very fact that when they directed a stream of electrons against the surface of a metallic crystal the reflected electrons distributed themselves according to a diffraction pattern precisely like that obtained when X-rays of wave-length $\lambda = \dfrac{h}{mv}$ are reflected from the same crystalline surface.[2]

Then, a little later in 1927, G. P. Thomson,[3] and after him many others, shot streams of electrons through thin crystal gratings and obtained in every case diffraction patterns precisely like those obtained by passing ether waves of wave-length $\lambda = \dfrac{h}{mv}$ through these same

[1] Oppenheimer, *Phys. Rev.*, XXXI (1928), 914. This is the first date of publication but it was in the fall of 1927 that these laws, experimental and theoretical, were being worked out at the Norman Bridge Laboratory.

[2] Davisson and Germer, *Phys. Rev.*, XXX (1927), 707.

[3] G. P. Thomson and A. Reif, *Nature*, CXIX (1927), 890.

gratings. The complete parallelism between the diffraction patterns obtained by passing ether waves through crystal gratings and streams of electrons through such gratings is beautifully illustrated in the reproductions shown on Figures 37, 38, 39, 40, 41, and 42.

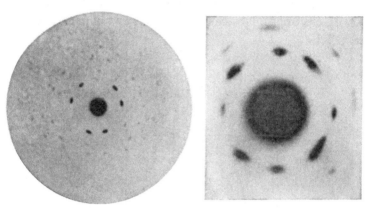

Fig. 37   Fig. 38

Fig. 37.—Diffraction pattern produced by passing a narrow beam of X-rays (ether waves) through a thin crystal of zinc blend parallel to the trigonal axis. (Friederich and Knipping.)

Fig. 38.—The quite similar diffraction pattern obtained by passing a narrow beam of electrons (particles) through a thin film of crystallized aluminum. (G. P. Thomson.)

Finally, Otto Stern of Hamburg and his collaborators performed similar experiments with streams of atoms of known velocities v and found in every case tried diffraction patterns which could be computed from the equation $\lambda = \dfrac{h}{mv}$ where m is now the mass of the *atom* in question.

In a word, then, experiments of this sort have shown with entire conclusiveness that streams of particles, when either passed through, or reflected from, crystal gratings, produce all the interference patterns which the 19th century had associated with and explained by the ether-wave theory. Inversely, as shown above in the photoelectric and Compton effects, we have found ether waves acting like localized bunches of energy which in their interplay with electrons follow consistently the laws of particle en-

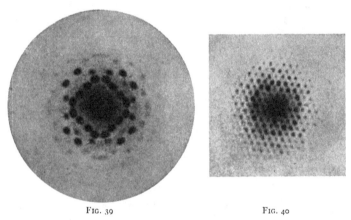

FIG. 39   FIG. 40

FIG. 39.—Diffraction pattern produced by passing a narrow beam of X-rays (ether waves) through zinc blend parallel to a cubic axis. (Friederich and Knipping.)

FIG. 40.—A quite similar diffraction pattern obtained by passing a beam of electrons (particles) through a thin mica crystal. (S. Kikuchi.)

counters. *In other words, particles actually are found experimentally to act like waves and waves to act like particles.* All these reciprocal wave-particle effects have been fitted into the equations of the so-called wave mechanics, that new field of theoretical physics which has been developed from the primitive beginnings made by de Broglie through the labors of Schrödinger and Heisenberg and Dirac and Oppenheimer and many others.

Whether the wave properties or the particle properties of light quants, electrons, atoms, and molecules are the more fundamental may not yet be altogether certain, but we are at least beginning to be able to formulate some of the rules hich Nature follows whenever and wherever we

 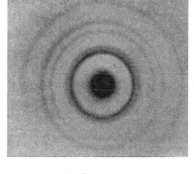

FIG. 41   FIG. 42

FIG. 41.—A typical X-ray (or ether wave) diffraction pattern produced by a powder or thin film in which the elementary minute crystals have been forced by the grinding and sifting, or else by the beating or rolling, to take a random distribution.

FIG. 42.—The corresponding diffraction pattern obtained when a beam of electrons (particles) passes through a thin gold foil in which the rolling or beating has given the elementary crystals again a random distribution.

find her playing this wave-particle game. These rules may be stated as follows:

Whenever, to take the simplest possible illustration, a single helium atom (alpha particle), a single electron, or a single light-dart shoots through a minute hole in a screen and impinges on another screen set up at some distance behind the first one, this second screen experiences *at one particular point* a bullet-like blow. About this there can be no doubt, for if the screen is made of a suitable fluorescent material an eye observing it actually sees a flash or scintillation where the bullet struck. In such individual *elementary processes* nature appears to exhibit only particle, or corpuscle, properties.

But when from a given very small source, commonly called a point source, *a large succession* of atoms, or electrons, or light-darts shoot through the aforesaid hole, no matter how large is the mean time interval between the successive shots, these are actually found to distribute themselves on the screen in the series of concentric circles which is the characteristic diffraction pattern for the case in which light of wave-length $\lambda = \dfrac{h}{mv}$ from a point source passes through a small hole and falls upon a screen set up parallel to the screen containing the hole. In the case of a light quant the m of the foregoing equation is given by the Einstein equation $E = mc^2$ or $h\nu = mc^2$, i.e., $m = \dfrac{h\nu}{c^2}$. In the case of an electron or an atom m is the mass of that electron or atom in grams. Just why the light quants, passing through holes, or slits, or lenses, or gratings of any kind, distribute themselves so as to produce in all cases precisely the patterns computed by the classical wave theory seems to transcend at present *physical* explanation. It is sometimes said that the light quants are "directed by" the "Poynting vector" or by the "field equations" or that the electrons or atoms "are guided by a train of waves." But this is merely to state in other words the aforementioned fact.

The new experimental discovery, made since the advent of the wave mechanics and stimulated by it, is that other kinds of particles besides photons, namely, electrons and atoms, also behave in the same way. We can now compute for these latter particles the wave-lengths corresponding to given masses and given velocities, and then actually get photographically and quantitatively, too, the predicted wave patterns. To state the newly discovered situation as generally and as briefly as possible, whenever we are dealing with a single *elementary process*, i.e., an individual "microscopic event," we find that we must use the particle conception in order to describe our experimental findings, but whenever we are dealing with a huge ensemble of elementary processes, i.e., a statistical array of microscopic events, we must use the wave conception in order to predict the experimental situation. From the standpoint of the thinking of the nineteenth century these two facts are contradictory.

Why have these contradictions arisen? Can we see even dimly a physical significance behind them? For, to the experimental physicist, at least, this world is at bottom more than a world of equations or even of ideas. Some external physical things are happening and we cannot rest indefinitely content with two types of physical interpretation of the same phenomena that seem to be mutually exclusive. The ultimate *elementary processes* which constitute light cannot be both waves and corpuscles. Which are they really? And what kind of legerdemain has nature played upon us to make them seem otherwise? How "did the rabbit actually get into the hat?" The only way I can see out of the contradiction is to assume that all microscopic or elementary processes, whether they are processes of matter physics or of ether physics, are at bottom discrete-particle processes, the four types of units involved being (1) elementary units of electrical charge, (2) elementary units of mass, (3) elementary units of radiant energy, and (4) elementary units of moment of momentum or "action" (called Planck's h units). Only when large numbers of these units are involved do we get over into the field of *continuous* processes of which waves constitute one of the best of examples. In other words, according to this view, all apparently continuous phenomena represent *statistical or mean behaviors* of elementary particles, in precisely the same way as the temperature of a mass is the mean kinetic energy of its particles, a fact which obviously does not imply that every individual molecule has that energy.

In the words of M. de Broglie, "the fundamental postulate of the wave mechanics is that with every independent particle, whether of matter or of radiation, there is associated the propagation of a wave, the *intensity* of this wave representing at each point and at each instant the *probability* that the associated particle will reveal itself at this point at this instant." The "wave" here spoken of as associated with the movement of the particle is then only an "artifice of calculation." It is a *probability* wave which describes accurately *how the particles do actually distribute themselves*, but it contains no information whatever as to why they do it. Yet this question of "why" will not down. Physicists are just now trying to find the reason for this behavior in the Heisenberg uncertainty principle, i.e., an uncertainty, or better an incorrectness in the assumption of an absolute length and an absolute time. The *probability function* which comes out of Heisenberg's equation has already succeeded in some cases in predicting the wave equation which yields the observed interference pattern.

At any rate, this recent particle-wave history that I have tried to recount shows at least that we are moving on and on continuously in our knowledge of the *actual behavior* of the physical world. We first discover the laws that govern the interaction of large bodies, Newton's laws, and these are just as valid now for large bodies as they ever were; then we discover the laws that hold in all the large-scale phenomena of ether physics, and they too are just as valid now as ever *for these large-scale phenomena.* Then we discover methods of pushing our researches farther into the field of individual, *elementary processes* and a whole group of new laws appears, not irreconcilable with the first but *containing the first as limiting cases as the number of units involved becomes large.* In other words, when we are buying sand by the ton we do not even need to know that it contains grains. And so long as we are merely building houses we get along satisfactorily without the knowledge of the granular structure. But fortunately the mind of man is not content with merely building houses. It seeks to "*know* so that life may be enriched."

CHAPTER XII

THE SPINNING ELECTRON

There is perhaps no chapter in the history of the development of modern electronic physics which better illustrates the usual steps in the process of discovery than does the chapter in which is told the story of the introduction of the idea of the spinning electron. This process here goes through the following sequence: (1) prediction on the basis of established findings and existing theory of results to be expected; (2) the search of the experimentalist to verify these predictions; (3) his discovery through this search of elements of inadequacy in the conception from which he started out; and (4) the subsequent extension of the theory so as to make it cover the new findings.

Up to the present the electron has been thought of as essentially a point charge. It is true that in the development of the idea of the electromagnetic origin of mass we were led to assign a radius to the sphere over which the electronic charge might be assumed to be uniformly distributed in order to obtain from our theory the rest-mass that the free electron is by experiment found to possess, but we were careful to call this the radius of "the equivalent spherical charge," since we had no real warrant for assuming that the electron itself is a charge distributed over such a sphere. Up to the present we have dealt with no properties of the electron which required of it anything more than (1) to attract or repel other charges from a location which might be looked upon as a point in space, and (2) to describe, in the case of the Bohr atom, an orbit about another essentially point-charge on the atomic nucleus. This was perhaps too simple a conception to be expected to last, for nature in the history of her encounters with the scientist has always shown herself to be a consummate poker player, often holding cards up her sleeve when he thought her hand was on the table. But at any rate the story of how her hand was forced in this case is somewhat as follows:

I. RELATIVISTIC INTERPRETATION OF FINE STRUCTURE

It will be recalled that Bohr's simple theory had considered only circular orbits, the radius r of a given orbit remaining constant and the azimuth φ alone varying as the electron moved about in its orbit.   In its inmost orbit the electron of the hydrogen atom had one unit of moment of momentum, the unit being actually $h/2\pi$, in its second orbit two units, in its third three, etc.   But this condition could be as well fulfilled with certain types of elliptical orbits as with circular orbits, but in that case two independent variables φ and r would have to be introduced to describe the motion in each ellipse.   Sommerfeld applied the same sort of quantum conditions to both of these variables as Bohr had applied to φ alone before, i.e., he made the very natural assumption that if it was necessary to quantize one of the independent variables which defined the system it was necessary to treat the other in the same way.   He thus made the total integral of the moment of momentum[1] of the system taken around a complete cycle

---

[1] More correctly "quantity of action" defined by $\int p_\varphi d\varphi + \int p_r dr$ for the Wilson-Sommerfeld rules for the quantization of the azimuthal co-ordinate φ and the radial co-ordinate r are

$$\int_{\varphi=0}^{\varphi=2\pi} p_\varphi d\varphi = kh \text{ and } \int_{\varphi=0}^{\varphi=2\pi} p_r dr = nh$$

p_φ representing "moment of momentum" with respect to φ and p_r repre-

—the so-called total quantum number—the sum of an azimuthal and a radial quantum number, the way in which the total integral of the moment of momentum was divided between the two determining the ellipticity of the orbit. Thus, when the total quantum number was 1, the azimuthal quantum number had to be 1 and the radial quantum number 0, or else the azimuthal 0 and the radial 1. The first condition meant physically a circular orbit, as in the simple Bohr theory, while the last meant a radial oscillation in which the electron had to pass through the nucleus, which seemed a physical impossibility (see, however, V, p. 300). Hence the inmost orbit, designated as a 1_1 orbit (the first integer denoting total, the last azimuthal quantum number, the radial being the difference between the two), had always in this theory to be a circle. (See Fig. 27, page 219.) For total quantum number 2, however, two different orbits were possible, namely, a 2_2 orbit, a circle (later on to be designated a p orbit), or a 2_1 orbit, an ellipse (later called an s orbit) having its major axis equal to the diameter of the circle and twice its minor axis. The orbits of total quantum number 3 might have the shapes 3_3, 3_2, or 3_1, just as shown in Figure 27, namely, a circle and two ellipses all having the same major axis but minor axes in the ratios 3, 2, 1.

So long as the Newtonian law of attraction governed the motions, all the orbits of a given total quantum number would have exactly the same energy (for this would be determined solely by the length of the major axis of the ellipse) and hence be spectroscopically indistinguishable,

---

senting momentum along r, so that, since $d\varphi$ has no dimensions while dr is a length, both quantities under the integral have the dimensions of "moment of momentum."

since the only directly observable quantity in spectroscopy is the emitted frequency, determined by the electron
jump from one orbit to another, i.e., by the difference in
the *energies* of two orbits.   But when the Einstein relativity corrections are applied to the Newtonian laws there
is introduced, as already indicated on page 218, a change
of mass with speed which makes the energy of binding of
the electron to the nucleus in the case of a series of orbits
of the same major axis (total quantum number) slightly
greater the greater the ellipticity of the orbit, and hence
makes the electron jumps into these orbits from a given
position correspond to a number of slightly different frequencies, thus introducing fine structure into spectral
lines.   The theoretical procedure here adopted is in part
the same as that used by Einstein in explaining precession
in the perihelion of Mercury, i.e., it is another instance of
the application of the laws of celestial mechanics to the
domain of atomic mechanics.   The quantitative prediction, however, of Mercury's precession involves the *general* relativity theory, whereas the foregoing has to do
only with "*special* relativity," since the *gravitational* force
between the nucleus and the electron is practically zero.

This Sommerfeld relativistic interpretation of fine
structure introduced new possibilities of spectroscopic prediction which from 1919 to 1924 had extraordinary successes.   Thus, all the lines of the so-called Balmer series
of hydrogen, which consist of jumps into the state of total
quantum number 2, a state possessing, according to the
foregoing theory, only two possible orbits, one a circle and
one an ellipse, should reveal a fine structure corresponding to the difference in the energies of these particular circular and elliptical electronic orbits.   To a first approxi-

mation they should actually all be doublet-lines with an accurately predictable difference in wave-length, or in frequency; and again *the prediction could be made wholly from the laws of orbital mechanics.* Sommerfeld's theoretical value of the frequency separation of the hypothetical hydrogen doublets of the Balmer series (total quantum number 2) came out .365 cm.$^{-1}$ frequency units. Now with instruments of high resolving power not only are these hydrogen lines *all* found actually to be doublets just as predicted, but the most accurate determination of this separation, obtained by Professor William V. Houston[1] in the Norman Bridge Laboratory from his measurements on the H_α, H_β, and H_γ doublets came out .36 cm.$^{-1}$—a value within 1 per cent of the computed value—an altogether outstanding and spectacular demonstration of the power of the physicist's newly discovered methods of spectroscopic prediction.

Again, the theoretical relativity formula made the frequency separation of a corresponding pair of lines vary with the *fourth power* of the atomic number so that the corresponding lines in ionized helium—like atomic hydrogen a simple nucleus-electron system, but with a charge on the nucleus twice that in hydrogen—should be sixteen times as far separated as in hydrogen. This prediction also was accurately verified, as early as 1916, by Paschen,[2] his observed separation divided by 16 yielding, with extraordinary accuracy, $0.365 \pm .0005$—another quantitative triumph of orbital theory.

But perhaps the most spectacular success of the relativity formula as applied to the interpretation of spectro-

[1] William V. Houston, *Astrophys. Jour.*, LXIV (1926), 81.

[2] Paschen, "Helium linien,"*Ann. der Physik*, L (1916), 901.

scopic fine structure came when this formula predicted with approximate correctness the frequency separation of the so-called L doublets in the X-ray spectra of the heavy elements like uranium. This, being a doublet corresponding, like the hydrogen doublets, to the electronic orbits of total quantum number 2, ought, if it were actually a relativity doublet as assumed by Sommerfeld, to have a frequency separation obtainable by multiplying the observed separation in the case of hydrogen by the huge factor of 71 million, for this represents the fourth power of the ratio of the atomic number of uranium (92) to that of hydrogen (1). The more accurate relation is

$$\Delta \nu_L = \Delta \nu_H (Z - s)^4$$

in which $\Delta \nu_L$ is the separation of the X-ray L doublet, $\Delta \nu_H$ that for hydrogen, viz., 0.365, Z the atomic number of the element in question, and s the screening constant for the L shell of electrons which Sommerfeld took as 3.5. This may be thought of as 2 for the two K electrons and 1.5 for the interaction of the electrons of the L shell upon one another, thus making $s = 3.5$. This prediction also checked with experiment.[1]

No theoretical formula in the history of physics has had more spectacular successes than has Sommerfeld's relativity formula as applied to spectroscopic fine structure.

II. THE SPECTRA OF THE ALKALI METALS

We have seen in the preceding section how Sommerfeld applied the relativity principle to the elementary Bohr theory so as to account for the fact that all the lines of the spectra of the atoms of both hydrogen and helium

---

[1] See Sommerfeld's *Atombau, etc.*, 4th Ger. ed., p. 445.

are *close* doublets, the elliptical, or 2_1, orbit corresponding to a slightly larger binding energy than the circular or 2_2 orbit.

The next logical application of the orbit theory as presented in Figure 27, page 219, was to the interpreta-

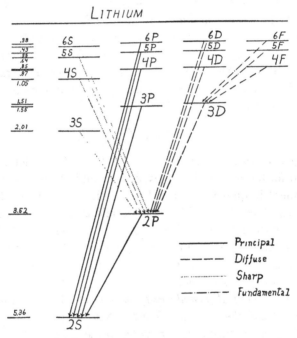

FIG. 43.—The diagrammatic representation of the spectra emitted by the atom of lithium.

tion of the spectra of the atoms of the alkali metals, of which lithium is the simplest example. Its spectrum is shown diagrammatically in Figure 43, the various possible orbits being here replaced simply by "energy levels" between which electrons jump in order to produce the observed spectral lines. The so-called "principal series" are produced by electron jumps into the so-called S levels, each of which in the Bohr theory, of course, corresponds to a particular orbit, indicated in the first column on the left. Since, according to the theory, there are in the second electronic shell (total quantum number 2) but two orbits, and since the lowest observed level corresponds to the highest numerical value of the binding energy, it will be seen that this lowest S level must be a 2_1 orbit, i.e., an ellipse. Similarly, the lines of the so-called sharp series are produced by jumps from the S levels into the P levels, or orbits, indicated in the second column from the left, and this P level in the second or L shell must according to the theory be a 2_2 orbit, i.e., a circle. Likewise, the so-called diffuse series correspond to electron jumps from the more remote D levels shown in the third column, into the P levels, and finally the so-called fundamental series to jumps into the D levels from the still more remote F levels of the fourth column.

If, now, it were possible for electrons to jump at will from any level of higher numerical potential energy to one of lower, lithium would show a spectrum having a very much greater number of lines than is actually found. Hence, so-called selection rules are set up so as to restrict the possible jumps to those corresponding to actually observed lines. Geometrically stated these rules limit the jumps to levels of lower numerical energy *in adjacent columns*, as shown in Figure 43, in which the lines connecting levels show all the electron jumps that are found to be normally possible.

But now let us see what these levels mean from the standpoint of the Bohr orbit theory. Lithium has the atomic number 3, and this means that it has three electrons in the region surrounding its nucleus, two in its K shell and one in its L shell. It is the jumping of the L electron from one level more remote from the nucleus to another less remote that produces all the lines indicated in Figure 43. The orbit in which this electron gets closest to the nucleus is of course the elliptical or 2_1 orbit. (See Fig. 27.) The other L orbit is a circle, designated by 2_2, and is represented by the lowest level in the second or p column. Now it will be remembered that Sommerfeld found the explanation of the very close doublet structure of the hydrogen and helium lines (so close that powerful spectroscopes are needed to separate them) in the fact that because of relativity the binding energy of the 2_1 orbit was just a trifle more than that of the 2_2 orbit. But how does it come that in lithium this same difference between a 2_1 and a 2_2 orbit produces forty thousand times that energy difference (for this is actually the order of the energy difference between the lowest levels shown in column 1 and column 2, Fig. 43)?

Bohr found the answer to this question in the fact, easily computable from the orbit theory, that the 2_1 orbit permits the electron describing it to cut inside the orbit in which are found the two K electrons. The necessary relations are all found merely in the *shapes* of the orbits, as these are fixed by the foregoing quantization rules. For suppose the single electron which belongs to the L shell of lithium is in the 2_1 orbit—an ellipse the major axis of which is twice the minor axis. Since the nucleus is at the focus of this ellipse, the revolving electron must at perihelion dip inside the circular K orbits of the two K electrons, as it is shown in Figure 27 to be doing, and there be relieved from the screening effect of these two K electrons for it. In other words, because of this *penetration of the elliptical orbit inside the K shell* the electron when in the 2_1 orbit gets into a field of force several times as intense as it would if it did not penetrate the K shell. It is thus *on the average* very much more tightly bound to the nucleus than is the electron in the 2_2 orbit, which does not thus dip inside the K shell. Hence, wherever there is such an inner shell to penetrate, the energy difference between the 2_1 and the 2_2 orbits will be very much larger than the relativity difference. The very beautiful way in which this sort of explanation fits the observed facts whenever there are inner shells to penetrate (see Fig. 27) constitutes another extraordinary triumph for the orbital conception. Indeed, these "interpenetration ideas" of Bohr, illustrated in Figures 28, 29, and 31, have been among the most fruitful of any of the orbital conceptions, illuminating in an extraordinary way the chemical properties of the elements as revealed particularly in the picture of the periodic table given in Figure 30.[1]

Up to this point, then, Sommerfeld's application of the relativity principle to the Bohr orbits has had altogether remarkable success, not only in explaining the fine structure of the spectra of the atoms of hydrogen and helium, but also in making intelligible both the similarities and the differences between the spectra of these simple atoms and those of the alkali metals.

[1] N. Bohr, *Three Lectures on Atomic Physics*, Vieweg: Braunschwerg, 1922; also *Ann. der Physik*, LXXI (1923), 228.

III. INNER QUANTUM NUMBERS

The only cause of spectroscopic fine structure thus far considered has been the relativity cause which operates by virtue of a difference in the shapes of orbits of the same total quantum number.   But it was found as early as 1920[1] that that was not enough to account for all the facts of fine structure.   In X-rays, for example, the so-called L orbits, or L levels, correspond to a total quantum number 2, and this permits of only two different orbits, one a circle designated as a 2_2 orbit, and one an ellipse designated as a 2_1 orbit; but X-ray absorption experiments, like those shown in Figures 22 and 23, brought to light *three* different levels or orbits, all close together in frequency, in which L electrons were actually found.   Two of these followed the relativity law, also known as "the *regular* doublet law" referred to in the last paragraph. The frequency difference of these two levels varied with the fourth power of the atomic number, as demanded by the relativity equation.   These two levels are represented by the two diverging lines marked L_1 L_2 in Figure 44. The third level, L_3, is seen to follow an entirely different law, for it runs everywhere parallel to L_2, and could therefore be accounted for by some difference in the apparent value of the nuclear charge such as might be produced by some difference in the screening of the nucleus from the two orbits in question. $L_2 L_3$ are therefore sometimes called "the screening doublet."   The frequency difference between L_2 and L_3 follows what has become known as the *irregular* doublet law, so that the geometrical representation in Figure 2 of the irregular or screening

[1] Sommerfeld, *Ann. der Physik*, LXIII (1920), 221. See also *Atombau u. Spectrallinien*, chap. viii.

doublet law is parallel lines, of the regular or relativity doublet law diverging lines.

Similarly, in the field of optics there are found *experimentally* three levels, or orbits, corresponding to the total quantum number 2, instead of merely the two permitted

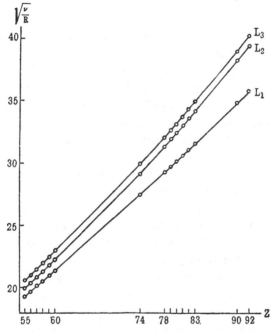

FIG. 44.—Regular and irregular X-ray doublets

by the relativity theory. Although they are not both shown in Figure 43, there are actually two *p* levels too close together to be distinguishable in this diagram, and not merely the one 2_2 orbit permitted thus far by the theory. It is these two *p* levels that are responsible for the familiar red doublet most characteristic of lithium.

Since, then, only two different *shapes* were possible, Sommerfeld introduced the idea that two orbits of the same shape, but of different *orientations*, might exist, and that some dissymmetry in the central force-field gave these two orbits slightly different energies. He introduced a so-called *inner quantum number* generally denoted by the symbol J to take care of this variation in *orientation*, just as changes in azimuthal quantum number took care of variations in *shape*.

The difference in the frequencies of the familiar doublet lines in lithium, for example, was supposed to be due to the fact that these two lines represented jumps into a common orbit called an *s* orbit from two orbits which differed only in inner quantum number, i.e., orbits of different orientations but the same shapes—in this case circles, or 2_2 orbits, known as the $p_1 p_2$ orbits. The *s* orbit, on the other hand, into which these two electrons jumped to form the lithium doublet, was the third possible orbit of the total quantum state 2, namely, the 2_1 orbit, of elliptical shape, the major axis being twice the minor. The two circular, or *p*, orbits differed only slightly in frequency, or energy, but the change in energy in going from either of the *p* orbits to the *s* orbit was relatively very large, as explained in the preceding section.

This was the situation with respect to both optical and X-ray doublets when I. S. Bowen and myself began our studies on the spectra of so-called "stripped atoms."

IV. THE SPECTRA OF STRIPPED ATOMS

These studies grew out of the development of the technique of what we termed "hot-spark vacuum spectrometry" with the aid of which it became possible to open up to exploration and careful study the whole range of wavelengths between the visible and the X-ray fields. Prior to 1920 Lyman[1] had succeeded in obtaining the spectrum of the gas helium down as far into the ultra-violet as to 600 Angstroms, but aside from this the ultra-violet spectrum below 1,000 A was a completely unexplored region and the available data between 1,800 A and 1,000 A was very meager. It was precisely in this region that lay the "stripped atom" spectra of greatest significance for the advances recorded in this chapter. For by 1923 with our high-vacuum, high-potential, "hot-spark" technique[2] we had carried our ultra-violet explorations down to a reported wave-length of 136 A using vertical incidence of the light upon the gratings, and as early as the spring of 1923 Bowen had developed and successfully used the grazing incidence technique and with it checked our wavelengths down to 124 A. He did not, however, publish this grazing incidence method since in his early work he did not reach much shorter wave-lengths than we had obtained using normal incidence. In 1926 this method was independently worked out and published by A. H. Compton, and has since yielded in many hands fine spectra down to as low as 1 A.

By the year 1924 Bowen and the author had succeeded with this "hot-spark" high-vacuum spectroscopy in stripping the entire outer or valence shell of electrons from each of the whole series of atoms Li, Be, B, C, N, and O. We thus obtained for the first time a long series

[1] Lyman, *Astrophys. Jour.*, XLIII (1916), 102.

[2] Millikan and Bowen, "Extreme Ultra-violet Spectra," *Phys. Rev.*, XXIII (1924), 1. See also Millikan, Sawyer, and Bowen, *Astrophys. Jour.*, LIII (1921), 150, for still earlier work on "hot sparks."

of light atoms having an identical electronic structure but a linearly increasing effective nuclear charge. For in view of the fact that the atomic numbers of these atoms are 3, 4, 5, 6, 7, 8, respectively, and each of them has two electons in its *K* shell, the effective nuclear charges for these stripped atoms are 1, 2, 3, 4, 5, 6. It is, of course, the return of one of these removed electrons through successive

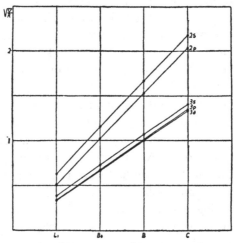

Fig. 45.—Moseley's law in the field of optics. Stripped atoms of the second row of the periodic table.

jumps from the more remote to the inner levels that produces, for example, the lithium spectrum shown diagrammatically in Figure 43, and precisely corresponding spectra pushed, however, farther and farther into the ultra-violet where our hot-spark spectroscopy could get at them, as the effective nuclear charge rose successively from 1 to 6 in going from stripped lithium to stripped oxygen. *But it is precisely this combination of identity of in-*

ternal electronic structure among heavy atoms with linearly increasing nuclear charge which is responsible for the existence of the Moseley law in the X-ray field—a law in which the square root frequency increases linearly with atomic number (Fig. 44).

When we tested this law with the foregoing sequence of stripped atoms, both in the levels for which the total

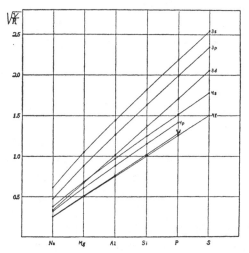

FIG. 46.—Moseley's law in the field of optics. Stripped atoms of the third row of the periodic table.

quantum number is 2 and those in which it is 3, we obtained the diagram shown in Figure 45. Similarly, we stripped the outer shell of electrons from the whole series of atoms of the third row of the periodic table, viz., Na, Mg, Al, Si, P, S, and Cl, and found the Moseley law again completely applicable (Fig. 46).

The new simplicity and orderliness introduced into the tangled empiricism of the field of optical spectra, by the opening to systematic exploration in the years 1920–25 of the ultra-violet region between 1,800 A and 150 A, is very strikingly illustrated by Figures 47 and 48. To understand what these plates mean consider first the familiar and prominent pair of red lines in the lithium spectrum. These may be taken as the *characteristic flag* of a one-electron system, if we here define a one-electron system as a stripped atom through whose series of empty quantum orbits the first electron is jumping back as the atom tries to restore itself to its normal condition. When then we found this characteristic flag in the beryllium spectrum pushed over toward the ultra-violet just the right amount to fit the Moseley law, and with just the right wave-length separation, we knew that there existed in our source stripped Be atoms and we then proceeded to look in the beryllium spectrum for all the other lines characteristic of a one-electron system as found in lithium. Then going farther toward the ultra-violet we sought first the characteristic doublet flag when boron was in our hot-spark electrodes, and found it, and afterward all the other expected lines. Then we did the same with carbon, then with nitrogen, then with oxygen, though not at that time with fluorine. Then going over to the next row of the periodic table we did the same with Na, Mg, Al, Si, P, S, and Cl, for every one-electron system always flies this doublet flag.

Precisely similarly every two-electron system, i.e., a nucleus with two electrons jumping through its otherwise empty quantum orbits, has its own characteristic spectrum, the presence of which can be easily recognized by some prominent group of lines that may be called the characteristic flag of a *two*-electron system. In just the

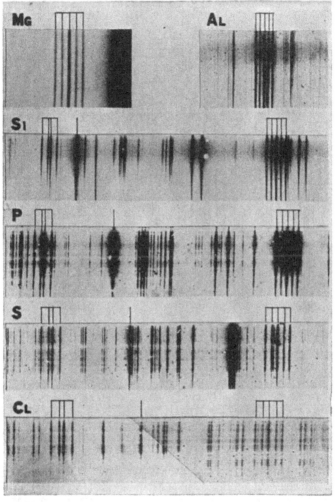

FIG. 47.—The five-band group of spectral lines (Figs. 47 and 48) is the characteristic flag of all two-valence electron systems. The four-band group is the corresponding flag of a three-valence system.

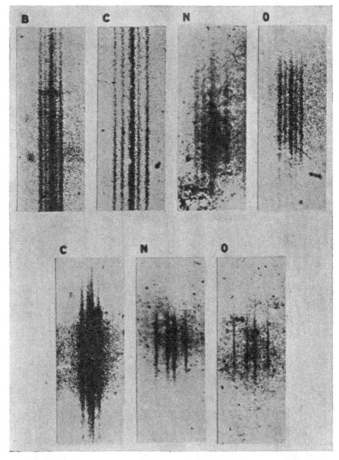

FIG. 48.—The five-band group of spectral lines (Figs. 47 and 48) is the characteristic flag of all two-valence electron systems. The four-band group is the corresponding flag of a three-valence system.

same way we can find a characteristic flag flown by a
three-electron system. Now Figures 47 and 48 show
very beautifully these characteristic flags of both two-
and three-electron systems, and hence tell us that any
source in which these flags appear contains such atomic
systems. Thus, the quintuplet of nearly equally spaced
lines (Figs. 47 and 48) has now been found for all the two-
valence electron spectra which have been studied by the
method of hot-spark spectroscopy, namely for Mg_I, Al_{II},
S_{III}, P_{IV}, S_V, Cl_{VI}, and for Be_I, B_{II}, C_{III}, N_{IV}, O_V, where the
roman subscripts now signify the number of valence
electrons that have been displaced from the normal atom
in order that this spectrum may appear. Similarly, the
quadruplet flag shown in the spectra of Si, P, S, Cl, C, N,
O is the characteristic flag for all three-valence electron
systems. Furthermore, these quintuplet and quadruplet
groups shown in Figures 47 and 48 are of exceptional
theoretical interest because their existence proves that
*two electrons, both of which are in unstable quantum states,
may simultaneously perform two definite quantum jumps
and integrate their combined energy-changes into a single
monochromatic radiation.*[1] The mechanism by which
such an integration takes place is at present entirely unknown.

But the results shown in Figures 45 and 46 had important consequences in still other directions. It will be
seen that the difference between the frequencies of the *s*
and *p* series is precisely like the difference between the
two upper lines of the X-ray series shown in Figure 44.
In other words, it is an irregular or screening doublet relation that exists between the *s* orbits and the p_2 orbits in

[1] Bowen and Millikan, *Phys. Rev.*, XXVI (1925), 150.

optics, just as between the L_2 and L_3 orbits in X-rays. We next found to our amazement, by careful measurements on the variation of the separations of the $p_1 p_2$ doublet lines in going through the stripped atom series Li, Be, B, C, N, O, and also through the series Na, Mg, Al, Si, P, S, Cl, that these separations increased according to a fourth-power law with effective nuclear charge, i.e., $p_1 p_2$ constituted a relativity doublet precisely as did L_1 L_2 in X-rays. In a word, there was complete correspondence between the s orbit in optics and the L_3 orbit in X-rays, the p_2 orbit in optics and the L_2 orbit in X-rays, the p_1 orbit in optics and the L_1 orbit in X-rays.

There had been indications of this relation earlier. Sommerfeld in the first edition of his book, 1919, had suggested it, but in later editions had discarded it as untenable because theoretically impossible and not experimentally justified. No unambiguous evidence could in fact be found until we had obtained a long series of atoms of identical electronic structure and varying nuclear charge to compare. It was work with these series that proved beyond a doubt that *the energy difference between the $p_1 p_2$ orbits followed in every respect the relativity equation, though it could not possibly be due to a relativity cause,* as this cause was then understood, since this cause required differences in *shapes* of orbits, while the $p_1 p_2$ orbits could not have differences in shapes but only differences in orientation. For the only difference in shapes of orbits permitted in the foregoing Bohr theory in the second or L quantum state was the $2_1 2_2$ difference required between the s and the p orbit in order to form the basis for the interpenetration ideas of Bohr which had had so manifold successes that they could not possibly be disregarded.

In papers published in 1924 and 1925[1] we therefore stated it as one of the most interesting problems of theoretical physics to retain relativity as a cause of fine structure in hydrogen and helium, and indeed in general, and yet to find another non-relativistic cause "magnetic, or magnetic and electrostatic combined," which would follow exactly the relativity equation. We concluded one of these articles with the following italicized statement:

"The evidence for a difference in kind between the fine structure of the lines of the two lightest elements, atomic hydrogen and ionized helium, and that of lithium and all the elements beyond it is so good, despite their apparent similarity in behavior in strong magnetic fields, that we propose to attribute the doublets of atomic hydrogen and ionized helium to a true relativity cause, and to introduce a new non-relativistic cause, which, however, obeys an equation almost exactly like the relativity equation, to account for the behavior of lithium and the elements of higher atomic number. This is the only possible way to retain both Bohr's interpenetration ideas and Sommerfeld's relativistic treatment of electron orbits; and both of them seem at present to demand retention. To find a new cause for the relativity-doublet-formula with only a little leeway in the value of the numerical constants is a problem worthy of the efforts of the theoretical physicist."

Probably never before in the history of physics had such an extraordinary—so well nigh impossible—a condition been imposed. And yet within a year of our state-

[1] *Phys. Rev.*, XXIV (1924), 209–28; *Proc. Nat. Acad. Sci.*, XI (1925), 119; *Phil. Mag.*, XLIX (1925), 923.

ment of the problem, two young Dutch physicists, Uhlenbeck and Goudsmit,[1] stimulated partly by our work, partly by other difficulties with existing theory, pointed out
especially by Landé, and having to do with the so-called
anomalous Zeeman effect, had found *in the assumption of
the spinning electron* another cause of fine structure which
followed exactly the same law in all respects as the relativity cause.   The incident furnishes as striking an illustration as the history of physics thus far affords of the
power of experimental and theoretical methods combined
for predicting new phenomena, interpreting old ones, and
thus slowly but inevitably, step by step, forcing open nature's hitherto fast-bolted doors.

The new physical conception introduced by Uhlenbeck and Goudsmit is that every electron within an atom
is not merely revolving in an orbit, but at the same time
rotating, just as does a planet, on its own axis.   There are
assumed to be but two possible directions of spin, $180°$
apart, but the moment of momentum of spin is assumed
to be always the same, namely, exactly one-half unit of
moment of momentum, i.e., $\frac{1}{2}h/2\pi$.   Such a conception
introduces precisely the right amount of energy-difference
between the $p_1 p_2$ circular orbits which is necessary to account for their observed spectroscopic frequency separation.   This effect is simply superposed upon the relativity effect, thus making the fine structure, even in hydrogen and ionized helium, somewhat more complex than
could be accounted for by the relativity effect alone.
*This newly predicted complexity was soon found by new and
more refined measurements by Professor Houston[2] to fit the*

[1] Uhlenbeck and Goudsmit, *Nature*, CXVII (1926), 264.

[2] W. V. Houston, *Astrophys. Jour.*, CXIV (1926), 81.

experimental facts much better than did the old theory.
Further, the introduction as above of half-units of moment of momentum due to electron-spin at once cleared
up difficulties which had appeared in interpreting the
facts of band spectra, so that the idea of the spinning
electron has definitely proved itself to be of very great
usefulness.   Whether the very notable results which have
flowed from its introduction could have been obtained in
any other way it is still too early to say.

V. THE NEW SPECTROSCOPIC RULES

With the aid of this new conception it has now become possible to make a nearer approach than heretofore
to presenting a physical interpretation of a group of remarkable spectroscopic rules developed, largely empir-
ically, between 1925 and 1927 by Russell,[1] Heisenberg,[2]
Pauli,[3] and Hund,[4] and embracing in an altogether remarkable way most of the facts of spectroscopy known up
to the present.   The success with which these empirical
rules describe the facts of spectroscopy is little less than
magical.   These rules naturally all start with, and grow
out of, the fundamental assumption underlying all quantum theory, namely, that all periodic motions must be
quantized, i.e., that periodic motion itself is unitary in its
nature.   As applied to atomic mechanics this means simply that all moments of momentum characteristic of the
periodic motions within the atom are to be assigned characteristic quantum numbers and are to be allowed to
change only by unit steps.

[1] Russell and Saunders, *ibid.*, LXI (1925), 38.

[2] Heisenberg, *Zeit. f. Phys.*, XXXII (1925), 841.

[3] Pauli, *ibid.*, XXXI (1925), 765.

[4] Hund, *ibid.*, XXXIII (1925), 345.

In the case of each individual electron there are just four sorts of such moments of momentum to consider. In other words, there are four elements necessary to the complete description of an electron's motion within the atom, namely (1) the size of its orbit, (2) the shape of its orbit, (3) the orientation of its orbit in space, and (4) the orientation, or direction, of its spin.

1. The total integral of the moment of momentum (quantity of action, see I, p. 271) of an electronic orbit is characterized by its total quantum number n introduced by Bohr. This fixes the size (or major axis) of the orbit.

2. The azimuthal quantum number which, with a given n or major axis, fixes the shape (minor axis) of the orbit has heretofore been characterized by the quantum number k. For some reason, perhaps not yet fully understood, but doubtless of profound physical significance (see below) in order to make the new spectroscopic rules fit the experimental facts it is found necessary to reduce by unity all values of k heretofore assigned. Since, however, we are not yet ready to discard entirely for the old purposes the old interpretations, this reduced value of k is for convenience denoted by a new symbol l, so that merely by definition $l = k - 1$. Thus for an s orbit $l = 0$, for a p orbit $l = 1$, for a d orbit $l = 2$, etc.

3. The projection of the orbital moment of momentum l upon any fixed direction of reference, which, in the consideration of the Zeeman effect, is the direction of the applied external magnetic field, is quantized and designated by the symbol m_l. This projection obviously fixes the orientation of the orbit in space. The physical significance of the fact that this projection is quantized is that only certain definite orientations of this orbit are possible

(such spacial quantization is directly proved by the so-called Stern and Gerlach experiments).

4. The projection of the moment of momentum of spin upon the fixed direction of reference is designated by the symbol m_r. As stated above, there are supposed to be in each atom but two possible directions of spin, 180° apart; so that m_r of course determines in which of these two directions a given electron is spinning. The quantities m_l and m_r are usually called magnetic quantum numbers merely because of their use in connection with magnetic fields.

Now one of the new and very illuminating spectrosopic rules known as the Pauli exclusion rule states that in a given atom no two electrons may be alike in all four of the above elements; in other words, two electrons cannot occupy one and the same electronic position within an atom.

This rule carries with it at once a whole group of conclusions which have been reached by piecing together evidence from many quarters. Thus it requires that the K shell of all atoms possess two electrons and no more; for since in this shell $n = 1$ and $l = 0$ and hence $m_l = 0$ it follows from the rule that in the fourth element of their motion, namely, the spin, the electrons must be different; and also that not more than two can exist without having two alike in all four elements, and hence violating Pauli's rule.

Similarly, and for precisely the same reasons, there can be but two electrons in s orbits in a shell of any total quantum number whatever, and this in turn requires that the eight electrons of the L shell shall be found two in s orbits and six in p orbits—a relation discovered by

Stoner[1] and Main-Smith[2] in England in 1924 by putting together evidence from a variety of sources. This relation underlies the whole of Table XIV, page 227.

Further, because of the opposite directions of spin of the two electrons of a K shell their joint or resultant moment of momentum and also their total magnetic moment must be zero, which in turn requires that helium be diamagnetic as it is found in common with all the noble gases actually to be.

Pauli's rule also requires that every closed shell, indeed every completely symmetrical electronic configuration, have a zero value of its resultant moment of momentum, and also of its magnetic moment. This accords with the fact that mercury, and other two-valence atoms, in their ground or unexcited states are diamagnetic.

Since every completed shell has zero moment of momentum, it follows at once that the total or resultant moment of momentum of an atom having a given electronic configuration[3]—and the number of different values that this moment can assume determines, of course, the number of terms in the fine structure corresponding to that configuration—must be made up of the combined moments of momentum of the electrons in the uncompleted or valence shell. This important relation was

[1] Stoner, *Phil. Mag.*, XLVIII (1924), 719.

[2] Main-Smith, *Jour. Soc. Chem. Ind.*, LXIV (1925), 944.

[3] The word configuration as above used means the definite distribution or assignment of the valence electrons to their proper types of orbit, e.g., s^2p^3 means two electrons in s orbits and three in p orbits. Again $s^2p^2 \cdot s$ means that one of the three p electrons, just referred to, has been pushed up into a higher state. Since there are a total of 6 p orbits, 3 electrons can be fitted into the 6 orbits in a considerable number of different ways and to each way corresponds a characteristic term value.

first perceived by Russell[1] who also first formulated the
new rules for the composition of the total moment of momentum of the atom from the foregoing components.
Thus, when more than one electron is present in the
valence shell the joint orbital moment of momentum L
of the whole group is obtained by taking the quantized
vector sum of the individual moments l. For example,
for two electrons in p orbits, for each of which $l = 1$, the
quantized vector sum is 0, 1, or 2. The physical significance of the fact that the vector sum of the ls is quantized
to obtain L is that the electrons are able to rotate only in
orbits of such orientations about the nucleus that this
vector sum is a whole number of units of moment of momentum.

The next step is to obtain the joint moment of momentum R of the spins of the individual electrons. Since
these spins are assumed to be in the same plane and each
of amount $r = \frac{1}{2}$, the quantized vector sum is here simply
the algebraic sum, i.e., in this case $R = \frac{1}{2} - \frac{1}{2} = 0$ or
$\frac{1}{2} + \frac{1}{2} = 1$.

Finally, to obtain the total moment of momentum of
the whole atom we take the quantized vector sum of the
total orbital moment L and the total spin moment R.
This is precisely the quantity which was originally called
by Sommerfeld the inner quantum number and designated by the letter J. Thus for the values $L = 2$, $R = 1$
the quantized vector sum is 1, 2, or 3.

The fundamental quantum condition is now that all
possible values of J constitute a series of which the successive steps differ by unity. If the value of R is an odd
number of half-units, i.e., if there are an odd number of

[1] Russell and Saunders, *Astrophys. Jour.*, LXI (1925), 38.

spinning electrons, then all the values of J are obviously half-integral ($\frac{1}{2}$, $\frac{3}{2}$, $\frac{5}{2}$, etc.) while if the value of R is integral, then all the values of J are likewise integral. The number of possible values of J obtained from a given pair of values of R and L gives the multiplicity, i.e., the number, of terms in the fine structure. The maximum of the multiplicity is actually $2R+1$.

The notation now in general use in the formulation of the new rules is as follows. When the value of L built up as above from the vectorial summation of the ls of the individual orbits is o, the term is by definition an S term. When $L=1$ the term is by definition a P term; when $L=2$ it is a D term; when $L=3$ it is an F term, etc.

The foregoing quantization rules predict a much larger number of terms than are actually obtained from a given configuration, but Pauli's exclusion rule succeeds in reducing the number of terms to those actually observed.

The new rules thus briefly outlined have had such altogether extraordinary success in predicting the character of the spectra emitted, not only by the simpler atoms, such as those which have been the subject of the studies of Dr. Bowen and myself, but by atoms like iron and titanium, thousands of whose lines have already been identified, that it seems as though we now have within our grasp the means of predicting the precise sorts of radiations which can be emitted by all the possible excited states into which any atom can be thrown. Indeed, the predictions even of complicated spectra have recently been so successfully made that one prominent spectroscopist has lately remarked that "the heroic age of spectroscopy is already past."

Nevertheless, these rules are still to a considerable extent empirical—a remarkable witness to the ingenuity of the physicist in picking out *rules of behavior*, and extrapolating from these rules from observed to unobserved phenomena. They are, however, little more than *rules*. They do not yet represent a completely logical and consistent scheme of interpretation, and they are only vaguely and very imperfectly translatable into physical pictures visualizable in terms of physical relationships. Thus, after the extraordinary and manifold successes detailed above which followed upon Sommerfeld's introduction of the idea of elliptical orbits—an idea definitely requiring that one unit of azimuthal moment of momentum be assigned to s orbits, two to p orbits, etc.—it is somewhat disconcerting to find that to fit the new spectroscopic rules s electrons can have no orbital momentum at all $(l = 0)$. This looks like a very fundamental contradiction and seems to spoil a whole group of interpretations which were thought to be quite definitely established. It is, however, just such contradictions which point the way to the next advance, as was so beautifully shown in the foregoing history of the development of the idea of the spinning electron. Furthermore, this advance is perhaps already dimly in sight, for the new wave mechanics actually does require that the s state (orbit) always represent a pure radial pulsation, centrally symmetric, and without any axial structure, and hence without any angular momentum. *The whole moment of momentum of electrons in the s state would then inhere in the spin.* This means that we have overdone somewhat *shape of orbit* in all the foregoing attempts at physical interpretation. Thus, for example, on page 272, we discarded $k = 0$ $n = 1$ as a possible case because it represented a linear oscillation or pulsation of the electron along a line containing the nucleus. The introduction of Heisenberg's principle of uncertainty, which means that while retaining as a first approximation the orbit conceptions of Bohr we can never specify the *precise* position of an electron in a particular orbit, makes it now possible to reintroduce this linear case $k = 0$ $n = 1$, which is precisely what we did in making $k = 0$ for the s orbit through the convention $l = (k - 1)$. This change also makes it possible to retain all that is essential in the interpenetration ideas of Bohr. In any case the foregoing series of possible values of moment of momentum of the electron must be retained and the introduction of the idea of the spinning electron has helped greatly in the attempt toward physical visualization of how these various moment of momenta come in. But just what influence this new group of ideas is to have upon the whole group of orbital conceptions developed in the early portion of this review is still somewhat uncertain.

The last twenty years of the history of spectroscopy, however, constitutes a remarkable illustration, first, of the rapidity of our modern rate of advance—a whole huge domain, a veritable dark continent—having been explored and reduced to order and civilization in a period of about twenty years; and, second, of the power of the physicist's two tools, analysis and experiment, when used properly together, for forcing open nature's most tightly barred and bolted doors and wresting her most jealously guarded secrets from her for the enrichment of the life of future generations. For every bit of added knowledge of nature adds so much to our control over her, that is, to our ability to turn her hidden forces to useful ends.

CHAPTER XIII

THE DISCOVERY AND ORIGIN OF THE COSMIC RAYS

I. THE DISCOVERY OF THE COSMIC RAYS

The term "cosmic rays" is not yet ten years old, and the branch of physics with which it deals first began to be opened up about 1910. It came about as follows: It will be remembered that X-rays were discovered in 1895 and that the characteristic that distinguished them from all "radiations" theretofore known was their very much greater penetrating power. Indeed, it was their ability to penetrate the flesh of the hand and leave shadow pictures of the bones upon photographic plates that first called their existence to the attention of Professor Roentgen[1] in Würzburg, Germany. It was Becquerel's search in Paris for other sources of such rays that led him to the discovery of the radioactivity of uranium in 1896. These rays were in their turn soon found to be very much more penetrating than were the X-rays, and after the separation from uranium of radium and its concentration by Mme Curie in 1898 careful studies of these penetrating powers began to be made.

It was 1902 when Rutherford and Soddy[2] succeeded in separating these rays into three groups, which they named α, β, and γ rays, the first two of which were found to be deflectable by a magnetic field and therefore to con-

[1] Roentgen, *Wied. Ann.*, LXIV (1896), 1.

[2] Rutherford and Soddy, *Phil. Mag.*, IV (1902), 370, 569, and V (1903), 576.

sist of electrically charged particles, while the last of the
three were not so deflectable and therefore were in this
respect like light. The β rays (electrons) were a hundred
times more penetrating than the α rays (helium nuclei),
and the γ rays a hundred times more penetrating than
the β rays. Indeed, the γ rays were found so penetrating
as to be detectable through as much as two inches of lead.
That these radioactive materials, giving off radiations of
just such great penetrating powers, were present in minute amounts in all kinds of rocks and soils was indicated
(later definitely proved) when it was shown in 1903 both
by Rutherford and Cook[1] and McLennan and Burton[2]
that the ordinary rate of discharge of an electroscope
could be somewhat reduced by surrounding it with lead
shields an inch or so thick, for these were experiments exactly like those that had already been performed with
concentrated samples of radium, uranium, and thorium.

Up to 1910 not a trace of evidence had appeared for
the existence of any rays of a penetrating power greater
than that of the γ rays of radium. Indeed, all the work
that had been done prior to 1910, even on rays capable of
discharging electroscopes through metal walls centimeters
or even inches thick, was interpreted in terms of such
earth rays, or of the radiations given off by the radioactive emanations getting from the earth into the lower atmosphere, and these are, in fact, responsible for much the
greater part of the observed electroscope-discharging effects found on the earth's surface. Also, prior to 1910,
not a trace of evidence had appeared that penetrating
rays entered the earth from outside. Such a hypothesis

[1] Rutherford and Cook, *Phys. Rev.*, XVI (1903), 183.
[2] McLennon and Burton, *ibid.*, p. 184

had not even been seriously proposed. Apart from a passing suggestion of Richardon's[1] in 1906 that electroscope-discharge effects observed at the earth's surface might possibly have something to do with solar influences —a suggestion quickly negatived by the fact that these effects are as strong in night time as in daytime—I can find no record of the existence anywhere up to 1910 of any ideas even remotely related to those that are now associated with the term "cosmic rays." Indeed, in 1909 all the work that had appeared in this field up to that date was reviewed by Kurz[2] and careful consideration given to each one of the only three possible origins of the observed electroscope-discharging effects, namely (1) the earth, (2) the atmosphere, and (3) the regions beyond the atmosphere. *The last two were definitely discarded* and the conclusion drawn that there was not the slightest evidence for the existence of any penetrating rays other than those produced by radioactive substances in the earth—this with full knowledge, too, dwelt upon at length in this article, that half a mile of the earth's atmosphere was sufficient to absorb all such radioactive radiations.

When, therefore, in 1910, the Swiss Gockel[3] took an electroscope three different times in a balloon to heights that reached 4,500 meters and found that *its rate of discharge was there even higher than on the earth, he had discovered something new and important*, namely, that although there are penetrating rays that do originate in the earth and are indeed abundantly given off from practically all kinds of rocks and soils in the earth's crust, as Kurz

[1] O. W. Richardson, *Nature*, LXXIII (1906), 607; LXXIV (1906), 55.

[2] K. Kurz, *Phys. Zeit.*, X (1909), 834.

[3] Gockel, *ibid.*, XI (1910), 280; also XII (1911), 597.

and the other workers prior to 1910 had rightly concluded, yet *there must be other rays abundant at high altitudes that come in from above* originating either (2) in the remoter regions of the atmosphere or else (3) coming in from outer space. Which one of these two possibilities corresponds to the correct origin it took a great deal of work by Hess, Kohlhörster, v. Schweidler, Bowen, Otis, Cameron, myself, and others, from 1910 to 1925 definitely to determine.[1]

The point of real significance in Gockel's experiments is simply the fact that the readings certainly did *not* fall at all to zero at altitudes above 1,000 meters, as they were obliged to do, as shown repeatedly before 1910, if the earth were the source of the observed effects. Instead of this, Gockel's experiments showed, to quote Hess's words, "a slight increase with altitude," thus definitely disproving the hypothesis generally favored up to this time and forcing a choice of one or the other of the hypotheses (No. 2 or No. 3) rejected by Kurz.

The next year, 1911, Hess, after repeating and checking Gockel's experiments, extending them to 5,200 meters, and making them more quantitative, spoke in favor of a return to Kurz' hypothesis No. 3, although at the same time suggesting the possibility of No. 2. He also pointed out the important fact that these discharging effects were apparently as strong in night time as in daytime. In 1913 and 1914 Kolhörster carried observations essentially like those of Gockel and Hess to 9,000 meters, and in interpreting his observed twelve- or thirteen-fold increase in discharge rate at that height over that at sea level, fa-

[1] Millikan, "History of Research on Cosmic Rays," *Nature*, CXXVI (July 5, 1930), 14.

vored hypothesis No. 3.   But, as late as 1924, Wigand[1] in
a comprehensive and admirable review summarizes the
situation as it existed at that date as follows: "The hypotheses as to the origin of the penetrating high altitude
rays look upon these sources as either outside the earth,
i.e., in the cosmos, or else in the higher layers of the atmosphere (Hess, 28, 33, v. Schweidler, Seeliger).   From
the many evidences for and against it is not possible as
yet to recognize a clear picture and one must, for the
present, reckon with several possibilities."

On account of the World War there were no observations significant for this history from 1914 to 1922 when,
in March and April of the latter year, Bowen and I made
from Kelly Field, San Antonio, Texas, the first ascent into
the stratosphere[2] (15,500 meters or 50,000 feet) with self-
registering electroscopes, barometers, and thermometers,
carried up in sounding balloons.   Figure 49 shows the
first type of electroscope to ascend so high.   It weighed
with all its recording and driving mechanisms about 7
ounces.   It was gone three hours and twenty minutes
and landed near Houston, a hundred miles from its starting point.   Our object in making this high flight was to
obtain what we hoped would be crucial evidence to differentiate between hypotheses No. 2 and No. 3.   For, we
argued, that if the rays originated outside the atmosphere
we should expect an exponential increase in ionization,
i.e., a geometrical progression of discharge rate, clear to
the top of the atmosphere [our instruments went .89 of

[1] Wigand, *Phys. Zeit.*, XXV (1924), 193.

[2] Millikan, *Carnegie Inst. of Wash. Year Book No. 21*, 1922, pp. 385–
86; also Millikan and Bowen, *Phys. Rev.*, XXII (1923). 198, and XXVII
(1926), 353.

the way to the top] with about the absorption coefficient
which had been computed from Kolhörster's highest

FIG. 49.—The first stratosphere flight with instruments for recording
cosmic rays was made from Kelly Field, San Antonio, Texas, in April,
1922. The electrometer with the recording mechanism for giving a
continuous record of temperature, pressure, and cosmic-ray intensity
weighed 7 oz. It was carried up to 15.5 Km. (50,000 ft.) by two balloons
filled with hydrogen to a diameter of 4 feet and brought back to earth
with a good record after $3\frac{1}{4}$ hours by the balloon that did not burst.
Skylight falling through the 0.001-inch slit at the right casts shadows
(diffraction images) of fibres on the photographic film placed just behind
a horizontal slit in the recording attachment to the left.

flight (to 9,000 meters), namely, =0.55 per meter of
water. Since our electroscope came down with a total
discharge very considerably less than that computed from

this coefficient, we concluded that *the apparent absorption coefficient went through a maximum before reaching the top*, a behavior at least consistent with hypothesis No. 2, so that these experiments left us still in doubt about the place of origin of the rays.

The difficulty was that up to this time no one had measured directly the penetrating power of these rays, but only the way their intensity varied with altitude. But very soft rays suitably distributed in origin in the atmosphere might show any desired distribution of intensity with height, while rays of external origin in order to be felt near the earth's surface would have to have a penetrating power sufficient to go through 10 meters of water—the equivalent in absorbing power of the atmosphere—while the rays from any known radioactive substances could not possibly penetrate more than, say, 2 meters of water.   So Otis, Cameron, and I, in the fall of 1922, set at the problem of directly measuring the penetrating power of the rays in question, and by the summer of 1925 had succeeded, through sinking our electroscopes meter by meter (down to 15 meters) in the high-altitude, snow-fed Muir Lake (11,800 feet) near Mt. Whitney, California, in proving that *there were here rays coming in exclusively from above of at least 18 times the penetrating power of the hardest known gamma rays*, a penetrating power amply sufficient to get through the atmosphere several times over.

We also proved by next sinking the same electroscopes in Arrowhead Lake, California (5,100 feet) and finding that each of the readings in this lower lake was identical with a reading in the upper lake 6 feet farther down (the 6 feet being just the water-equivalent of the amount of atmosphere between the levels of the two lakes), that *the atmosphere between these two levels acted simply as an absorbing blanket and contributed not a bit to the intensity of radiation found at the lower level.*

These two experiments seemed to us to settle the question that the rays were of cosmic origin, for the first showed that they had ample penetrating power to come from outside, and the second showed that the sources were not distributed throughout the atmosphere. Further, we could not conceive of any type of events, of the energies here involved, taking place in our atmosphere which would be entirely absent from the sun's atmosphere, as we could definitely prove the sources of these rays to be (see Section II of this chapter). Therefore, late in 1925, we wrote the first article in which we called these rays "cosmic rays."[1] In the matter of absorption coefficient in 1923 Kolhörster[2] too had gone at the problem of trying to measure directly the penetrating power of the rays in question. To do this he took readings, first on top of a glacier in the Alps and second in a crevasse in the glacier where if local rays from the surrounding mountains and from glacial dirt were assumed to be absent all the ionizing rays had to pass through a known thickness of ice. He thus obtained an absorption coefficient of the same order of magnitude as that which we found in 1925. He obtained results of this order of magnitude too in shallow bodies of water, and drew the conclusion that he had at least "established the existence of a hard gamma ray having an absorption coefficient about $\frac{1}{10}$ that of the

[1] *Proc. Nat. Acad. Sci.*, XII (1926), 48–55; *Phys. Rev.*, XXVIII (1926), 851.

[2] Kolhörster, *Sitz. Ber. Preuss. Akad. Wiss.*, XXXIV (1923), 366.

hardest known gamma ray." However, at the Volta Centenary held at Lake Como, as late as 1927 one of the most distinguished of living physicists declared himself still a believer in an upper atmospheric origin of these radiations, so that up to that time neither our work nor Kolhörster's had carried complete conviction.

Today, however, I think the cosmic origin has been practically universally conceded and with that concession it follows from the measurements of intensities which we carried very close to the top of the atmosphere in 1922 and carried up to still higher altitudes in 1932[1] and 1933,[2] as did also Regener, Piccard and some others (see below), that the total cosmic-ray energy falling into the earth is approximately one-half of the total energy coming in in the form of radiant heat and light from the stars. We get this total of the inflowing cosmic-ray energy simply by counting the total number of ions produced and multiplying this by the known mean energy required to ionize air, viz., 32 electron-volts. Again, since the energy per cubic centimeter in the form of heat and light inside our galaxy must be much greater than it is in intergalactic space, while the cosmic-ray energy is just as dense outside as inside (for Cameron and I found the cosmic rays quite as intense when the Milky Way was out of sight as when it was overhead), it follows that *there is in the universe much more energy in the form of cosmic rays than in the form of heat and light. From the astonomical estimates of the distribution of the nebulae we conclude that the total radiant energy in the universe existing in the form of cosmic rays is from 30 to 300 times greater than that existing in all other forms of radiant energy combined.*

[1] Bowen and Millikan, *Phys. Rev.*, XLIII (1933), 695.
[2] Bowen, Millikan, and Neher, *ibid.*, XLVI (1934), 641.

II. EVIDENCE AS TO THE PLACE OF ORIGIN OF THE COSMIC RAYS

In the last paragraph I have touched upon the most amazing property of the cosmic rays, namely, *the uniformity of their distribution over the celestial dome.* In spite of an enormous amount of experimenting by Hoffmann, Steinke, Corlin, Lindholm, Hess, and many other[1] observers, *no one has yet brought to light any certain, direct influence of the sun on the intensity of the cosmic rays.* Within the limits of my own observational uncertainty they are just as strong at night time as in daytime. In my own careful tests upon this point,[2] covering years of time, I found small daily changes amounting to 1 or 2 per cent, as did other observers, but the phases of these changes did not coincide with the times of appearance and disappearance of the sun, nor with the time of its passage across the meridian where any influences from it must encounter the minimum of atmospheric absorption, so that any direct solar influence should show symmetrical variations about the noon hour, which they did not do. All the very small observed variations appear to me consistent with the view that they are due to daily convective changes in the thickness of the atmospheric blanket through which the rays must pass to reach the earth's surface, and, despite much controversy, I think this view is now quite generally accepted. The exceedingly careful and exact measurements of Hoffman,[3] in Halle, are quite convincing on this point. Nevertheless, if the cosmic

[1] See review by Axel Corlin, Dissertation. Lund, Sweden, 1934; also Summary by Hoffmann, *Phys. Zeit.*, XXXIII (1923), 633.

[2] *Phys. Rev.*, XXXIX (1932), 391.

[3] Hoffman, *Zeit. f. Physik*, XLIX (1931), 704.

rays are shooting uniformly in all directions through space, *secondaries released in the sun's atmosphere might produce slightly greater day-time than night-time effects* such as Hess finds (1934). This effect, if it exists, is well masked by the other causes of fluctuations, and has no bearing anyway on the following conclusions.

The processes that give rise to these rays do not then take place in appreciable amount in the earth's crust, else the rays would not all come in from above, as the measurement of intensities by Cameron and Millikan, clear down to within a few feet of the bottom of lakes, hundreds of feet deep, definitely proved them to do. The equality of day-time and night-time effects shows equally definitely that these rays do not come to us in appreciable amount from the sun, and the entire lack of effect of the presence overhead of the Milky Way—a point which in 1926 we tested[1] carefully in South America, where we could get entirely out of sight of any portion of the Milky Way for several hours at a time—shows convincingly that these rays do not come from any of the stars of our galaxy. Positively stated, these experiments seem to show altogether definitely that *these rays come from beyond the Milky Way.* If, again, they originate in any kind of atomic or nuclear transformations, the normal conditions existing in the earth, the sun, and the stars are apparently unfavorable for these transformations. In a word, the portions of the universe in which the great bulk of the matter of the universe is concentrated are certainly not the portions in which the cosmic rays originate.

If, then, these rays are being formed *now* anywhere,

[1] Millikan and Cameron, *Phys. Rev.*, XXXI (1928), 169.

we must conclude that the conditions of temperature and pressure existing wherever matter is largely concentrated are unfavorable to such formation, and we seem to be forced, then, to seek their origin in the conditions of exceedingly low densities, temperatures, and pressures existing in interstellar space if we would account for the uniformity, by night and by day, of their distribution over the celestial dome. This possibility is a most stimulating one, for it would mean that there is some heretofore undreamed-of kind of activity taking place, more or less uniformly, throughout the depths of space—an activity which manifests itself in continuously raining enormously energetic bullets of some kind (photons, electrons, or both) from all directions upon the heads of us mortals who live on the surface of the earth.

III. HYPOTHESES AS TO THE MODE OF ORIGIN OF THE COSMIC RAYS

The only kind of activity that I have been able to think of that can furnish bullets of the requisite energy to account at least for the least penetrating, and by far the most powerfully ionizing part of the cosmic rays, is the occasional sudden building up, out in the depths of space, of one or another of the heavier elements out of the hydrogen nuclei from which our atomic weight table tells us that all the heavier atoms have at some time or other actually been built up. This atom-building hypothesis will presently be shown to be a special case of the so-called mass-annihilation hypothesis as to the origin of the cosmic rays. The only other source than matter-annihilation that has been suggested for these observed cosmic-ray bullets is a cosmic electric field, essentially symmetrical with respect to the earth, and of such strength and sign as to drive electrons or ions continually into the earth with energies which must rise as high as the highest energies observed in the cosmic rays, namely, several billions (I use a billion as a thousand million) of electron-volts. The difficulties with a hypothesis of this latter kind are so great that I have, for the present, discarded it altogether as completely untenable.[1]

But in any case, so far as we can now see, one has only these two alternatives between which to choose as to the origin of the observed energies, namely, cosmical electric fields, symmetric with respect to the earth and rising to the values indicated, or else some kind of atomic transformation which is capable of releasing the observed energies. These latter must be one or the other of two types of processes, namely, atom-building processes or atom-annihilating processes, or both, and these are, in their most fundamental aspect, not essentially different, for in accordance with Einstein's equation, $E = mc^2$, every process which releases radiant energy of any kind must be accompanied by a corresponding decrease in the mass of the radiating system, i.e., it must correspond to the transformation of mass into radiant energy. As a matter of fact, the mass of every known atom is less than the sum of the masses of the hydrogen nuclei which, at some time, have come together to form it, so that, as stated above, *atom-building processes represent merely a partial stage in the atom-annihilating process.*

From the measured masses of the various atoms it is possible to compute the energy released if an atom of helium is formed by the sudden union of four atoms of hydro-

[1] Bowen, Millikan, and Neher, *Phys. Rev.*, XLVI (1934), 641.

gen. It comes out 28 million electron-volts; that released by the sudden formation of an oxygen atom out of 16 hydrogen atoms comes out 116 million electron-volts; that released in the formation of silicon out of 28 atoms of hydrogen is 216 million volts, of iron out of 56 atoms of hydrogen 460 million volts, and of uranium, the heaviest known atom, about 1,800 million volts. On the other hand, *complete* sudden annihilation of an atom of hydrogen would yield about 1,000 million electron-volts, of helium 4,000 million, of lithium 7,000 million, of carbon 12,000 million, of oxygen 16,000 million, etc.

We shall see later that the great majority of the cosmic rays appear to have energies within the ranges indicated above as corresponding to the sudden building of the common elements, oxygen, silicon, iron, possibly also helium, out of hydrogen. It is interesting to observe in this connection that Henry Norris Russell concludes from his astronomical studies that more than 90 per cent of the universe is still in the form of hydrogen.

However, there are a few of the cosmic rays that have energies as high as 10 billion of electron-volts, so that these fall into the range to be expected from the sudden and *complete* transformation of the whole mass of some of the lighter atoms into radiant energy. How these atom-building and atom-annihilating processes can both be taking place throughout the depths of space we can only conjecture. My own tentative hypothesis has been that under the exceedingly low temperatures existing in interstellar space there are formed aggregates or clusters of hydrogen atoms, just as clouds consisting of aggregates of water molecules are formed at ordinary temperatures in our atmosphere. Only at temperatures near absolute zero, such as exist in interstellar space, could such condensation be expected from hydrogen, the lightest of the elements. As a matter of fact, the modern astronomer thinks he has evidence for the existence of a vast amount of cosmic dust of some kind floating about in space. If we may assume, then, the existence of vast quantities of "hydrogen dust" scattered throughout the universe, it is then to be expected that in accord with the postulates of the modern quantum theory a time will come sooner or later when some such cluster of hydrogen atoms finds itself in just the right condition to jump over the "potential wall" that has thus far kept it a loose cluster of separate atoms of hydrogen and to clamp itself together into the nucleus of a helium atom, or perchance an oxygen atom or one of iron. This act, because of what is known as the "packing effect," destroys a part of the mass existing when the hydrogen atoms were still in the form of hydrogen, and could release a ray of one of the smaller energies listed above.

But again, we may also assume that in rare instances one of these clusters, having got started on the road to ruin, instead of stopping at the stage of the partial annihilation of its mass required to produce say an atom of helium or of oxygen, completes the catastrophe and transforms its whole mass into radiant energy. We could thus account for the uniformity of distribution of both the hard and the soft components of the cosmic rays.

A slight modification of this hypothesis is suggested by Baade and Zwicky,[1] who instead of letting these clustering and subsequent partial or complete annihilation processes take place under the influence of the intense cold of

[1] Baade and Zwicky, *Proc. Nat. Acad. Sci.*, XX (1934), 259.

316 THE ELECTRON

interstellar space, have assumed that the catastrophic processes which result in the sudden flashing-up of temporary stars, called "novae," may create another type of extreme conditions that facilitate these atom-building and atom-annihilating processes. If novae are assumed to be uniformly distributed throughout space the uniformity of distribution of the cosmic rays would thus be accounted for. Still another hypothesis to account for uniformity of distribution is to assume that the cosmic rays were created in bygone ages, when the universe was in a different state from that existing now. All these, however, are only variants of the partial or complete atom-annihilating hypothesis, for *our present knowledge of the universe reveals no other source of such enormous energies as are found in the cosmic rays.*

IV. THE BANDED CHARACTER OF THE COSMIC RAYS

We can gain some rough notion of the relative energies of the various components of the cosmic rays from measuring their relative penetrating powers. The first evidence that they consist not at all of a single band or group of rays of one particular penetrating power, but rather of at least two (possibly more) such bands, came as a result of the 1925 measurements[1] on intensities as a function of depth beneath the surfaces of Muir and Arrowhead lakes; for the absorption coefficient of the rays found near the surface of Muir Lake came out 0.30 per meter of water, while at a depth of 50 feet it was less than half as much, so that it was considered that a definite inhomogeneity, or a "spectral" distribution of the cosmic

[1] Millikan and Cameron, *Phys. Rev.*, XXVIII (1926), 851; XXXII (1928), 533; XXXVII (1930), 235.

rays had been established. More accurate, and more elaborate, studies of this spectral distribution were carried out in the summers of 1927, 1928, and 1929, the depth-ionization curve being followed accurately down to a depth of 70 meters (235 feet) beneath the surface of Gem Lake (California), and cosmic rays found at that depth had a penetrating power fully ten times that of those most abundant at the surface. This meant rays capable of penetrating more than 20 feet of lead. Indeed, Regener, repeating these under-water experiments[1] later in Europe, found traces of these most penetrating cosmic rays at depths as low as 230 meters. When it is remembered that the most penetrating gamma rays from radioactive substances which correspond to an energy of 2.6 million electron-volts can only penetrate say 2 meters of water at most, it will be seen that the energies of the most penetrating band of cosmic rays must be exceedingly high. The most complete study, however, of the relative penetrating powers of the different cosmic-ray bands has come from combining the foregoing under-water results with corresponding high-altitude studies made by sending automatic recording electroscopes up to 29,000 feet in airplanes and up to 60,000 feet in balloons.[2] Figure 50 shows the type of very sensitive high-pressure electroscope —30 atmospheres of argon—that we developed for the under-water and the airplane measurements. For the airplane work, however, it was redesigned internally and made self-recording (p. 412). The four most important conclusions from these studies are: (1) that it is in no way

---

[1] Regener, *Phys. Zeit.*, XXXIV (1933), 306.

[2] Bowen and Millikan, *Phys. Rev.*, XLIII (1933), 695, and Bowen, Millikan, and Neher, *ibid.*, XLIV (1933), 246.

possible to build up the curve found in equatorial latitudes without the assumption of at least three cosmic-ray bands of widely different penetrating power, namely, about $\mu = 0.5$, $\mu = .07$, $\mu = .015$ per meter of water, though four bands $\mu = 0.55$, $\mu = 0.12$, $\mu = .03$, $\mu = .0075$ will do equally well; (2) that though there is no unique solution for such an absorption curve and though the smaller co-

Fig. 50.—The sensitive electroscope, filled with air to a pressure of 450 lbs., with which the first cosmic-ray intensity curve was obtained in high-altitude, snow-fed lakes down to depths of 300 feet. This curve assisted greatly in the solution of the problem of the nature of the cosmic rays.

efficients can be adjusted considerably, yet all possible solutions give absorption bands, or regions, bearing a rather close resemblance to the foregoing solution, the most notable characteristic of which is that the softest component possesses a coefficient which is from four to seven times that of its nearest neighbor; (3) that the softest band, namely $\mu = 0.5$, is capable of very little adjustment; that it takes care of the great bulk of the ionization of the atmosphere above say 14,000 feet; that it, indeed, is responsible for more than 90 per cent of the whole ionization found within the atmosphere; (4) that this softest component is only six to seven times more penetrating than the most penetrating gamma rays, those having an energy of 2.6 million electron-volts. This banded character of the cosmic rays is at least consistent with the atom-building and atom-annihilating theory as to their origin. It is, however, to be remembered that the experimental fact that one of these bands, for example the least penetrating one, throughout a certain range of altitudes, *acts like* a homogeneous beam does not necessarily mean that it is such. Indeed, it is quite certain that in temperate latitudes the shape of the depth-ionization curve at high altitudes, from which the coefficient of absorption of the least penetrating band is determined, is due to the joint action of two quite different causes, one of which is tending to push down the apparent value of μ, the other to raise it. Further facts bearing upon this theory will be presented in chapter xvi.

CHAPTER XIV

THE DIRECT MEASUREMENT OF THE ENERGY OF COSMIC RAYS AND THE DISCOVERY OF THE FREE POSITIVE ELECTRON

Prior to the night of August 2, 1932, the fundamental building-stones of the physical world had been universally supposed to be simply protons and negative electrons. Out of these two primordial entities all of the ninety-two elements had been formed. The proton was itself the positive unit of charge, or the positive electron, exactly like the negative unit so far as charge alone was concerned, but by its very nature different from the negative unit in that it was always associated with a mass 1,834 times that of the rest-mass of the negative electron. Thus, we assumed an intrinsic difference between the natures of positive and negative electricity, and explained the fact that not only was the whole positive charge of an atom concentrated in its nucleus, but also practically the whole of its mass as well. To fit all this into the electromagnetic theory of the origin of mass, most of us, in view of J. J. Thomson's famous equation for the mass of a spherical charge, namely $m = \frac{2}{3}\frac{e^2}{a}$ (Appendix D), conceived the positive unit of charge as having a radius only $\frac{1}{1834}$ of the radius of the negative unit, as this equation seemed to require.

A cosmic-ray photograph taken in the Norman Bridge Laboratory on the afternoon of August 2, 1932, put a stop to our complacency with the foregoing picture.

Carl D. Anderson had taken the photograph and developed the film. He at once realized its importance, and, with Seth Neddermeyer, spent the whole night trying in vain to see if there were not some way of looking at it that would save the old point of view. It could not be done! The photograph revealed indubitably the track of a *free positive electron* having precisely the properties of the free negative electron, save for the sign of its charge—not at all the properties of the proton. The name of the proton might still remain, but its character had changed, for it was no longer itself the ultimate positive entity. A cosmic ray had apparently picked off from it the disembodied positive unit of electrical charge, the identical twin of the free negative electron. The way this discovery came about is as follows.

I. THE DIRECT MEASUREMENT OF THE ENERGIES INVOLVED IN COSMIC RAYS

Up to the year 1931 no particle energies higher than 15 million electron-volts had been measured. From the enormous penetrating power of the cosmic rays, as well as from the known energies freed in the building of the common elements out of hydrogen, I had estimated as early as 1928[1] that the most penetrating of the observed cosmic rays had an energy of as much as 500 million electron-volts, and even if these came to the earth as photons they would be expected to release electrons of about this same energy from the atoms of the atmosphere. It was therefore of the utmost importance to find some way of *directly measuring* electron energies of this order of magnitude, since the estimates of energies made in other ways

[1] Millikan and Cameron, *Phys. Rev.*, XXXII (1928), 533.

were most uncertain. In the summer of 1929, therefore, Dr. Carl D. Anderson and I went at the design of a *vertical* Wilson cloud chamber—the first of this type—set in the midst of an exceedingly powerful and widely extended horizontal magnetic field.

FIG. 51a.—The photograph of the huge electromagnet with which the measurement of charged-particle energies has been pushed up from about 15 million electron-volts, the limit up to 1930, to about 6,000 million electron-volts. Cosmic-ray particles of this energy have actually been found.

Figure 51a shows a photograph of the resulting cosmic-ray energy-measuring device, and Figure 51b a schematic diagram of it. Two thousand amperes of current from a thousand-horse-power motor-generator create a 24,000 gauss field of nearly uniform strength over a space measuring 17 cm. × 17 cm. × 4 cm. In this apparatus the

electron tracks 15 cm. long permit of the measurement of
their curvatures up to energies of as much as 6 billion
(10^9) electron-volts, so that the highest electron energies
that are being measured now are 400 times the highest
energies that were accessible to measurement up to 1931.

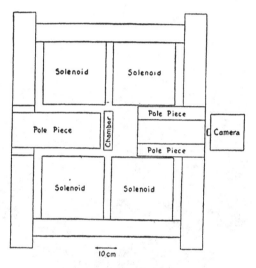

Fig. 51b.—Diagram of cosmic-ray magnet shown in 51a. A nearly uniform magnetic field in the cylindrical chamber 17 cm. in diameter, 4 cm.
thick, is created by a current of 2,000 amperes through the solenoids.
The magnetic field throughout this large volume is 24,000 gauss.

To show strikingly how this apparatus is capable of
differentiating between the effects of gamma rays and
cosmic rays, two of the first photographs taken in the
summer of 1931 are shown in Figures 52 and 53. When
gamma rays from ThC″, maximum energy 2.6 million
electron-volts, are passing through the expansion chamber, the electrons jerked out of the atoms of the gases in

the chamber are forced by the powerful field to move in spirals about the lines of force, and since the camera is looking along these lines the photographs should ideally reveal only circles. Since the light from some parts of the field comes at a slight angle with the lines of force, the

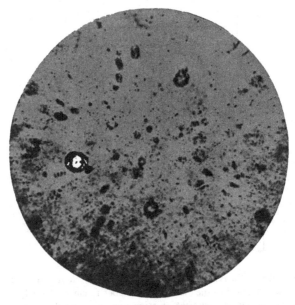

Fig. 52.—12,000 gauss. A test photograph showing tracks in air of secondary electrons ejected by gamma rays from radium filtered through 2.5 cm. of steel. The energies of the electrons range from 2×10^6 electron-volts down.

spiral form of the tracks can actually be seen in Figure 52. From the strength of the field, here 12,000 gauss, the curvature of the path and the known mass of the electron, the energy can be at once calculated and it comes out for the largest circle in Figure 52 about 2 million electron-volts, as it should. Figure 53 shows a cosmic-ray track.

The curvature shows that the energy with which this electron was jerked out of its parent atom was about 8 million electron-volts, more than three times the energy that the most penetrating gamma rays from ThC'' could furnish.

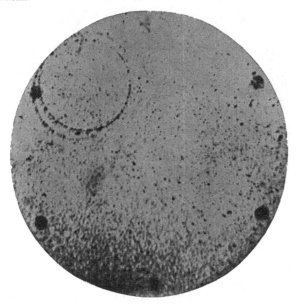

Fig. 53.—Field strength was 12,000 gauss. Track of an 8-million-volt cosmic-ray electron which makes over 1½ complete revolutions in the chamber.

But the next photographs, Figures 54 and 55, show not only much higher energies but something much more important, namely, that *the incoming cosmic rays have here encountered the nucleus of an atom, and as a result of that encounter thrown out from that nucleus both a positive and a negative particle*, for no other interpretation of the opposite curvatures of the two tracks was possible. Since positives could not possibly come from extra-nuclear encounters, this was direct, unambiguous proof

Fig. 54.—17,000 gauss. A pair of associated tracks; at the left an electron of 120 million volts energy, at the right apparently a proton of 130 million volts energy. Since protons and free electrons for a given velocity have equal specific ionization, then from the measurements of Williams and Terroux (*Proc. Roy. Soc.*, CXXVI [1929–30], 300) on the ionization by β particles a 130-million-volt proton is expected to ionize about 3 times as heavily as the electron. It appears to be doing so.

first, that the nucleus plays an important rôle in cosmic-ray absorption, and, second, that both positive and nega-tive particles can fly out from a nucleus when it is hit by a cosmic ray. The nearest approach to anything like this that had been observed before was found in the discovery by Rutherford in 1919 that the impact of alpha particles upon the nuclei of most of the light atoms could knock protons out of them. But cosmic rays were certainly not alpha particles, so that here was an entirely new phenomenon.

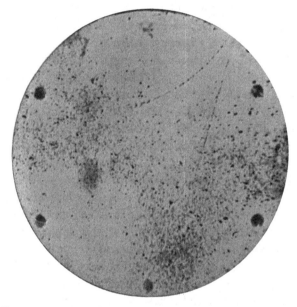

Fig. 55.—12,000 gauss. A negative electron of 27 million volts energy and a positive of about 450 million volts energy.

I presented these photographs in lectures in both Cambridge, England, and Paris at the Institute Poincare in November, 1931, and interpreted Figure 54 as showing that a cosmic ray had knocked out of a nucleus a proton, the track curving downward and to the right, and a negative electron, the track curving downward and to the left. From the curvatures the computed energy of the negative electron was 120 million electron-volts, of the proton 130 million, so that the energy of the incoming cosmic ray had to be at least 250 million electron-volts. This particular photograph was first published on December 18, 1931, by Science Service, along with Dr. Anderson's photograph, under the title "Cosmic Rays Disrupt Atomic Hearts."   I am giving it some attention in this narrative partly because it brought to light the first *direct* proof of the importance of nuclear encounters in the interpretation of cosmic-ray phenomena, but also because it was this photograph that actually delayed the discovery of the free positive electron for at least eight months, and that for the following reason.   A proton of the foregoing energy should ionize the gas much more copiously than should a 120-million-volt electron, and this is just what the positive particle of Figure 54 appears to be doing.   Hence, we had no doubt whatever that it was a proton.   This checked completely with the then universal conception of the composition of the nucleus.

The photograph shown in Figure 55, however, taken at about the same time, bothered us very much, and Dr. Anderson and I discussed it at great length.   The positive particle, to the right, has only a little curvature, and from this curvature we concluded that it was a proton of 450 million electron-volts of energy, while the track to the left was computed to be an electron of 27 million electron-volts.   Now we knew that protons of more than a billion (10^9) electron-volts of energy would give an ionization along its track scarcely distinguishable from the ionization produced by a free electron, but at 450 million electron-volts the ionization due to a proton should be from $1\frac{1}{2}$ to 2 times that due to an electron of 27 million electron-volts, and yet *no trace of a difference in ionization between the right- and the left-hand tracks of Figure 55 was discernible.* We saw no way out of this difficulty save in the assumption that in this enormously high-energy range, in which no one had worked before, perhaps there was something wrong with the theory of the way the ionization of protons varied with energy.   In the lower energy range represented by Figure 54 everything seemed all right.   Contenting ourselves temporarily, at least, with the foregoing excuses, though we definitely promised ourselves to come back to this point when time permitted and find out what it was that was wrong, we continued to call the particles that produced all our positive tracks "protons," and it so happened that for some eight months we found no positives of so large curvature as to make the foregoing explanation impossible.   It is to be remembered that only about one-eighth of our tracks were "associated" in the sense of showing at least two tracks branching downward from a common center,[1] and these alone were able to be unquestionably identified as a positive and a negative. Any isolated single track of small ionizing power that curved rapidly to the right, and therefore ought to be a positive going down, might also be considered a negative going up, so that a crucial test of the proton hypothesis was at this date (fall of 1931) not so easy a matter as it might seem, and furthermore the idea that the proton was itself the fundamental unit of positive electricity was so deeply rooted that almost any other kind of a hy-

[1] Skobelzyn, *Zeit. f. Physik*, LIV (1929), 686, had first called attention to these branching tracks. He had found no evidence, however, nor did he suggest, that one of them consisted of a positive. Indeed his field was too weak to enable him to obtain information on this point, which is the point that ties the phenomenon definitely to the nucleus.

pothesis to get out of a difficulty would at that time have seemed preferable to that of abandoning it.

II. THE DISCOVERY OF THE FREE POSITIVE ELECTRON

But on August 2 Dr. Anderson took the photograph shown in Figure 56. In the middle of the chamber is seen a 6 mm. lead plate that had been inserted for the very purpose of finding the loss in energy of the charged particles in traversing a known thickness of lead. The curvature below the plate corresponds to an energy of 63 million electron-volts in a body having the rest-mass of the electron. The curvature above the plate corresponds to an energy of the same body of but 23 million electron-volts. Since it was quite impossible that it could have *gained* the difference in these two energies in traversing the lead, the direction of motion of the particle had to be upward, not downward.[1] This, with the direction of curvature in the field of known direction, fixed the sign of the charge of the particle as positive. But not only was the thinness of the track precisely like that shown by negative electrons of this curvature, but its length above the lead was at least ten times greater than the possible length of a proton path of this curvature. It seemed to Dr. Anderson, therefore, after his all-night vigil with it, that the

---

[1] The upward motion of an electron is sometimes due to the effect of the magnetic field in reversing its direction. Indeed, without a field so few secondary electrons move upward that Dr. Neher and the author found no measurable difference in the readings of an electroscope when it was taken up in an airplane to an altitude of 15,000 feet, first when resting on a light wooden frame, and second when resting on a heavy mass of lead 10 cm. thick, in spite of the fact that lead is known to stimulate very many new secondaries. This shows that the great majority of all these secondaries move downward in the direction of the incoming beam.

track simply had to be that of a free positive electron. In early September I had the pleasure of showing this track to Dr. Aston of the Cavendish Laboratory, who was on a

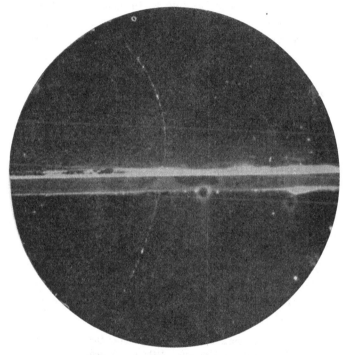

FIG. 56.—A 63-million-volt positron ($H\rho = 2.1 \times 10^5$ gauss-cm.) passing through a 6 mm. lead plate and emerging as a 23-million-volt positron ($H\rho = 7.5 \times 10^4$ gauss-cm.). The length of this latter path is at least ten times greater than the possible length of a proton path of this curvature.

visit to the Bridge Laboratory, and of assuring him that we should withhold publication until we had found other unambiguous cases. These came very soon, and in September[1] Dr. Anderson published his discovery.

[1] Anderson, *Science*, LXXVI (1932), 238.

Even then, however, and for some months thereafter, we regarded the appearance of the free positive electron as a rare event, and kept on interpreting most of our high-energy positive tracks as due to protons. For reasons that will appear below, we later came to regard this view as entirely erroneous. Indeed, after studying thousands of track photographs, the only one among them all that we now see no way of interpreting save as the track of a proton is the original one shown in Figure 54, so that our present conclusion is that if protons, or any other sort of nuclei, ever appear in an ion chamber or an electroscope as a result of the passage of cosmic rays through the chamber, their number is so small as to be inappreciable in the tests we have made to discover them. Since among electrons of energies above say 100 million electron-volts as many positive tracks appear as negatives, indeed a few more in the highest energy ranges (above 3 billion volts), it is clear that approximately half of the ionization due to cosmic rays of energies above these values is due to the passage through the gas of free positive electrons or "positrons"[1] released in cosmic-ray encounters with the nuclei of atoms.

[1] To remove the ambiguity in the definition of the term "electron" existing at the present time because of the double sense in which it is used in the literature, namely, to denote on the one hand—as for example in the universally used expression electron-volts—*the magnitude of the elementary quantity of electric charge*, and on the other hand, the name of a *particle* of a particular mass, the terms "negatron" and "positron" are here used. These terms are used merely as convenient contractions for the fully descriptive *particle* designations, "free negative electron" and "free positive electron." The term electron then retains its historical, derivative, and logical meaning as the name of the elementary unit of charge, and the present ambiguity no longer remains. It is pointed out that this suggestion is not at all in conflict with the tradition and usage

Between September, 1932, and March, 1933, a large number of confirmatory cosmic-ray photographs revealing unambiguously the existence of positrons was taken and a second report was published by Dr. Anderson[1] in March, 1933, in which fifteen of these photographs were discussed.   Figure 57 is one of the most illuminating of these photographs.   The track at the left of the central group of four tracks is made by an 18-million-volt free negative electron (which from symmetry we may now contract to negatron[2] whenever a differentiation in the sign of charge is the important consideration, though this is not often necessary), while the track at the right is that of a positron of an energy of about 20 million volts.   This whole group represents tracks all of which are *"associated in time,"*[2] i.e., represent ionizations necessarily occurring at the same instant, since they all have the same amount

of the term electron.   Even today probably nine-tenths of the usage has reference in the mind of the author to charge rather than to mass, as, to take but a single example, in all cases in which the number of electrons going to a given electrode is under consideration.   The usage we are suggesting is merely for the sake of removing the ambiguity, the bad effects of which are becoming increasingly felt since the discovery of the "free positive electron," and since the discussion of nuclear processes has become more common.   In this usage there is no difficulty in speaking of electrons as existing in the nucleus, since one has then in mind only the number of units of electric charge.

[1] Anderson, *Phys. Rev.*, XLIII (1933), 491.

[2] The tracks in Fig. 55 are geometrically associated, i.e., spring from a common center, as well as being associated in time. In general geometrically associated tracks are also necessarily time-associated.   For with the infrequency with which in the fall of 1931 we got cosmic-ray tracks at all—about one in thirty exposures—the chance that two curved tracks coming from a common center should have originated in independent events is practically nil. This is why we felt so sure that the positive track in Fig. 54 corresponded to a proton. But during the year 1932 in our discussions we adopted the term "associated in time" to describe the num-

of diffuseness. This diffuseness is here due to the fact
that the expansion occurred say half a second after the
passage of the ionizing particles, so that the ions had dif-

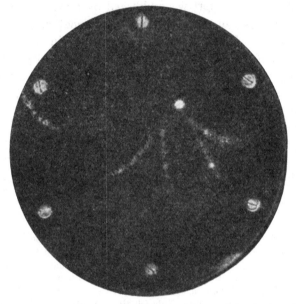

FIG. 57.—A group of six particles projected from a region in the wall of
the chamber. The track at the left of the central group of four tracks is a
negatron of about 18 million volts energy (Hρ = 6.2 × 10^4 gauss-cm.) and
that at the right a positron of about 20 million volts energy (Hρ = 7.0 × 10^4
gauss-cm.). Identification of the two tracks in the center is not possible.
A negatron of about 15 million volts is shown at the left. This group represents early tracks which were broadened by the diffusion of the ions.
The uniformity of this broadening for all the tracks shows that the particles entered the chamber at the same time.

bers of photographs taken on which appeared tracks, like some of those
in Fig. 57, which show identity in diffuseness without geometrical association. Later when we got many of these on a given photograph we used
the term "a photon spray." The phenomenon is what Blackett and
Occhialini describe by the term "non-ionizing links."

fused away from their point of origin before their motion was stopped by the condensation of water vapor upon them. This makes it possible to count *accurately* under a microscope the number of ions per cm. of path, as well as to measure the curvatures. These two elements taken together, when applied to a number of such tracks, fixed both the charges and the masses of the two electrons as the same within 10 per cent and 20 per cent, respectively [1] so that within these limits positrons and negatrons have been definitely proved to be "identical twins" with a difference only in sign of charge. Thus far they had appeared only in cosmic-ray photographs taken at the Bridge Laboratory. In March, 1933, beautiful confirmatory evidence for the existence of positrons resulting from cosmic-ray encounters was presented by Blackett and Occhialini[2] of the Cavendish Laboratory based on similar experiments with a vertical cloud chamber in a magnetic field of 3,000 gauss and actuated by the responses of Geiger-Müller counters.

III. POSITRONS DUE TO GAMMA RAYS

At once search began for these positive tracks from other sources. In April, 1933, Chadwick, Blackett, and Occhialini,[3] Curie and Joliot,[4] and Meitner and Phillipp[5] all reported that the bombardment of beryllium by the alpha rays of polonium is able to produce radiation that results in the production of positrons, though in none of

[1] *Ibid.*, XLIV (1933), 406.

[2] Blackett and Occhialini, *Proc. Roy Soc. A*, CXXXIX (1933), 699.

[3] Chadwick, Blackett, and Occhialini, *Nature*, CXXXI (April 1, 1933), 473.

[4] Curie and Joliot, *Compt. Rend.*, CXCVI (1933), 1105

[5] Meitner and Phillipp, *Naturwiss.*, XXI (1933), 286.

these experiments was it possible definitely to identify the nature of the radiations producing the positrons. Nevertheless, Curie and Joliot showed that the yield of positrons decreased approximately as was to be expected if gamma rays resulting from the alpha-ray bombardment of the beryllium in their turn produced the positrons. The first experiments proving directly, however, that a gamma-ray photon impinging upon a nucleus gives rise to positrons were made by Anderson at the Norman Bridge Laboratory, also in April, 1933, using the gamma rays from thorium C″, and reported to the National Academy of Sciences in Washington[1] on April 24, 1933. In this paper the fact that free electrons of both positive and negative sign are produced simultaneously by the impact of a single gamma-ray photon, an observation of considerable theoretical importance, was first presented. Figure 58 shows one of many such cases photographed. A very narrow beam of photons from Th C″, that was found experimentally to eject electrons so infrequently that the possibility of two such ejections occurring exactly simultaneously was practically nil, comes in from above. Out of the 2 mm. lead sheet in the upper part of the chamber a single photon is then seen to knock two negatives and a positive. The latter is seen to shoot through the thin aluminum plate 0.5 mm. thick in the middle of the field of view, to lose energy in so doing, and to come out below with increased curvature. The curvature below the middle plate actually corresponds to an energy of 520,000 volts, while above the middle plate the curvature corresponds to 820,000 volts.

[1] Carl D. Anderson, *Nat. Acad. Sci.* (meeting April 24, 1933), and *Science*, LXXVII (1933), 432.

Measurements of the energies of these positives produced by the γ rays of ThC″ have been made by Anderson and Neddermeyer,[1] Curie and Joliot,[2] Meitner and

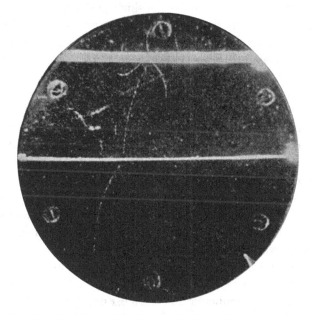

Fig. 58.—A positron ejected from a lead plate by gamma rays and
passing through a 0.5 mm. aluminum plate. The energy above the aluminum is 820,000 volts, below 520,000 volts.

Phillipp,[3] and Chadwick, Blackett, and Occhialini.[4] Anderson and Neddermeyer were the first to make a prolonged statistical study of as many as 2,500 tracks of
single electrons, both positive and negative, and positive-

[1] Anderson and Neddermeyer, *Phys. Rev.*, XLIII (1933), 1034.

[2] Curie and Joliot, *op. cit.*, p. 1581.

[3] Meitner and Philipp, *Naturwiss.*, XXIV (1933), 468.

[4] Chadwick, Blackett, and Occhialini, *op. cit.*

negative pairs ejected from plates of lead, aluminum, and carbon by γ rays from ThC'' filtered through 2.5 cm. of lead. Their purpose was to determine the frequency of occurrence both of pairs and of single positrons along with their energy-distribution for absorbing materials of different atomic numbers. This ejection of particles was observed from lead plates of 0.25 mm. thickness, aluminum plates of 0.5 mm. thickness, and a graphite plate of 1.4 cm. thickness. For the measurement of these low energies the magnetic field was here adjusted to 825 gauss.

Both the single positives and the pairs (the sum of the energies of the positive and the negative components being taken) ejected from the lead plate showed a maximum energy of 1.6 million electron-volts, though 80 per cent of the single positives had an energy of less than half this value. Similarly, in the case of the positives and the pairs ejected from the plates of aluminum the maximum energy was also 1.6 million electron-volts.

On the other hand, the maximum energy of the single negatives was in all cases about 2.6 million electron-volts. When it is remembered that the most penetrating and the most intense constituent of the gamma radiation incident upon the lead and the aluminum from ThC'' has an energy of but 2.62 million electron-volts, it will be seen that these experiments seem to say quite definitely that about a million volts more energy is required to get a positive electron out of an atom than to eject from it the least tightly bound (and of course extra-nuclear) negative electron, and that, further, this difference is the same for both lead and aluminum. So far as experiments with lead alone are concerned, these results have recently been checked even more accurately at the Cavendish Laboratory.[1]

This result is of peculiar significance if for no other reason than because of how it came to be obtained, which was as follows:

While the discovery of the positive electron and its appearance under the impact of both cosmic rays and gamma rays was made without the guidance of any theory whatever, as was also the discovery of the frequent occurrence of positive-negative pairs of tracks, like those of Figures 54, 55, and 58, when Blackett and Occhialini published in March, 1933, their check of Anderson's discovery of the positron they first suggested an interpretation of this phenomenon in terms of the so-called Dirac theory of the creation of positive-negative electron pairs through the impact of photons on the nuclei of atoms. Also, when Anderson and Neddermeyer in April, 1933, had proved that gamma rays, as well as cosmic rays, eject apparently both single positives and positive-negative pairs from the nuclei of both lead and aluminum, Oppenheimer and Plesset[2] at the Norman Bridge Laboratory set about working out in detail the consequences of the Dirac theory for the absorption of the photons of ThC″ by the nuclei of atoms. It was also this theory of Dirac's and Blackett and Occhialini's use of it that first suggested to Anderson and Neddermeyer that the maximum energy associated with a positron produced by ThC″ should be a million volts less than the energy of the incident photon,

[1] Chadwick, Blackett, and Occhialini, *Proc. Roy. Soc. A.*, CXLIV (1934), 235.

[2] Oppenheimer and Plesset, *Phys. Rev.*, XLIV (1933), 53.

i.e., that it should be 1.6 million electron-volts, as they very soon found it in fact to be.

This million volts is obtained from Dirac's theory and Einstein's relation, $E = mc^2$, as follows: It will be remembered that the rest-mass of the electron is about one two-thousandth that of a hydrogen atom which, as we have seen on page 314, chapter xiii, has a mass equivalent to a billion electron-volts, so that the energy equivalent of the mass of the electron is roughly a billion divided by two thousand, or a half-million electron-volts. Hence, to create a positive-negative electron pair would require an expenditure of a million electron-volts of energy and leave of the total 2.6 million volts contained in a quantum of ThC'', 1.6 million to be divided between the positive and negative of the pair in a way which the theory had nothing to say about.

To find experimentally this distribution was one of the objects of Anderson and Neddermeyer's statistical study both of the numbers of single negatives, of single positives and of pairs, as well as of the energies of each. Out of a total of 1,542 electrons ejected from the 0.25 mm. lead plate by the γ rays of ThC'' filtered through 2.5 cm. of lead, 1,387 were found to be single negatives, 96 single positives, and 59 pairs. From an aluminum plate 0.5 mm. thick and ejected by the same radiation there were, out of a total of 943 electron tracks, 916 single negatives, 20 single positives, and 7 pairs. Of these apparent pairs some may not be pairs at all since a negative going up and bounding back from a lower surface of the lead plate is indistinguishable from a pair.

The foregoing negatives may be assumed to have arisen in general from Compton and photoelectric encounters with extra-nuclear electrons, and this is why their maximum energy reached practically 2.6 million electron-volts. But all the single positives and probably most of the pairs correspond to nuclear encounters. If we assume that on the average an equal number of positives and negatives results from nuclear impacts, we can calculate the ratio of the nuclear to the extra-nuclear absorption. This amounts to about 20 per cent for lead and 5 per cent for aluminum—values in reasonably good agreement with those obtained by Chao,[1] Meitner,[2] and Gray and Tarrant[3] by entirely different methods. That the nuclear absorption in carbon for ThC'' rays is still smaller in carbon than in aluminum, as is to be expected from the above relations of lead and aluminum, is shown by the fact that as compared with 415 negatives knocked out of a carbon plate 1.4 cm. thick there appeared only two pairs and six single positives.

On the whole, then, the energy relations of the positives and pairs from both the aluminum and the lead appear to be quite consistent with the pair-creation hypothesis, as are also the approximate values of the excess absorption in lead and aluminum as calculated by Oppenheimer on this pair-creation assumption. But the fact that all told there are nearly twice as many single positives as pairs, and possibly more, and in aluminum and carbon at least three times as many, present some difficulties to the pair theory. Further, Anderson has observed one case in which two negatives and two positives were all

[1] Chao, *Proc. Nat. Acad. Sci.*, XVI (1930), 431, and *Phys. Rev.*, XXXVI (1930), 1519.

[2] Meitner and Hapfield, *Naturwiss.*, XIX (1931), 775.

[3] Gray and Tarrant, *Proc. Roy. Soc. A.*, CXXXVI (1932), 662

observed to originate at one point in the lead plate. The possibility that this can represent two pairs accidentally associated in time and position is so remote that the experiment may be taken as evidence that photons of energy even so low as those of ThC″ can occasionally give rise to "showers" such as are common with cosmic rays (see § IV below), a phenomenon that according to Oppenheimer cannot as yet be handled satisfactorily by the pair-formation theory.

IV. COSMIC-RAY SHOWERS

Whatever may be the mechanism by which gamma-ray photons of from two to three million volts of energy produce positive and negative electrons through their encounters with the nuclei of atoms, a study of cosmic-ray showers with the aid of our big magnet leaves little doubt that a billion-volt photon can cause to emerge one or many electrons, both positive and negative, from the nucleus of a heavy atom with which it collides.

Some 88 per cent of all the tracks that we obtain when expansions and exposures are taken at random are the tracks of single electrons of energies varying from say 50 million to 6,000 million electron-volts. The other 12 per cent are showers, a shower being defined as two or more associated tracks. Tracks are considered "associated" if they are produced by electrons that pass through the chamber at the same time, as indicated by the identity in the sharpness or in the diffuseness of the tracks they leave. The distribution of these tracks in both energy and sign of charge has been carefully studied by Anderson[1] with the result that between 100 million and 3,000

[1] Carl D. Anderson, *Phys. Rev.*, XLIV (1933), 411; also Anderson, Millikan, Neddermeyer, and Pickering, *ibid.*, XLV (1934), 352.

million volts, positives and negatives appear in about equal numbers for all energies. Of the 12 per cent of associated tracks perhaps a tenth show more than five tracks, but by placing a bar of lead across the middle of the expansion chamber the chance of bringing a "many-track-shower" into evidence is of course increased. Further, an arrangement in which two Geiger-counters are placed one just above and one just below the expansion chamber, and the circuits so adjusted that the passing of an electron through them both produces both the expansion and the exposure and thus forces the cosmic ray to take its own photograph, is one that has a marked selective action on showers for a reason that will presently appear.

The photographs shown in Figures 59–64 were taken with such automatic Geiger-counter controlled expansions and exposures, first used by Blackett and Occhialini. As will be seen from these photographs, with their legends, a new fact is here strikingly brought to light, namely, that *in the absorption of the cosmic rays there are produced in addition to electron showers in some instances sprays of large numbers of secondary photons of relatively low energies.* These photons are presumably produced by the same mechanism that gives rise to the general X-radiation when cathode rays are shot into a target of some dense metal, for in this latter case we think it is the sudden impact of the cathode-ray electron against the nucleus of a heavy atom that generates the impulse radiation (bremsstrahlung) of which the general X-radiation consists. So here when the powerful incoming cosmic-ray is absorbed by the electrons (+ and −) within the nucleus, or, if one prefers the Dirac theory, creates electron pairs by its impact with the nucleus, these electrons receive very large kinetic energies and colliding then several or many times with the dense elements of the nucleus in

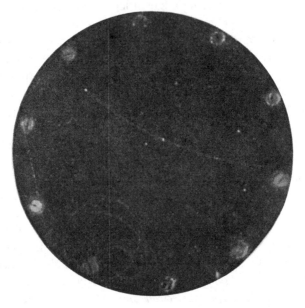

FIG. 59.—17,000 gauss field. An electron pair, positron 75 million electron-volts, negatron 290 million, ejected presumably from the nucleus of a lead atom above the upper counter. These electrons in getting through or out of the nucleus presumably collided with its mass, and produced thereby "bremsstrahlung." This photon spray shot downward and its absorption in the gas of the chamber, or the surface layer or its wall, produced the four secondaries seen between the electron tracks of energies in millions of electron-volts 9, 9, 4, 1.

their endeavor to get through or out of it they throw out from these collisions a spray of soft X-rays. Figures 59 and 60 show, in addition to the positive and negative electrons ejected by the incident cosmic ray, low-energy circular tracks, presumably due to the absorption in the gas of the chamber of these soft photons of the general

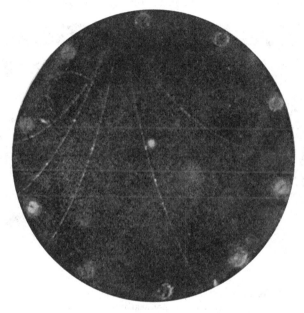

FIG. 60.—17,000 gauss field. A shower presumably originating in the impact of a cosmic-ray photon upon the nucleus of an atom of lead in the bar just above the top counter. Energies in millions of electronvolts; positrons, 145, 38, negatrons 104, 65, 28; sum of all 380. Again the presence of secondary photons is demonstrated by the tracks of low-energy particles at the left; their energies are: 6.7, 4, 2, 0.1.

X-ray type. Figure 61 shows more than 80 of these low-energy tracks. It is this spray of relatively low-energy photons that sets off both counters even though the original cosmic-ray photon passed through neither of them.

This is, then, why this Geiger-counter control exerts so selective an action upon the appearance of "showers."

That pair production or shower formation by a fast electron (positive or negative) as distinct from a photon

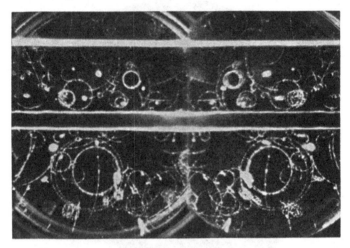

FIG. 61.—17,000 gauss field. More than 80 low-energy tracks. A stereoscopic study with the aid of the direct image (left) and the reflected one (right) of the orientations and directions of these tracks shows that in most instances their motions are nearly in the plane of the chamber so they could not have originated except from a considerable number of separate centers, hence indicating a large number of secondary low-energy photons (100,000 to 10 million electron-volts) presumably resulting from the collision of a primary photon with a lead nucleus above the upper counter. It is this shower that must set off both counters, as well as produce the cloud-chamber effects.

is a relatively rare event is shown by the fact that more than a thousand fast electrons have been observed to traverse a 1 cm. lead plate, and only in one instance has a definite pair (+ and −) been projected from the lead by such a fast electron, while a very large number of secondary negative electron tracks have appeared as the result of close encounters with the extra-nuclear electrons within the lead. Figure 62 shows one of the rare cases in which a negatron (curvature to left) seems simply to make a nuclear hit within the lead and transfer its energy

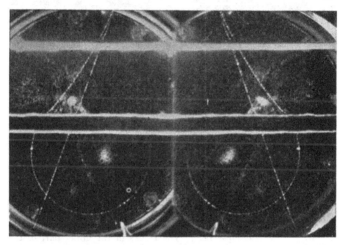

FIG. 62.—17,000 gauss field. At the left an electron passes into the middle lead plate and either transfers its energy to a positron or else forms a pair, both the negatron-component of the pair and the original negatron being absorbed in the lead. The former interpretation seems more likely. The difference in energy above and below the plate is consistent with observed values of the specific energy loss for electrons in lead, inasmuch as the fluctuations are rather large. Energies: $(-)$ above, 90; $(+)$ below 26. (Specific energy loss 49×10^6 e.v./cm.)

to the positron (curvature to right) that emerges from the bottom of the plate.

Because of the very powerful magnetic field we are using, it is possible to deflect all but a very small number of the electrons projected in the showers by the photon impacts. In general, in a shower a pronounced asymmetry is noted in the numbers of positive as compared with negative electrons emerging from the lead plates, in one instance 7 positives and 15 negatives (Fig. 63), and in a second case 15 positives and 10 negatives (Fig. 64).

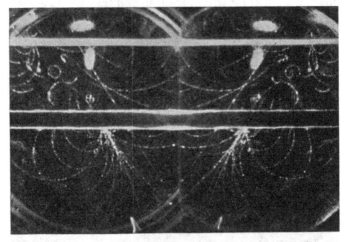

FIG. 63.—17,000 gauss field. The tracks in the upper part of the photograph, at least three of which converge to a region in front of the chamber, indicate the occurrence of a shower above the chamber. A second shower of 22 particles, 7 positives, 15 negatives, is seen to originate in the lead plate, the initial directions of the particles indicating that the photon (or photons) producing it passed through a point very close to the origin of the upper shower. The high-energy positron (520 million electron-volts) passing through both lead plates probably has its origin at the point above the chamber through which passed the photon which gave rise to the showers. There are other tracks not in line with the main shower, as, for example, a group of three tracks from the upper lead plate at the left, which we attribute to the absorption there of secondary photons. The two heavy white patches just above and just below the upper lead plate cannot be associated in time with the shower. Their diffuse appearance may be explained by the assumption that they are due to recoil nuclei released *before* the expansion. The total energy of all the tracks is about a billion volts. All this suggests that a high-energy photon may knock out one or many electrons from several nuclei which it may encounter along its path.

These effects are only with some difficulty reconciled with the Dirac theory of the creation of pairs out of the incident photon. Rather might they indicate the existence of a nuclear reaction of a type in which the nucleus plays a more active rôle than merely that of a catalyst, as for example the ejection from it of positive and negative

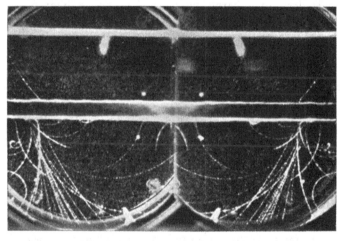

Fig. 64.—17,000 gauss field. Shower of 28 electron tracks resulting presumably from the absorption of a very high-energy primary photon in the central lead bar. From one main center at the left there diverge 15 positrons and 10 negatrons, while the three remaining tracks may arise from the photon spray. The total energy is about 2.5×10^9 volts, slightly less than that of the highest energy single tracks we have observed.

charges which then appear in the showers as free positive and negative electrons. The essential difference, however, between these two points of view may be merely that in one case the nucleus may change its charge, and in the other it does not do so.

To study nuclear absorption in a light element more than four hundred successful photographs were taken in which a carbon plate of 1.4 cm. thickness replaced the lead plate. Many of these showed showers originating in a block of lead placed above the chamber, but in no instance was a secondary shower observed in the carbon plate. This indicates, in agreement with data obtained with ThC″, the relatively small probability in comparison with lead of a carbon-nucleus absorbing a photon by shower production.

One other striking and significant result of these studies is that the total energy available in the incident ray seems to be able to appear as a single positive, a single negative, a pair, or a multiple-track shower, but in general the larger the number of tracks between which the energy is divided the smaller the energy in the individual tracks. Thus, we have measured both individual positives and negatives of energies above three billion electron-volts, but the sum of the energies in all the tracks of the most energetic shower, Figure 64, is just under three billion electron-volts. Similarly, the sum of the energies of the many-track shower of Figure 63 is between one and two billion electron-volts.

V. POSITRONS FROM ARTIFICIALLY ACTIVATED SUBSTANCES

From the time of its discovery by Becquerel in 1896 up to January 15, 1934, the phenomenon of radioactivity had been found to be beyond the control of man. Certain kinds of atoms were discovered to be *spontaneously* disintegrating with the emission of α, β, or γ rays, or of two or three of them simultaneously, but the rate at which these radiation-processes were going on was fixed by the nature of the atom and was independent of its physical or chemical environment.

On the aforementioned date, however, Mme Irene Curie and her husband, M. F. Joliot, read a paper[1] before the French Academy in which they showed (1) that by bombarding boron, magnesium, or aluminum with the alpha rays of polonium all of these substances could be rendered radioactive; (2) that the radioactive rays emitted by them consisted of positrons; (3) that the half-decay time for the positron radioactivity of aluminum was 3 minutes, 15 seconds; of boron 14 minutes; and of magnesium 2 minutes and 30 seconds, while the average energy of the positrons from aluminum, estimated from absorption measurements, was about 2.2 million electron-volts, and that from boron and magnesium 0.7 million electron-volts. Curie and Joliot also predicted from their view of the nature of this new kind of radioactivity that if carbon could be bombarded by heavy hydrogen (hydrogen containing 2 protons per nucleus instead of 1) it, too, would show positron radioactivity.

Meanwhile, at the California Institute of Technology, Professor C. C. Lauritsen had developed as early as 1929 his million-volt X-ray tube which, in addition to its continuous use for four years for therapeutic purposes, had been modified in 1932–33 so as to adapt it to the problem of bombarding targets with ions of the lighter elements after they had fallen through million-volt fields,[2] and more. (The Lauritsen tube has actually been run continuously at 1,200,000 volts.) With this arrangement

[1] Curie and Joliot, *Compt. Rend.* CXCVIII (1934), 254.

[2] Crane and Lauritsen, *Rev. Sci. Inst.*, IV (1933), 118; also Crane, Lauritsen, and Soltan, *Phys. Rev.*, XLIV(1933), 692 and XLV (1934), 507.

they found in February, 1934, that both boron and carbon
could be endowed with positron radioactivity by bombarding them with deutons (heavy hydrogen) of $0.9 \times 10_9$
electron-volts of energy.   The carbon activity had a
half-period of 10.3 minutes, the boron of about 20 minutes,[1] quite consistently with the well-known behavior of
ordinary negatron or beta radioactivity.   Figures 65 and
66 are photographs of the tracks of these positive electrons taken by Anderson and Neddermeyer[2] with the aid
of our big magnet.   That the rays consist of positives is
shown by the fact that they all bend to the right in the
"direct" image.   Figure 65 was taken immediately after
the carbon plate had been activated and placed in the upper part of the cloud chamber, where it can be seen in the
photographs. Figure 66 was taken some minutes later
when the induced radioactivity had fallen to about a third
of its initial intensity. These experiments confirm in a
most satisfactory way the findings of Curie-Joliot and
extend them somewhat in that the radioactivity is now
produced wholly artificially, since here the bombardment
is done not by the pre-existent and uncontrollable alpha-
ray activity of polonium but by artificially produced ions
of any desired kind freed in a specially designed discharge
tube and accelerated to any desired energy by a man-
made field. This wholly artificial radioactivity was of the
order of a thousand times the intensity of that before obtained. This step added notably to the possibilities open
for the study of different kinds and conditions of artificially stimulated radioactivity. In fact, Professor Laurit-

[1] Crane, Lauritsen, and Harper, *Science*, LXXIX (March, 1934).
234; also Crane and Lauritsen, *Phys. Rev.*, XLV (March 15, 1934), 430.

[2] Anderson and Neddermeyer, *Phys. Rev.*, XLV (1934), 653.

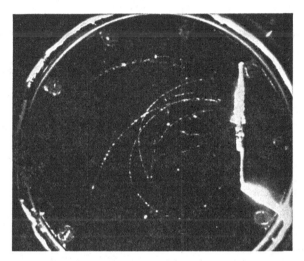

Fig. 65.—The carbon plate in the upper part (here right side) of the chamber is seen to be emitting positrons. It has been bombarded by 900,-000-electron-volt protons, then immediately placed in the cloud chamber.

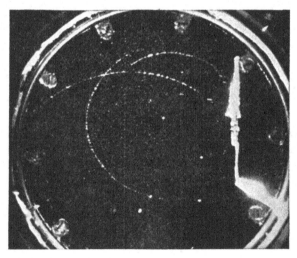

Fig. 66.—The same as Fig. 65 after the induced positron radioactivity has been in the chamber for some 20 minutes.

sen[1] and his collaborators have obtained similar results
with the use of protons and of helium nuclei used as bom-
barders, as well as with deutons. They have also obtained
most interesting results with a series of other targets,
notably lithium and fluorine. Also a little later Professor
Lawrence[2] of Berkeley got similar results with his unique
ion-accelerating device.[3] The kind of nuclear transformations involved in the generation of this artificial radioactivity will be considered more fully in the next chapter.

There is, however, one further very important particular in which Professor Lauritsen's group at the Kellogg
Radiation Laboratory of the California Institute has
checked and extended the work of other investigators on
the properties of the positive electron, as follows.

VI. THE ULTIMATE FATE OF THE POSITIVE ELECTRON

From the time of the first discovery of the positive electron in the summer of 1932 the question of what became
of it had been the subject of prolonged discussion within
the Norman Bridge Laboratory. According both to all
experimental findings and all theory up to the time of its
discovery all the positive electricity that was demanded
to neutralize the negative electrons that swarm about the
nucleus of every atom and by their occasional detachment
from the outer shells of these atoms produce the phenomena of conductivity, all this neutralizing positive
electricity was imprisoned within the nuclei of the atoms

---

[1] Crane, Lauritsen, and Soltan, *Phys. Rev.*, XLV (February 1, 1934),
226; Crane and Lauritsen, *ibid.* (March 1, 1934), 344; *ibid.* (April 1,
1934), 493 and 497; *ibid.* (April 15, 1934), 550.

[2] Henderson, Livingston, and Lawrence, *ibid.* (March 15, 1934), 428.

[3] Lawrence and Livingston, *ibid.* (1934), 608.

and the total number of positive electrons there found was of necessity the same as the total number of extra-nuclear negative electrons in all neutral systems. Anderson's discovery showed that a cosmic-ray photon, possibly also a high-energy electron, colliding with a nucleus, could eject positive electrons out into the inter-atomic world which swarmed everywhere with negative electrons. That it could not live there long after its energy of ejection had been spent was certain, both from experiment and theory. For the former showed that these positive electrons play no part in conduction—gaseous, electrolytic, or metallic—and hence that they do not remain free. Theory, on the other hand, said that free positive and negative electrons existing side by side would of necessity rush together and destroy each other unless in some way through the presence of the free positive electron a new proton could be formed which by then capturing from the surroundings a new negative electron would give rise to a new atom of hydrogen. For in all nuclei, including that of hydrogen, the mutual destruction of positive and negative electricity is avoided through the usual quantum conditions holding within the atom. But no matter how the positron disappears, the region in which such disappearances are taking place must be a region from which radiant energy is emerging, for with either alternative as to mode of escape, according to Einstein's equation, the disappearing potential energy of two separated attracting systems, corresponding to the m in $E = mc^2$, must of necessity appear in radiant form.

Now, before it was known that positrons were produced by the more energetic of the gamma rays, Bowen, Cameron, and I had obtained and presented to the National Academy of Sciences in November, 1929,[1] convincing evidence that the nucleus of the atom of lead plays an appreciable rôle in the absorption of cosmic rays, and very early in 1929 we asked Mr. Chao to test this point carefully with respect to the rays from ThC″, using a much narrower and more sharply defined beam than had before been employed, for we wished to improve the accuracy in the measurement of scattering, secondary radiation, and absorption. With this arrangement Chao[1] first found definite evidence of nuclear absorption in addition to Compton scattering, and he brought to light a secondary, as distinct from scattered, radiation of wavelength of 22.5 XU (0.55×10^6 electron-volts). He also found that this radiation was emitted in about the same intensity in a number of directions making angles of between 30° and 150° with the primary beam; that, further, this secondary radiation was stimulated at a threshold value of the incident photons of about 2.0 million electron-volts.

Meanwhile Tarrant,[2] who had been working at the Cavendish Laboratory on the absorption of ThC″ rays in lead, published in July, 1930, evidence in agreement with Chao's in suggesting nuclear absorption, and in 1932 Gray and Tarrant[3] checked most satisfactorily Chao's 1930 discovery of the stimulation in lead by ThC″ rays of a secondary isotropic radiation of an energy of about half a million volts. Again in 1934,[4] they checked both

[1] Chao, *Proc. Nat. Acad. Sci.*, XVI (1930), 431 and *Phys. Rev.*, XXXVI (1930), 1519; also *Proc. Roy Soc. A.*, CXXXV (1931), 206. Chao's results were first presented before the National Academy of Sciences on April 29 (see *Proc.*, XVI [1930], 421).

[2] Tarrant, *Proc. Roy Soc. A.*, CXXVIII (1930), 345.

[3] Gray and Tarrant, *Proc. Roy Soc. A.*, CXXXVI (1932), 662.

[4] Gray and Tarrant, *ibid.*, CXLIII (1934), 681.

themselves and Chao in the determination of the properties of this radiation, though they do not succeed in fixing
its energy with a certainty greater than perhaps 25 per
cent. They think they find also a weaker isotropic radiation of an energy of a million volts, but Lauritsen and
Oppenheimer[1] in a critical analysis of all the already ex-
tensive work in this field reach the conclusion that the
addition of Chao's isotropic half-million-volt radiation to
the various types of scattered rays to be expected is sufficient to account for all the facts thus far observed.

This half-million-volt isotropic radiation thus brought
to light by Chao is obviously just what is to be expected
if the positrons which are produced by the absorption of
the 2.6-million-volt rays of ThC″ in lead, etc., are ultimately destroyed through combining with negatives, for
though such mutual annihilation of two electrons should
release a million volts of radiant energy the Newtonian
momentum law cannot be satisfied unless two half-million-
volt photons go off in opposite directions from the scene
of the catastrophe.

This annihilation hypothesis as to the fate of the positive electrons has then the advantage over its only possible rival, namely, the hypothesis that the positron's end is
the formation by some sort of mechanism of a new proton
in that the energy released by such formation would in
this latter case have to be *arbitrarily* assumed to be about
half a million volts, whereas the annihilation hypothesis
predicts the value actually found. The greatest weakness
in the position of this hypothesis so far lies in the fact that
none of the observers using the foregoing method have

[1] Lauritsen and Oppenheimer, *Phys. Rev.*, XLVI (1934), 80.

been able to determine with sufficient accuracy the energy of the photons resulting from the annihilation.

The discovery of artificial radioactivity by Curie-Joliot assisted, however, in obtaining greater precision, since this made it possible to obtain a pure positron source of large intensity. But just prior to this discovery Thibaud[1] and Joliot[2] presented, independently, papers in the same issue of the *Comptes Rendus* (December 18, 1933) in which they both had separated by means of a suitably arranged magnetic field the positrons from the negatrons when both were produced by the bombardment of aluminum by the alpha rays of polonium. When the magnetic field was thrown on in one direction the positives were caused to fall upon a thin sheet of platinum (Thibaud) or lead or aluminum (Joliot), and when the field was thrown on in the other direction it was only negatives that fell on the thin sheet. From the region of the positives there emerged gamma rays of much greater intensity than when the field was reversed, and the energy of these gamma rays was found, by interposing absorbing sheets, to correspond to about half a million volts. This constitutes a step somewhat in advance of that taken by Chao and Gray and Tarrant, since the positron beam that creates the gamma rays is completely under control and directing it upon a given spot causes the gamma rays to emerge from that spot.

The next step was taken by Lauritsen and Crane, who with their powerful million-volt tube produced in plates of boron, carbon, aluminum, etc., wholly artificial positron radioactivity of a thousand times the intensity of that theretofore available. Ordinary electroscopes, in-

[1] J. Thibaud, *Compt. Rend.*, CXCVII (1933), 1629.

[2] F. Joliot, *ibid.*, p. 1622.

stead of Geiger-Müller counters, could be used for its measurement.   For example, one experiment was as follows: Dr. Lauritsen activated strongly a small plate of carbon, placed it activated side up on the top plate of an electroscope which was too thick to let any of the positrons through, and observed the rate of discharge of the electroscope due to the gamma rays produced by the disappearance of the positrons that shot downward from the activated layer of carbon.   Then he simply laid a piece of aluminum on top of the activated carbon.   At once the electroscope practically doubled its rate of discharge, for the half of the positrons that shot upward had now all to disappear in the aluminum, where they could throw their annihilation gamma rays into the electroscope, while prior to the time the aluminum plate was placed upon the activated carbon they had shot off to the walls of the room whence practically none of their progeny of annihilation-rays could get back to the electroscope. This constitutes a very simple and direct way of showing that these gamma rays are indeed generated by the disappearance of the positrons.

Lauritsen's and Crane's measurement of the absorption coefficient[1] of these rays agreed with the theoretical value to within some 4 or 5 per cent—a very considerable improvement in accuracy—so that it now seems to be fairly well established experimentally that the positive electron lives outside the nucleus only until it has lost the bulk of the kinetic energy that has been imparted to it— this will perhaps be of the order of 5×10^{-10} seconds—when it disappears somewhere, probably by committing suicide with the first negative that comes near it.

[1] Crane and Lauritsen, *Phys. Rev.*, XLV (1934), 430.

CHAPTER XV

THE NEUTRON AND THE TRANSMUTATION
OF THE ELEMENTS

I. EARLY EVIDENCES OF TRANSMUTATION

The first suggestion, after the days of alchemy, of the transmutability of the elements came with the appearance in 1815 of Prout's hypothesis; for if, as Prout supposed, all atoms had weights that were exact multiples of the weight of the atom of hydrogen it is at once to be inferred that they have been built up at one time or another out of hydrogen, and this of course means that they are ideally transmutable. But for three-quarters of a century nothing further happened to suggest that atomic transformations ever actually take place.

Then in 1896 came the discovery of radioactivity. It was this discovery that definitely destroyed the idea of the independence and permanence of the elements. But while through this radioactive process a few of the elements were universally recognized to be spontaneously transforming themselves into other elements, no definite evidence appeared until 1919 that atomic transformations could be in any way controlled by man. In that year Rutherford,[1] by bombarding nitrogen with swift a particles, definitely proved that protons, or hydrogen nuclei, were in some instances knocked out of the atoms of nitrogen by this bombardment. Also the fact that this proton shot out sometimes in a forward direction, some-

[1] Rutherford, *Nature*, CIII (1919), 415; *Phil. Mag.*, XXXVII (1919), 537, 571, 581.

times a backward one, and with an energy that enabled it to make a cloud-chamber track more than 50 cm. long—greater than could possibly be due to energy immediately transferred to it from the α particle—meant that some kind of change capable of supplying energy had taken place in the nucleus. The inference was that the α particle had been captured by the nucleus, thus forming out of nitrogen (N^{14}) a rare isotope of oxygen (O^{17}), though no definite proof of this could be brought forward. But in any case the production of an atom of hydrogen through the bombardment of nitrogen by α rays was an unquestioned fact. Here, then, was man-controlled transmutation, though natural radioactive substances were needed to bring it about.

By 1930 this type of semi-artificial transmutation of elements had been studied with a good deal of care,[1] and it had been shown that all the elements from boron to potassium, save only carbon and oxygen, can be disintegrated by such bombardment with α particles. The result of the disintegration was in every case the emission of a hydrogen nucleus or proton. The second step in artificial disintegration came in 1930 with the discovery of Bothe and Becker, treated at length in the next section, of a new, highly penetrating radiation excited in beryllium by bombarding it with alpha rays from polonium. Bothe could see no source of such energy as seemed to be required for such penetration without the assumption of a nuclear transformation within the beryllium atom that would release the requisite energy in accordance with the Einstein equation.

[1] See Rutherford, Chadwick, and Ellis, *Radiations from Radioactive Substances*, 1930, chap. x.

The next step in artificial nuclear disintegration came when, in the fall of 1931 at the Norman Bridge Laboratory, we got the definite proof already presented in chapter xiv that the tracks of both positive and negative particles appeared on our photographic plates as a result of a cosmic-ray photon encounter with the nucleus of a heavy atom. But like all our predecessors we could only conjecture what sort of nuclear changes had been occasioned by this kind of bombardment.

All three of the foregoing types of atomic changes are semi-artificial in the sense that all that man does is to place substances in the way of projectiles which nature has herself provided, namely, alpha rays of naturally radioactive substances, and cosmic-ray photons. But men were at work in many laboratories trying to find ways of producing artificial projectiles of high enough energy to disrupt atomic nuclei. The first successful device of this kind was built at the California Institute of Technology by Professors C. C. Lauritsen and R. D. Bennett.[1] In 1928 they succeeded in building and continuously operating a giant X-ray tube in which streams of either electrons or positive ions could be accelerated up to potentials of 750,000 volts; and four years later, 1932, Professor Lauritsen was able to report to the Physical Society that he had been able to run a modification of this tube continuously at 1,200,000 volts. These Lauritsen tubes, which have now been introduced into the regular equipment of quite a number of hospitals, made it possible for the first time to produce wholly artificially rays of essentially the same quality as the gamma rays of radi-

[1] Lauritsen and R. D. Bennett, *Phys. Rev.*, XXXII (1928), 850; Lauritsen and Cassen, *ibid.*, XXXVI (1930), 988.

um, but in an intensity that could not be equaled by 100 grams (or at current prices seven million dollars' worth) of radium.

Mr. W. K. Kellogg of Battle Creek, Michigan, and Mrs. Seeley W. Mudd of Los Angeles at once provided funds for carrying out a five-year experimental program designed for the purpose of developing the technique of the proper therapeutic use of these powerful artificial gamma rays primarily in the treatment of cancer, and although Dr. Lauritsen and his associates have for four years been daily occupied with this humanitarian program—a program which prevents the use of this particular tube for the production of an artificial beam of atomic projectiles—they have yet found time to build a second similar tube for this specific purpose (Fig. 67) and in the summer of 1933[1] succeeded in producing with it a wholly artificial a ray beam of greatly increased intensity, which in its turn produced atomic transmutations of precisely the sort that Irene Curie and F. Joliot, as well as Chadwick, had previously produced with the a rays from radioactive substances (see § II below).

Simultaneously and quite independently Professor E. O. Lawrence[2] at Berkeley had been developing a most ingenious ion-accelerating device with which he and his colleagues have reached ion-bombarding energies as high as two million volts, though their currents are as yet very much smaller than Lauritsen's. Also Tuve[3] of the Carnegie Institution of Washington and Van de Graaff of the

[1] Crane, Lauritsen, and Soltan, *ibid.*, XLIV (1933), 514; also XLV (1934), 507.

[2] Lawrence, Livingston, and Lewis, *ibid.*, XLIV (1933), 35.

[3] M. A. Tuve, *Journal of the Franklin Institute*, CCXVI (July, 1933), 1.

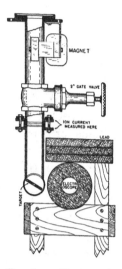

FIG. 67a.—Part-sectional view of the lower end of the tube, showing the lead shielding and the position of the electroscope with respect to the target.

A million-volt (root-mean-square value) cascade transformer is used as the source of high potential. The top of the upper tube is connected, through a suitable protective resistance, to the high potential end of the transformer set, and the midpoint between the two tubes is connected to the half-potential point of the transformer set. The apparatus therefore operates as two separate tubes, each giving the ions half their total acceleration.

The source of the ions is located in the end of the inner electrode of the upper tube. A metal ring in the shape of a doughnut, 3 in. (7.5 cm) in diameter and having a 1 in. (2.5 cm) hole is supported from the bottom plate of the upper tube, about 5 in. (12.5 cm) from the end of the electrode. This modifies the field in a way that is favorable to concentration of the ion beam. From here the ions pass down the hollow central electrode, and receive the second half of their acceleration in the gap of the lower tube. The target consists of a 2 in. (5 cm) brass disk mounted on a shaft, so that either side can be exposed to the ion beam. One side of the disk is covered with the material to be disintegrated, and the other side is covered with some material which gives no effect, such as brass or aluminum for the purpose of comparison.

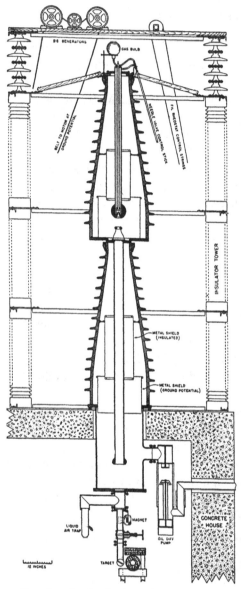

FIG. 67.—Sectional view of Lauritsen-Crane million-volt tube.

Massachusetts Institute of Technology had joined in the race of producing artificially very high-energy ionic projectiles. But before any of them had got their outfits into practical working order, Cockroft and Walton[1] of Cambridge, England, discovered that certain types of atomic disintegrations could be produced *wholly artificially* with the use of comparatively small accelerating voltages—those of the order of 100,000 volts. The present explanation of this important phenomenon is as follows.

Before the advent of the wave mechanics it had been thought quite impossible that a proton could force its way into the nucleus of an atom like lithium against the field created by the three positive charges on that nucleus unless it were endowed with an energy of more than a million electron-volts. According to the principles of the wave mechanics, however, there is a certain *probability* that it can "leak through" the potential barrier about the nucleus in quite the same way in which electrons leak through the potential barrier at the surface of the metal in Oppenheimer's wave mechanical interpretation (see p. 261) of the phenomena of cold emission. At any rate, the experimental fact now is that protons, or hydrogen ions, endowed with energies of as low as 30,000 electron-volts do sometimes get into the nucleus of the atom of lithium and produce the following transformation first discovered by Cockroft and Walton.

The hydrogen nucleus H^1 forces itself into the lithium nucleus Li^7 and the resulting nucleus then breaks up into two a particles which are ejected in opposite directions as they must be to satisfy the momentum equations.

[1] Cockroft and Walton, *Proc. Roy. Soc. A.*, CXXXVII (1932), 229.

Kirchner,[1] and Dee and Walton,[2] have made very beautiful cloud chamber photographs, samples of which are
shown in Figure 69, in which the bombarding ions come
in from above, as shown in Figure 68, hit a lithium target,
and the two oppositely directed α particles shoot out

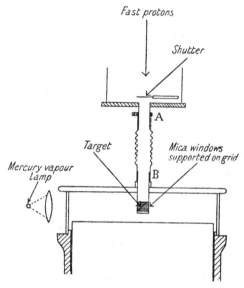

FIG. 68.—Diagram of arrangement for observing tracks of α particles
made by bombarding lithium by protons (after Dee and Walton).

through very thin windows in the end of the tube in which
the target is placed, and are thus seen shooting in opposite directions through the cloud chamber. The reaction
is

$$Li^7 + H^1 \rightarrow He^4 + He^4 .$$

[1] F. Kirchner, *Sitz. Ber. Bayerischen Akdwis. Sitz.*, März 4, 1933,
p. 129.

[2] Dee and Walton, *Proc. Roy. Soc.*, CXLI (1933), 733.

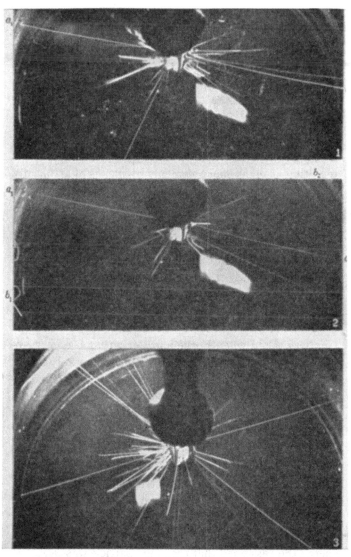

FIG. 69.—(1).—Disintegration of lithium by ions of the heavy isotope of hydrogen, showing the emission of two particles (a_1, a_2) in opposite directions passing out of the expansion chamber and with ranges therefore greater than 10 cm. ($Li^6 + H^2 \rightarrow 2He^4$). The thin long track ($b$) is probably a fast proton.

(2).—This photograph shows the type of disintegration described above, (1), and also a disintegration with the emission of a pair of opposite 8.4 cm. particles (b_1, b_2), probably due to the presence of protons in the positive ion beam.

(3).—This shows several cases of particles with range greater than 10 cm. lying in opposite directions. There are also particles with different ranges less than 8 cm. ending in the chamber. (Photographs by Dee and Walton.)

Here was then a case of the *completely* artificial transmutation of lithium into helium; and further, the synthesis of atoms was a part of the process, too, since the proton had to be incorporated into one of the resultant atoms of helium. Following Cockroft and Walton, Lewis Livingston and Lawrence obtained evidence for the disintegration of many more elements by means of artificially accelerated deutons (H^2) and Crane, Lauritsen, and Soltan first succeeded in producing by means of *artificially* accelerated ions ($_2He^4$ and $_1H^2$) transformations yielding "neutrons." But in order to follow further the history of artificial transmutation it is necessary to tell the story of how this new body, the neutron, comes into the picture.

II. THE DISCOVERY OF THE NEUTRON

The first step that led up to the discovery of the neutron was taken when Bothe and Becker[1] at Giesen, Germany, found that if certain light elements, notably beryllium, and in lesser degree boron and lithium, are exposed to the bombardment of the alpha rays of polonium they emit rays of several times the penetrating power of the most energetic gamma rays from known radioactive substances, namely those of ThC″ (2.62 million electron-volts). They used the responses of Geiger counters to detect these rays and to estimate the value of their penetrating power. They interpreted their experiments as bringing to light gamma rays of an energy of as much as 14 million electron-volts, and to account for so high an energy they assumed that the alpha particle was captured by the nucleus of the bombarded atom of beryllium and

[1] Bothe and Becker, *Zeit. f. Physik*, LXVI (1930), 289, and *Naturwiss.*, XIX (1931), 753.

that the energy released in the radiant form came from the known excess in the mass of $Be^9 + He^4$ over that of the atom (C^{13}) assumed to be formed by their union, i.e., they adopted the atom-building hypothesis and used the Einstein equation $E = mc^2$ (see p. 313) to supply the necessary energy.

The second important step was taken by Irene Curie and her husband, F. Joliot,[1] in Paris, who, in repeating these experiments used ionization chambers for measuring the radiation and found that when they interposed in the path of the rays to this chamber sheets of carbon, aluminum, copper, silver, and lead, the ionization currents in the electroscope were practically unaffected, but that when they interposed a sheet of paraffin, the current was doubled. Similar, though smaller, increases were observed when water, cellophane, or other substances *containing hydrogen* were used in place of the paraffin. Curie-Joliot then showed that this increased current was due to the fact that Bothe's rays in passing through hydrogen compounds knocked out of them protons (hydrogen nuclei). They measured the length of these hydrogen tracks in a cloud chamber and concluded that these protons had received energies as high as 4.5 million electron-volts. They also presented convincing evidence that Bothe's rays could impart kinetic energies to the nuclei of helium and carbon, as well as hydrogen, and thought they had shown by these experiments that electromagnetic rays of high frequency are capable of imparting high kinetic energies to the nuclei of hydrogen and other light atoms.

[1] Curie-Joliot, *Compt. Rend.*, CXCIV (1932) (January 18). 273; (February 22), 708.

But such a transfer of momentum and energy from a photon of the energy in question to a proton or other nucleus is impossible without a violation of the laws of momentum and energy that have been found to hold everywhere thus far, even in encounters between photons and electrons (Compton effect).

It was Chadwick,[1] of Cambridge, England, who resolved the difficulty by showing not only that the principle of the conservation of momentum could be retained, but that the momenta all came out consistently for the different kinds of nuclei to which both his own and Curie-Joliot's experiments showed that motion was imparted by the Bothe rays *if a non-ionizing particle of the mass of a proton were the bombarding agent.* In a word, he brought forward convincing evidence that when the alpha particles from polonium bombard atoms of beryllium, boron, etc., they knock out of the nuclei of these atoms particles of about the mass of the proton but *devoid of charge* and hence properly called *neutrons.* It is these neutrons then, that because of their lack of charge have a very high penetrating power, and that also because of their mass are able to transfer their energy and momentum to the hydrogen nuclei that Curie-Joliot found shooting through their cloud chamber (see Fig. 70) when they interposed hydrogen-bearing substances between it and the beryllium-polonium combination. Chadwick assumed that the capture of the alpha particle by the nucleus of the atom of beryllium must release "a neutron" through the following reaction, the atomic numbers of the elements be-

[1] J. Chadwick, *Nature,* CXXIX (1934), 312; *Proc. Royal Soc. A.,* CXXXVI (1932), 692; *ibid.,* CXLII (1933), 1.

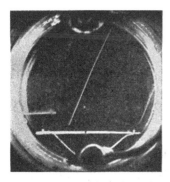

(a) A neutron entering the chamber from below traverses a plate of paraffin extending across the lower part and ejects from it a proton (H^1) which shoots clear across the chamber.

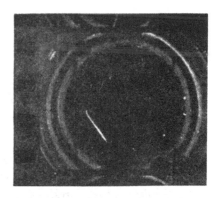

(b) The chamber is here filled with helium A neutron entering from below collides with a He nucleus four times its own mass and can only impart to it a range of about 5 mm

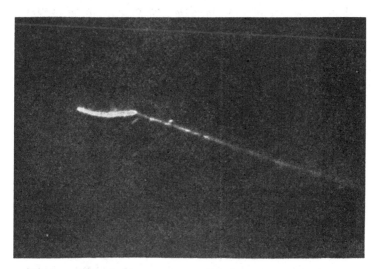

(c) The neutron here causes the transmutation of a nitrogen nucleus which it enters and by so doing occasions the ejection of an α particle while the heavy short trajectory reveals the recoil of the remainder of the nucleus. The pressure was here low and the magnification large.

FIG. 70.—Photographs of the effects of the collision of neutrons with the atoms of hydrogen (a), helium (b), and nitrogen (c) (after Curie-Joliot).

ing written as preceding subscripts, the atomic weights as superscripts:

$$_4\text{Be}^9 + _2\text{He}^4 \rightarrow _6\text{C}^{12} + _0\text{n}^1 \,.$$

Chadwick's first suggestion of the "Possible Existence of a Neutron" was published in a brief note to *Nature* on February 19, 1932. His elaboration of his views with the supporting evidence appeared in the June 1 issue of the *Proceedings of the Royal Society* three months before Anderson's announcement of the existence of the positive electron.

Chadwick advanced the view that the neutron consists of a proton and a negative electron in close combination. This view required no *fundamental* change in the usual conception of the nucleus as a number of protons equal to the atomic weight held together by a number of negative electrons equal to the atomic weight minus the atomic number. The new view merely went a bit farther in specifying the locations within the nucleus of these two familiar constituents. Thus on this view all the negative electrons within the nucleus simply became closely attached to protons, thus making as many neutrons as there were negative electrons, and the excess protons then provided the free positive charge on the nucleus, or, the atomic number.

This view required that the mass of the neutron be precisely that of the hydrogen atom, namely 1.00777 less the change in the binding energy of the negative electron to the proton that has taken place because of the transfer of the electron from the innermost hydrogen orbit to the extremely close association that it is assumed to have with the proton in the neutron. By applying the Einstein equation to the transformation involved in the capture of an α particle of known energy by the nucleus of an atom of boron, and the emission of a neutron of measured energy, Chadwick obtained an experimental determination of the mass of the neutron as 1.0067 and hence concluded that the binding energy of the proton and negative electron was 0.001 atomic weight units or about a million electron-volts.

Chadwick's reasoning in arriving at the foregoing important conclusion as to the mass of the neutron was as follows. He assumed that when alpha particles, i.e., nuclei of helium, bombard boron and in the process unite with the nucleus of the boron atom of atomic weight 11, atomic number 5 ($_5B^{11}$) they form thus an atom of nitrogen ($_7N^{14}$) and throw out the observed neutron ($_0n^1$) the equation of transformation being

$$_5B^{11} + _2He^4 \rightarrow _7N^{14} + _0n^1 .$$

The kinetic energy with which the neutron was ejected was obtained by letting these neutrons pass through paraffin and measuring the maximum range communicated to the protons that were found to emerge from the paraffin. This was found to be 16 cm. in air, a range that corresponds, according to a carefully worked out empirical curve, to a velocity of 2.5×10^9 cm. per second. But, since the mass of the neutron is practically the same as that of the proton ejected by it from the paraffin, in the most favorable case the whole momentum of the colliding body is transferred to the struck body so that the foregoing number is the maximum velocity of the neutron liberated from boron by an α particle of polonium, the velocity of which is known to be 1.59×10^9 cm. per second. Again assuming that the momentum is conserved in the inelastic collision of the a particle with the B^{11} to form N^{14} we can at once calculate the velocity communicated to the N^{14} nucleus and we then know the masses and the kinetic energies of all the particles concerned in the nitrogen-building process and can at once insert the numerical values of all the masses and all the kinetic energies involved in the Einstein equation, namely, mass of $B^{11}+$ mass of He^4+Kin. En. of $He^4 =$ mass of $N^{14}+$mass of n^1+Kin. En. of $N^{14}+$Kin. En. of n^1.

The masses, according to Aston, are $B^{11} = 11.00825\pm$ .0016, $He^4 = 4.00106\pm0.0006$, $N^{14} = 14.0042\pm0.0028$ and the above kinetic energies reduced to mass units[1] by $E = mc^2$ are for the a particle 0.00565; for the neutron 0.0035; and for the nitrogen nucleus 0.00061. From these figures the mass of the neutron comes out 1.0067. The errors quoted for the mass measurements are those given by Aston. Chadwick thinks that making adequate allowance for all the errors in the foregoing the mass of the neutron cannot be less than 1.003 and "that it probably lies between 1.005 and 1.008."

In order that the reader may have the data with the aid of which all such computations are made there is listed herewith the masses of thirteen of the lighter atoms as determined with much precision by Aston and Bainbridge in their mass spectrographs, the mass of the O^{16} atom being taken as exactly 16. The figures given represent the masses of the neutral atom with all the

[1] If we desire to express m in grams and E in electron-volts, then $E = \frac{mc^2}{e} = \frac{c^2}{e/m} = \frac{(2.9979\times10^{10})^2}{1.7570\times10^7\times10^7} = 512,000$ electron-volts and the mass of the hydrogen atom $= 1.0078\times1835\times512,000 = 946,000,000$ electron-volts. This gives .001 atomic weight units $= 939,000$ electron-volts.

extra-nuclear electrons in place. To obtain the masses of the nuclei alone one must subtract the product of the number of extra-nuclear electrons by 0.00055, which is the rest-mass of a single electron. The listed uncertainties are "probable errors," which are only about one-third of the maximum uncertainties quoted above from Aston.

H^1	$1.007775 \pm .000035$	C^{12}	$12.0036 \pm .0004$
H^2	$2.01363 \pm .00008$	N^{14}	$14.008 \pm .001$
He^4	$4.00216 \pm .00013$	O^{16}	16.0000 (standard)
Li^6	$6.0145 \pm .0003$	F^{19}	$19.000 \pm .002$
Li^7	$7.0146 \pm .0006$	Ne^{20}	$19.9967 \pm .0009$
Be^9	$9.0155 \pm .0006$	Ne^{22}	$21.99473 \pm .00088$
B^{10}	$10.0135 \pm .0005$	Cl^{35}	$34.9796 \pm .0012$
B^{11}	$11.0110 \pm .0005$	Cl^{37}	$36.9777 \pm .0019$

III. THE NATURE OF THE NEUTRON

The foregoing mass is about what is to be expected if the neutron consists of a closely bound proton and a negative electron the sum of the masses of which in the hydrogen atom is 1.0078, thus leaving a binding energy, or a mass defect, of about 0.001 or roughly a million electron-volts due to the much closer association, or approach, of the two charges in the neutron than in the hydrogen atom. Up to this point, then, everything fitted nicely into the old conception of a universe made up of the two fundamental primordial entities—negative electrons and protons, neutrons representing merely a closer union of the two than there had been any particular reason for postulating before Chadwick's discovery of the neutron.

All this simplicity vanished with Anderson's discovery of the existence of the free positive electron, which showed that *the positive unit charge could exist entirely detached from the mass of the proton.* This made it necessary to postulate the existence of at least three fundamental entities to account for all the phenomena that had before been taken care of by the assumption of the two fundamental entities, negative electrons and protons. The three now required were (1) positive electrons, negative electrons, and protons, or else (2) positive electrons, negative electrons, and neutrons. In this last conception the proton loses its rank as a fundamental entity, being now regarded merely the combination of a neutron and a positive electron, while the neutron rises to the rank of one of the independent fundamental building-stones of the universe. The whole conception of the electromagnetic origin of mass as such would then be abandoned. If one does not wish to give this up he can retain it by taking the first alternative and making the positive electron, the negative electron, and the proton the three fundamental entities. He may then assume further, if he so wishes, that the fundamental unit of positive electricity can exist both in the proton form and the free positive electron form and then, in order to get into consistency with the Einstein equation, he may also assume that under suitable conditions the proton can blow itself up into the free positive electron, for instance by expanding its radius about 2,000 times. By such an artifice one might be able to retain the conception of the electromagnetic origin of mass and with it Chadwick's conception of the neutron as a mere combination of a proton and a negative electron.

It is conceivable that experiment may lead to a choice between these two hypotheses. Thus, if the act of the transformation of a proton into a positive electron ever takes place, then according to Einstein's equation radiant energy of a total value of a billion volts must escape from the scene of the explosion; for from the energy standpoint this is practically the same as the annihilation of an atom of hydrogen. By analogy with the event of the mutual annihilation of a positive and negative electron with the emission in opposite directions of two half-million-volt photons this transformation of a proton into a positron might be expected to produce two half-billion-volt photons.

Events of this sort are apparently not taking place on the earth or in the atmospheres of the stars, since the cosmic rays of this order of energy are actually not found to come from these sources. As already indicated, however, such annihilation processes are needed somewhere in the universe in order to account for the observed existence and energies of the cosmic rays. Indeed, it is not impossible that 500-million-volt photons do actually constitute the strongest component of the cosmic rays, though this is somewhat higher than my estimate in the last chapter of the energy of the least-penetrating photon beam. However, if this should be proved to be the case it would merely be evidence in favor of the annihilation of hydrogen atoms throughout the heavens, and would not decisively differentiate between the relative claims of the neutron and proton as primordial entities, though the retention of the electromagnetic theory of the origin of mass is a point in favor of the proton.

There is another line of experimental approach that already has yielded a bit of evidence and bids fair to yield still more. Chadwick's determination of the mass of the neutron placed it, according to his estimate, between the limits 1.003 and 1.008. Now, we know that

the mass of the hydrogen atom in atomic weight units is
1.00777 ± .00002; that of the electron 1/1,835 as much or
0.00055 mass units. This makes the mass of the proton
alone quite definitely 1.00722, for the binding energy of
the electron to the proton in the hydrogen atom is negligible—only about 14 volts. If, then, the neutron is a fundamental thing to which a positive electron of mass
0.00055 becomes attached to form a proton, then its mass,
M, must be given by $M + 0.00055 -$ binding energy $=$
1.00722 or

$$M = 1.0067 + \text{binding energy.}$$

This equation says, then, that if the neutron is a primordial thing and the proton a complex thing, the neutron
mass cannot possibly be lower than 1.0067, but it may be
as much larger than 1.0067 as you please, the difference,
whatever it is, determining the binding energy, which the
apparent stability of the neutron indicates should be
high. Any experimental neutron mass lower than, or
even close to, 1.0067 speaks in favor of the *proton* as a
primordial thing.

On the other hand, if the neutron is a proton and a
negative electron in close association, then the mass of
this neutron system can be anything you please *less* than
$(1.00722 + .00055) = 1.0078$, the difference, whatever it is,
in this case representing the binding energy. In other
words, for this case

$$M = 1.0078 - \text{binding energy.}$$

Any experimental neutron-mass higher than, or even
close to, 1.0078 speaks in favor of the *neutron* as a primordial thing. If Chadwick's actual measurement of the

mass of the neutron (viz., 1.0067) were accurate to one part in a thousand, then on the hypothesis of a primordial neutron there would be nothing left to serve as the binding energy of the positive electron to the neutron, whereas if the neutron is a closely coupled proton and a negative electron then, as Chadwick pointed out above, there would be about a million volts left to serve as the binding energy of the negative electron to the proton. This looks like evidence in favor of the proton as a fundamental entity.

But this evidence must not be taken too seriously for the reason that Chadwick's accuracy is not at all adequate for the definite drawing of the foregoing conclusion, as the following work of Lauritsen and Crane makes clear.

IV. LAURITSEN AND CRANE'S WORK ON THE GAMMA RAYS EMITTED IN THE PROCESS OF ARTIFICIAL TRANSMUTATION

In setting up the foregoing equation Chadwick tacitly assumed that the capture of an alpha ray by boron and the ejection of a neutron in accordance with the foregoing reaction and computation was an operation in which no energy was lost through the emission of a gamma ray. But in the case of the bombardment both of beryllium and boron by α rays Bothe and Becker[1] and Rasetti[2] as well as Curie-Joliot[3] had proved that weak gamma rays, as well as neutrons, are actually given off. To measure the energy of these very weak but very penetrating gamma rays Bothe and Becker adopted the following most ingenious technique. They set up two Geiger counters so that the

[1] Bothe and Becker, *Zeit. f. Physik*, LXXVI (1932), 421.

[2] Rasetti, *ibid.*, LXXVIII (1932), 165.

[3] Curie-Joliot, *Jour. d. Phys. et le Radium*, IV (1933), 21.

electrons ejected from matter placed in the path of these gamma rays could produce audible responses at an observed rate by their successive passage through both counters. Then they interposed increasing thicknesses of aluminum, or other substances, between the two counters until the responses ceased—a condition reached when the ejected electrons had just insufficient energy to traverse the intervening sheets of matter. From this thickness they estimated that the gamma rays produced when boron is bombarded by α rays have an energy of about 3 million electron-volts, while when beryllium replaces the boron the gamma rays now produced have an energy of about 5 million electron-volts. *These were the highest gamma-ray energies yet obtained from terrestrial sources.*

If radiations of anything like such energies were involved in the reaction assumed by Chadwick, his estimate of the mass of the neutron needed notable reduction, so that the figure 1.0067, even if derived from very precise measurement of masses and energies, was rather an upper limit than an exact determination, at least until it became definitely established that no radiative losses had been involved in the assumed reaction. However, Chadwick's estimated range of uncertainty, viz., 1.003 to 1.008, was sufficient to take care of the escape even of a 5-million-volt gamma ray. It was the great intensity of the stream of bombarding particles that Lauritsen and Crane[1] were able to obtain in their wholly artificial production of neutron beams— intensities a thousand times those used by any of their predecessors—that made possible the bringing to light and quantitative study of the accompanying gam-

[1] Crane, Lauritsen, and Soltan, *Phys. Rev.*, XLIV (1933), 514, 783.

ma radiations in the case of most reactions giving rise to neutrons.

Their method of separating the gamma rays from the neutron is illustrated in Figure 71,[1] in which the curves all apply to the radiations produced when deutons, i.e., heavy

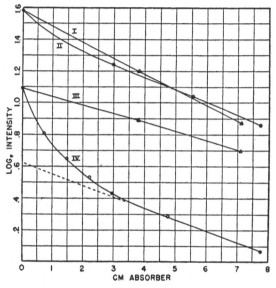

FIG. 71.—Absorption of the beryllium radiation. I. Paraffin-lined chamber, paraffin absorber; II. Paraffin-lined chamber, lead absorber; III. Lead-lined chamber, paraffin absorber; IV. Lead-lined chamber, lead absorber.

hydrogen atoms, are made to bombard a beryllium target in the Lauritsen million-volt tube. The absorption in both lead and paraffin of the radiations stimulated in the beryllium by the deuton bombardment is measured with the aid of two electroscopes, the first lined with lead, the

[1] Crane and Lauritsen, *ibid.*, XLV (1934), 226.

second with paraffin. The paraffin-lined chamber is much more sensitive to neutrons than the lead-lined chamber, since the ionization within it is increased by the protons ejected from its walls by the neutrons. On the other hand the lead-lined chamber is very much more sensitive to gamma rays than is the paraffin-lined one, since gamma rays eject many more electrons from lead than from paraffin. The figure shows four absorption curves obtained by using the two kinds of absorbers and the two chambers. Curves II and IV are taken in the paraffin and lead lined chambers, respectively, but the beam was successively weakened in both cases by introducing before the chambers a series of *lead* absorbers. It will be seen that at thicknesses greater than 4 cm. the slope is the same for the lead and the paraffin chambers. This means that the radiation is of a single type, either entirely neutrons or entirely gamma rays, since but one absorption coefficient is involved. But the large excess in total intensity of Curve II (paraffin chamber) over Curve IV (lead chamber) shows clearly that the radiation responsible for these curves consists of *neutrons*, since passing through the paraffin walls increases its intensity so largely. Hence, *the slope of Curves II and IV beyond 4 cm. is taken as the absorption coefficient of neutrons in lead.*

At the thicknesses of the absorber less than 4 cm., curve IV shows a steep rise which clearly indicates the presence of a component of radiation that is more absorbable than is the neutron radiation and that is practically entirely absorbed in 4 cm. of lead, as shown in both Curves II and IV. That this component consists of gamma rays is indicated by the fact that it occurs strongly in Curve IV (lead-lined chamber) but only weakly in Curve II (paraffin-lined chamber), for it is well known that gamma rays stimulate secondaries much more strongly in lead than in paraffin. Curves I and III show that the paraffin absorbers cut down but slowly the joint effects of the two types of rays, while the difference in the slopes of these two lines shows that the proportions of the different types of rays are not the same in the lead-lined chamber as in the paraffin-lined one, a result obviously to be expected.

By extending the straight part of Curve IV backward as indicated by the dotted line in Figure 71 we can determine the intensity contributed by neutrons alone for absorbers less than 4 cm. Then the difference between the total intensity and the intensity represented by the dotted line should be just the intensity due to the gamma rays alone. The intensity of gamma rays thus obtained, as a function of thickness of absorber, is plotted on a log scale in Figure 72, along with a similar curve taken with the gamma rays from radium under identical conditions. This shows that the gamma rays from beryllium are quite monochromatic with an absorption coefficient considerably greater in lead than the rays of radium after the latter have been filtered through 1-2 cm. of lead. Since the strength of the radium was accurately known, the number of gamma rays emitted from beryllium could be calculated and it came out equal to the number of neutrons to well within the accuracy with which the latter could be determined. Lauritsen and Crane therefore drew the important conclusion that in beryllium the neutrons and gamma rays are produced in the same reaction, which they wrote down as follows:

$$_4Be^9 + _1H^2 \rightarrow _5B^{10} + _0n^1 + \gamma .$$

The foregoing illustrates the extreme importance of knowing in the case of every reaction under consideration

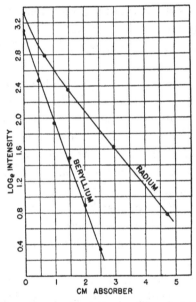

FIG. 72.—Absorption curve of the beryllium-deuton gamma rays, with an absorption curve of radium gamma rays, made under the same experimental conditions, for comparison.

whether or not there is a gamma ray released in the reaction, and if so just how great is its energy. Lauritsen and Crane have supplied this knowledge for a very considerable number of transmutation reactions, and have thereby notably pushed forward our insight into nuclear effects.

Perhaps the most interesting reaction that they have studied with great care is that produced by the bombardment of lithium by high-speed protons, in which, as already indicated (see p. 365) Cockroft and Walton had first shown that the proton is simply captured by the lithium isotope of mass 7 and the resulting nucleus of mass 8 then breaks up into two alpha particles that are ejected in nearly opposite directions. The energy released in this reaction, easily computed from the masses of Li^7, H^1 and He^4, given on page 375, comes out quite accurately 17.0 million electron-volts. If the whole of this energy plus the 200,000 electron-volts with which the protons were accelerated in the Cavendish Laboratory experiments went into the pair of oppositely flying a particles (Fig. 69) each would have had an energy of $\frac{17.2}{2} = 8.6$ million electron-volts, and when this is translated into a particle-range with the aid of the empirical curve relating a ray range and energy, now accurately known from twenty years of experimenting on the ranges and the magnetic deflectabilities of a rays emitted by the uranium and thorium families, the result comes out 8.2 cm.

Now Oliphant, Kinsey, and Rutherford,[1] by interposing in the path of these, same a rays, emitted by lithium bombarded by protons, successive layers of mica each of accurately known air-equivalence, have obtained the relation shown in Figure 73 between the air-equivalent range and the number of a particles having any particular range.

Their method of taking the data for such a curve is as follows: They let the alpha rays shoot into an ionization

[1] *Proc. Roy. Soc.*, CXLI (1933), 722.

chamber a centimeter or two in diameter and a few mm. deep. Each α ray entering this chamber produces a practically instantaneous ion current which is amplified by an ordinary tube amplifier, and the resulting "oscillograph deflection" photographed on a movie film so that the number of kicks appearing on the film is the same as the

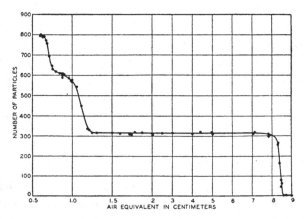

FIG. 73.—Distribution-in-range curve of the α particles resulting from bombardment of lithium by protons. (Oliphant, Kinsey, and Rutherford.)

number of α particles that produce them, as shown in Figure 74. By interposing between the source and this chamber a sheet of mica of known air-equivalence the number of particles having energies at least sufficient to get through that thickness will be recorded. Figure 73 shows the number of particles that get through varying thicknesses of mica, these numbers being plotted as ordinates, the air equivalent thicknesses being plotted as abscissae. The broad plateau of Figure 73 means, then, that all the points taken on this plateau correspond to the

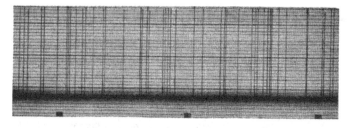

(a)

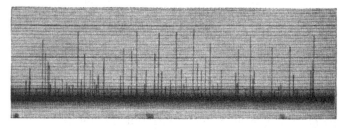

(b)

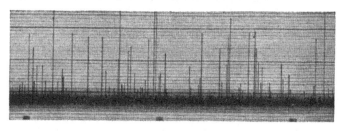

(c)

Fig. 74.—Three records of the ionization produced by individual particles in a shallow ionization-chamber: (a) alpha-particles of nearly the same speed, (b) protons of various speeds, (c) particles of several kinds which had been set into motion by impacts of neutrons. The fogging along the base-lines is much fainter in the original records than in these reproductions. (J. R. Dunning of Columbia University.) (Taken from an article by Carl Darrow.)

large group of tracks all of which have a range in air of 8.4 cm., which is seen to be the *maximum* distance to which the tracks reach when the mica sheets are all withdrawn.

But it will have been noticed that the range here accurately observed is very close to that computed above with the aid of the assumption that the whole of the energy released in the aforementioned transmutation of lithium into helium went into the kinetic energy of the flying helium nuclei or α particles. This excellent quantitative agreement is convincing evidence of the correctness of the assumptions of Cockroft and Walton and the whole Cavendish group as to the nature of the transmutation-reaction which produced these α rays of 8.4 cm. range in air. *In this particular reaction there is certainly no energy whatever left to go off as a gamma ray,* the whole of the librated energy being used to impart velocity to the two α particles.

How, then, did Lauritsen and Crane find strong gamma rays given off by the bombardment of lithium by protons? In order to find the answer to this riddle we must first know how they tested for the presence of gamma rays and what sort of energies they found. From electroscope tests of the kind already described in the case of beryllium rays they got an absorption coefficient in lead close to that found for γ rays from radium filtered through 2 cm. of lead, and hence at first concluded that the energy involved in these two radiations, one from bombarded lithium the other from radium, was about the same, namely, about 1.6 million electron-volts.

But having begun to suspect the reliability of these absorption formulas for the determination of energies, at least in this region of frequencies, with the assistance of

Messrs. Fowler and Delsasso, they let these same gamma rays from bombarded lithium pass into a cloud chamber set in the midst of a 1,200 gauss magnetic field and measured directly the energies of the electrons shot into the cloud chamber by the absorption of the gamma rays in thin sheets of lead and other metals also placed inside the cloud chamber. This was precisely what Anderson and Neddermeyer had first done when they found the energy distribution of positives and negatives released from lead and aluminum by the gamma rays of ThC″—experiments, it will be remembered, in which they discovered that the maximum energy of the negatives was very close to the 2.6 million electron-volts which is the long-known energy of the incident gamma rays, while the maximum energy of the positives was 1.6 million volts, the difference representing, according to the Dirac theory, the million volts necessary to create a positron-negatron pair.

A large number (about 1,000) of cloud-chamber measurements of this kind, made by Lauritsen, Crane, Fowler and Delsasso, on the electrons released by the gamma rays from lithium bombarded by protons yielded a positron energy distribution curve that ran into the energy axis at about 12 million volts (see Fig. 75), while the distribution curve for the negatives intercepted the axis at 13 million electron-volts. In other words, the energy of this gamma-ray beam produced by the bombardment of lithium by protons was not about 1.8 million volts, as previously estimated from the absorption measurements in lead, but *it was as here directly determined about 13 million volts and it was the highest energy gamma rays as yet found from terrestrial sources.* Not only was this high-energy gamma ray of itself a discovery of im-

portance, but even more so was the discovery of so high
an energy combined with so high an absorption coefficient
in lead (see § V, p. 402).

Before attempting to interpret the latter, let it be
noted that both the curve corresponding to the positive

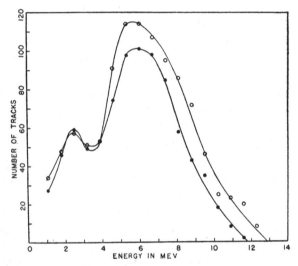

FIG. 75.—Energy spectra of the negative and positive electrons ejected
from a thick lead plate by the gamma radiation from lithium bombarded
with protons. Circles indicate negative electrons and dots indicate positive electrons. Each point represents the number of electron tracks in a
1.4 million electron-volt energy interval.

electrons and that corresponding to negative electrons
show a subsidiary maximum at about 4 million volts, thus
indicating that there are two gamma rays given out in this
reaction, one having an energy of about 4 million electron-volts and the other of about 13 million. *In other
words, as much as 17 million electron-volts of energy sometimes go off from this reaction in the form of gamma rays.*

But this is practically all the energy that was available, for though Lauritsen and Crane's bombarding protons had 900,000 volts behind them we have here found about 17 million out of 17.9 million in gamma rays alone. A comparison, then, of Figure 73 with Lauritsen and Crane's curves, Figure 75, seems to show that the event of the capture of a proton by Li^7 may take place in such a way that all of the 17 million electron-volts of released energy may appear in the kinetic energies with which the two normal alpha particles fly apart, or on the other hand, it may take place in such a way that one of the newly born alpha articles may be thrown at birth into an excited state, of energy-value about 17 million volts, from which excited state it falls back to normal by two stages, of energy values 4 and 13 volts, respectively. This process of course requires that the α particles fly apart in this particular instance with but the small amount of the total available energy (17.9 million volts) left over after about 17 million of it has gone off in the form of gamma rays.

This begins to show the complexity of the events that can happen in a nuclear transformation. Thus the proton can get into the nucleus of lithium in such a way as to leave in their normal state the two α particles formed, in which case these two particles must take the whole energy released; they then have the 8.4 cm. range. Or the entering proton may do its job unskilfully and get one of the α particle twins badly excited by pushing it up to a 17-million-volt level, in which case it falls back to the unexcited state in two jumps, stopping at the 13-million-volt level and letting out a 4-million-volt gamma ray, then falling back to normal with the emission of a 13-million-volt gamma ray. This particular transmutation-reaction, too, is probably one of the very simplest possible. In some other reactions protons of varying speeds are released, in some neutrons, in some gamma rays or photons, in some positrons, and in some negatrons.

Especially because of its relation to the very important subject of the mass of the neutron, Lauritsen and Crane also analyzed very carefully with the great intensities obtainable with their apparatus the mass-energy relations in the reaction produced when lithium is bombarded by "deutons" instead of as above by protons. This they had before proved to give rise to neutrons through the reaction

$$_3L^7 + _1H^2 \rightarrow _2 _2He^4 + n^1 .$$

Although there was some justification for assuming that gamma rays are not a product of this reaction an experimental test of this point had not been available so that calculations of the mass of the neutron made with the aid of this equation and without certain knowledge as to the presence of gamma rays were to be considered strictly valid only as an upper limit.

By analyzing the rays produced in this reaction by exactly the method indicated above for Be, Lauritsen and Crane[1] showed that no gamma rays at all come out of this reaction, but only α rays and neutron rays. The distribution of the ranges of the α rays had already been very beautifully analyzed by Oliphant, Kinsey, and Rutherford[2] with the results shown in Figure 76. It will be seen from this figure that out of this reaction there comes a

[1] Lauritsen and Crane, *Phys. Rev.*, XLV (1934), 550.

[2] Oliphant, Kinsey, and Rutherford, *Proc. Roy. Soc. A.*, CXLI (1933), 722.

large group of α particles which have a uniform range of
13.2 cm.—a range that corresponds to an energy of 11.5
million electron-volts.

Now this 13.2 cm. range actually corresponds remarkably closely with the total energy released in the reaction

$$_3Li^6 + _1H^2 \rightarrow 2_2He^4 + E$$

in which E is the total kinetic energy of the α particles.
Indeed, the mere differences in the masses of the atoms on
the left and right sides of this equation amounts to 22.2×10^6 electron-volts and if we add the 200,000 electron-volts
for the bombarding energy of the deutons we see that the
energy of each of the two ejected α particles should be
$\frac{22.4}{2} = 11.2 \times 10^6$ electron-volts while it was seen that the
range, 13.2, reduced to electron-volts is 11.5.   This excellent agreement is then fairly conclusive evidence, even
without Lauritsen and Crane's proof that gamma rays are
not produced in this reaction, that the deuton is here
captured by Li^6 in such a way that the whole of the released energy is transformed into the kinetic energy of the
two oppositely flying α particles.   It is agreements like
this that gives us confidence in the correctness of these
transmutation equations.

But what is then the explanation of the sloping line of
Figure 76 which shows clearly that this bombardment
of lithium by deutons also gives rise to α particles the
ranges of which vary continuously from 1 cm. up to a maximum of 7.8 cm?   If no gamma rays are produced there is
nothing left save the observed neutron rays to divide the
energy with the α particles so as to cause the latter to have

all sorts of ranges up to 7.8, which is taken as the range
when all the energy goes into the α particles and none into

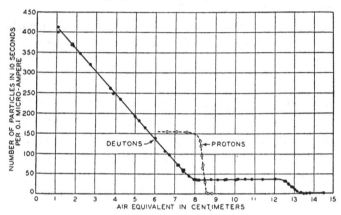

FIG. 76.—Distribution-in-range curve of the α particles resulting from
bombardment of lithium by deutons. (Oliphant, Kinsey, and Rutherford.)

the neutron. But from the masses involved it is at once
obvious that neutron rays can only come from the reaction

$$_3Li^7 + _1H^2 \rightarrow 2He^4 + _1n^1 + E_1 ,$$

in which E_1 is the total kinetic energy carried away by
both the α particles and the neutron. After combining
this with the other reaction, discussed above, in which
lithium is bombarded by protons, viz.,

$$_3L^7 + _1H^1 \rightarrow 2_2He^4 + E_2$$

Lauritsen and Crane point out that one obtains by sub-
traction

$$n^1 = H^2 - H^1 + E_2 - E_1$$

in which the only atomic masses involved are H^1 and H^2
which are known with considerable precision. The kinet-

ic energy of the bombarding protons (H^1) and deutons (H^2) as used by Oliphant, Kinsey, and Rutherford was the same (about 0.2×10^6 electron-volts) and hence cancels. The ranges of the α particles resulting from the two reactions are nearly the same (7.8 cm. and 8.4 cm., respectively) and since only the difference in energy is made use of any systematic error in the measurements, or in the conversion from range to energy, tends to cancel out. Using $E_1 = 2 \times 8.3 \times 10^6$ and $E_2 = 2 \times 8.75 \times 10^4$ electron-volts the difference is very closely equivalent to 0.001 atomic-weight units, and adding this to $H^2 - H^1 = 1.00586$ *we obtain for the mass of the neutron 1.0068.* On account of the accuracy with which ($H^2 - H^1$) is known *this value seemed to be the most accurate yet obtained.* Oliphant, Kinsey, and Rutherford estimate the errors in the ranges 7.8 cm. and 8.4 cm. as ± 0.2 cm., but even if two such errors should chance to add their effects the result could scarcely be affected by more than .0003 of an atomic-weight unit, so that the value of the neutron appeared to be found to lie between 1.0065 and 1.0071. Had this work yielded a value definitely lower than 1.0067 it would have spoken strongly against the neutron and in favor of the proton as the elementary particle. As it is, both possibilities are still open, though if the choice be in favor of the neutron it must be admitted that the binding energy between it and the positive electron is somehow strangely small.

In August, 1934, Chadwick[1] published a preliminary report on a closely related method. He exposed a chamber filled with deuterium, heavy hydrogen gas, to the 2.62×10^6-volt gamma radiation of ThC''. The chamber was connected to a linear amplifier and oscillograph in the

[1] *Nature*, CXXXIV (August 18, 1934), 237.

usual way. A number of kicks were recorded by the oscillograph which had to be attributed to protons resulting
from the splitting up of the deutons into protons and
neutrons. When radium rays of 1.8×10^6 volts were tried
the kicks were greatly reduced and Chadwick concludes
that the binding energy of a proton and a neutron to form
a deuton must be more than 1.8×10^6 volts. He estimates it at 2.1×10^6. The atom of heavy hydrogen is simply an atom of ordinary hydrogen after a neutron has
been added to its nucleus. Hence

$$H^1 + n^1 - \text{binding energy} = H^2,$$

or $\quad n^1 = H^2 - H^1 + \text{binding energy} = 1.0058 + \dfrac{2.1 \times 10^6}{946 \times 10^6} =$

$\pm 1.0080.$

If these figures should turn out to be correct the mass of
the neutron would be above the limit (1.0078) that is consistent with the idea that the proton is an elementary particle. The two last modes of approach to the mass of the
neutron did not seem as yet brought into consistency, and
whether the neutron or the proton was to be regarded as
the elementary particle waited future determination.

At the meeting of the International Union of Pure and
Applied Physics on October 24, 1934, Dr. Oliphant pointed out that Chadwick and Goldhaber had just made more
accurately by Chadwick's latest method the determination of the mass of the neutron and obtained the value
1.0080 ± 0.0005. Oliphant preferred, also, to make the
assumption discussed at length in Lauritsen and Crane's
paper and discarded by them as having too small a probability to be the determining factor in fixing the 7.8 maximum range. This assumption is that this range corre-

sponds not to the case in which the neutron takes none of
the energy and the two α particles divide it equally between them, but rather to the case in which the neutron
and one α particle receive together the same momentum
in one direction that the other α particle receives in the
opposite direction. This makes the latter alpha particle
receive five-ninths of the total energy released, and thus
gives the value of the total energy corresponding to the
7.8 cm. range 14.9 million electron-volts, and brings the
mass of the neutrons obtained by Lauritsen and Crane's
method up to 1.0083, in good agreement with Chadwick
and Goldhaber's[1] value. It is obvious that the uncertainty in this assumption still makes it pertinent to retain
the spirit of caution of the foregoing paragraph.

V. THE LAWS OF ABSORPTION OF HIGH-ENERGY PHOTONS

The discovery, first made independently and in wholly
different ways by Chao,[2] and by Bothe and Becker,[3] then
greatly extended as shown above by Lauritsen and Crane,
that artificially produced nuclear changes are capable of
emitting new types of gamma-ray photons, some of them
of higher energies than any heretofore originating on
earth, has opened up a large new field of both experimental and theoretical advance in our knowledge of the interaction between radiation and matter.

Prior to the work of Chao the theoretical physicist had
developed his laws of the absorption, scattering, and

[1] Chadwick and Goldhaber, *Nature*, CXXXIV (1934), 237.

[2] Chao, *Proc. Nat. Acad.*, XVI (1930), 431; *Phys. Rev.*, XXXVI (1930), 1519.

[3] Bothe and Becker, *Zeit. f. Physik*, LXVI (1930), 289; *Naturwiss.*, XIX (1931), 753.

energy degradation of photons on the assumption that the sole agents conditioning these processes were extra-nuclear electrons. Up to the time of the experiments of Chao, the Klein-Nishina formula, the most theoretically sound of our absorption formulas, had been assumed by all of us, Millikan, Jeans, Regener, and all the rest, though with varying degrees of assurance, to be applicable to all frequencies, even those of the most penetrating of the cosmic rays. Chao first showed sharply that the nucleus had its own laws of absorption and that they were wholly different from those described by the Klein-Nishina formula. Gray and Tarrant, Meitner and Phillip (see p. 337) confirmed with much elegance and precision these findings of Chao with respect to the nuclear absorption and re-emission in a new form of the energies of ThC″ photons, then Anderson discovered the positron and proved conclusively that the encounters of ThC″ photons with the nuclei of both lead and aluminum produce such positrons. Meanwhile Blackett and Occhialini had suggested the interpretation of cosmic-ray showers in terms of the creation of positron-negatron pairs, Dirac's negative energy states being the basis of this conception, and immediately thereafter Anderson and Neddermeyer found the maximum energy imparted to positive electrons by ThC″ rays just a million volts less than that imparted to extra-nuclear negatives, thus bringing to light the first new fact predicted by the pair-creation theory.

There followed the very elegant proof first by Thibaud and Joliot independently, then with much precision by Lauritsen and Crane, that Chao's half-million-volt isotropic gamma radiation excited by the impact of ThC″ photons on the nuclei of both heavy and light atoms is in all probability nothing but the ether signals sent out from the spot at which a positive electron disappears, presumably by committing suicide with a negative. However, this "annihilation ray" theory must be regarded as resting on evidence that is quite independent of all hypotheses as to where the observed positrons come from, whether from pair creation without any alteration in the charge on a nucleus, or by the extraction from a nucleus with a consequent diminution by one unit of its positive charge. One or the other or both of these origins may exist and both are entirely consistent with the annihilation-ray hypothesis. For in any case we know that these positives appear in our cloud chambers when photons of sufficient energy encounter nuclei, and we know that they disappear somehow, and also that when they disappear half-million-volt photons emerge from the scene of the disaster. We know, too, simply from the Einstein equation, that the radiation equivalent of the mass of two electrons is a million volts—more accurately $2 \times 512,000$ (see p. 374)—and also that the principle of conservation of momentum requires that this million volts emerge from the point of origin as two oppositely directed half-million-volt photons.

But meanwhile the theorist had been at work. In particular at the Norman Bridge Laboratory Oppenheimer and Plesset[1] had been working out from the pair-creation theory (1) how nuclear absorption should vary for a given nucleus as the energy of the approaching photon increased from the two-million-volt limit where Chao had found it first beginning to set in, and (2) how it should depend upon the atomic number of the nucleus.

[1] Oppenheimer and Plesset, *Phys. Rev.*, XLIV (1933), 53.

Fortunately Lauritsen and Crane's intense sources of artificial gamma rays made it possible to obtain a comparison between experiment and theory in the energy interval included between 2 million and 13 million electron-volts, and in the atomic-number interval copper (29) and

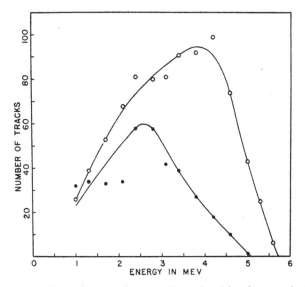

FIG. 77.—Energy spectra of the negative and positive electrons ejected from a thick lead plate by the gamma radiation from fluorine bombarded with protons. Circles indicate negative electrons and dots indicate positive electrons. Each point represents the number of electron tracks in a 0.7 million electron-volt energy interval.

lead (82). By precisely the method used in obtaining Figure 75 they measured the energy distribution curves for both the positive and negative electrons ejected into a cloud chamber from a 3 mm. lead plate by the gamma radiation from CaF$_2$ when it was bombarded by 800,000 volt protons. The results are shown in Figure 77 which it is most interesting to compare with Figure 75. Both figures show nicely the million-volt difference between the maximum energy of ejection of the negatives and the positives discovered by Anderson and Neddermeyer for ThC'' rays and accurately checked by Chadwick, Blackett, and Occhialini,[1] this being, according to the pair theory, the energy lost by the incident photon in the creation of an electron-pair.

But the comparison of Figures 75 and 77 shows something more, namely, that the difference between the number of negatives and the number of positives is very much greater with the 5.4-million-volt fluorine rays than with the 13-million-volt lithium rays. This of course means merely that, as is to be expected, the nuclear absorption is a much smaller part of the whole absorption in the case of the 5.4 rays, for, as Chao showed, the threshold at which nuclear effects begin to appear at all is at about 2 million electron-volts. On the assumption that the nuclear absorption shown in these experiments is due solely to pair-formation, the difference between the positive and negative curves should give the effect of Compton encounters with extra-nuclear electrons, in other words the part of the absorption that is taken care of by the Klein-Nishina formula. This is seen to be much less at 13 million volts than at 5.4 million—*indeed the difference is about as predicted by that formula.* This is then another success for the pair theory.

A still further success is seen in Figure 78, in which is shown Professor Oppenheimer's computation of the pair absorption in lead and in copper as a function of quantum

[1] Chadwick, Blackett, and Occhialini, *Proc. Roy. Soc. A.*, CXLIV (1934), 253.

energy. The curves of total absorption represent merely Klein-Nishina absorption added to this theoretical pair absorption as computed by Oppenheimer. The difference between the lead and copper curves is due to the fact that according to the pair theory the absorption is proportional to the square of the atomic number. The result is that after pair absorption begins, in the case of a heavy atom like lead, the absorption actually *rises* with increas-

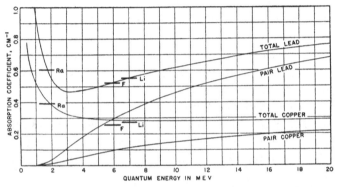

FIG. 78.—Curves giving the total absorption coefficients of gamma rays in lead and in copper as a function of their quantum energy, and also curves giving the absorption due to the formation of electron pairs alone.

ing energy so that there are two values of the energy such, for example, as 2 million volts and 7 million volts, or 1.6 million and 13 million (see p. 388) that have the same absorption coefficient. In confirmation of this double-valued coefficient both MacMillan at Berkeley and Lauritsen and Crane at Pasadena have made absorption measurements with different kinds of absorbers on the gamma rays from fluorine bombarded with protons, and the latter observers have done the same with the gamma rays from lithium bombarded with protons with the results shown in Figure 78. The observed absorption coefficients are indicated by horizontal lines intersecting the theoretical curves. Although the absorption coefficients *in lead* of these gamma rays are seen to be not very different from that of radium gamma rays it is clear from the corresponding coefficients *in copper* that their quantum energy is such as to place them on the high-energy side of the minimum, namely, at about 6.9 million electron-volts for lithium and 5.8 for fluorine, rather than on the low-energy side as the authors originally inferred. This result for flourine is completely and beautifully confirmed by the direct measurements, shown in Figure 77. In lithium the 6.9 value obtained from absorption measurements is a reasonable mean between the 13-million- and the 4-million-volt components shown in Figure 75, for Lauritsen and Crane[1] show that the 4-million component is quite as intense as the 13-million-volt one.

All this is in notable confirmation of the demands of the pair theory for this small range of frequencies herein investigated. That, however, a Z^2 absorption law cannot hold for higher energies seems to be definitely established by our cosmic-ray measurements, as well as by those of Steinke and his collaborators at Königsberg. These measurements show that when different absorbers are used for the same cosmic-ray photons the absorption is roughly proportional to the first power of the atomic number rather than to the second power. The whole mechanism of nuclear absorption may therefore change as these higher energies are reached. This is an interesting and important field in which both our experimental and our theoretical knowledge is still very incomplete but in which new data are rapidly appearing, so that the question here raised bids fair to find an early solution.

[1] *Proc. International Union of Physics.* London, Roy. Soc., 1934.

CHAPTER XVI

THE NATURE OF THE COSMIC RAYS

Up to the present no assumption whatever has been made as to the nature of the cosmic rays. The energy considerations presented in chapter xii are equally applicable to all theories as to their nature. Their very high penetrating power, however, suggests at once their kinship with the gamma rays instead of with the beta rays since at the highest *measured* energies (2.6×10^6 electon-volts) the former have, for a given energy, a penetrating power of the order of a hundred times that of the latter. Something like this ratio seems to hold also in the cosmic-ray field, for it will presently be shown quite definitely that it takes an energy of about 6 billion (10^9) electron-volts to enable an electron to penetrate the atmosphere. To find, then, primary incoming charged particles which have penetrated down to depths beneath the surface of a lake equivalent to a thickness of from 10 to 30 atmospheres, where our experiments and Regener's revealed traces of the rays, is well-nigh unthinkable. From the start the photon hypothesis seemed required to get such enormous penetration without calling upon impossibly large energies, namely, up to 30×6 or 180 billion electron-volts. But with a penetrating power of the order of a hundred times that of electrons, the two- or three-billion-volt photons, such as must have produced the showers of

Figures 63 and 64, would be able to reach down as far as could 200-billion-volt electrons, i.e., down to the depths at which we actually find traces of these rays. The great penetrating power, then, of the cosmic rays seemed at once to ally them to the gamma rays.

But there is another and a still more important parallelism between cosmic rays and gamma rays which was brought to light by the work of Hoffmann,[1] Millikan[2] and Cameron, and Compton, Bennett, and Stearns[3] between 1927 and 1931. The observation as made by Hoffmann and Millikan and Cameron in 1927 and 1928 was that when the pressure in a given electroscope was raised from 1 atmosphere of air to 30 atmospheres, although the ionization current was thus multiplied but some 14 times instead of 30 times, this multiplying factor of 14 *was the same for the gamma rays of radium as for the cosmic rays.* Bowen and Millikan[2] then proved that the reason the multiplying factor was but 14 instead of 30 was that with the high pressure 16/30 of the ions formed had recombined before the electric field could separate them and carry them to the electrodes; and the fact that this same factor held for the gamma rays as well as the cosmic rays meant that *the mechanism of ionization had to be the same in both cases.* But since gamma rays are known to ionize only through the mechanism of the secondary high-speed electrons or beta rays which they send across the chamber, the conclusion could be drawn with

[1] Hoffmann and Lindholm, *Gerlands Beitr.*, XX (1928), 12; also *Zeit. f. Physik.*, LXIX (1931), 704.

[2] Millikan and Bowen, *Nature*, CXXVIII (1931), 582; Millikan. *Phys. Rev.*, XXXIX (1932), 401.

[3] Compton, Bennett, and Stearns, *Phys. Rev.*, XXXVIII (1931). 1565.

considerable confidence that *the cosmic rays also ionize only through the mechanism of free "electron shots" which they cause to pass through the chamber.* Very important consequences will later be drawn from this type of parallelism between gamma rays and cosmic rays—a parallelism that Neher and I have roughly tested directly or indirectly clear up to altitudes of 60,000 feet.

But the most direct evidence that the immediate ionizing agent in all cosmic-ray ionization is either a free positive or a free negative electron and not in general a proton or any other heavy nucleus comes from a study of the thousands of photographs of cosmic-ray tracks that we have taken. Only electrons and photons can possibly produce the type of effects shown in Figures 59, 60, 61, 62, 63, and 64. These show a continuous gradation in energy as measured by curvature from a few million electron-volts up to a billion without the slightest trace of a difference in ionization between positives and negatives, such as in this range of curvatures is very pronounced between electrons and heavier charged bodies, such as nuclei of any sort. Nothing but electrons (+ and −) can have the combination of curvature and range, to say nothing of ionization, here shown. Further, if the Newtonian laws of encounter hold, protons, neutrons, or other heavy particles cannot transfer energies of millions of electron-volts to electrons such as we find in a great many of our photographs. In the study of all of our plates we have found only two tracks that seem to be those of protons (Fig. 54). Whatever then the nature of the original cosmic rays it will henceforth be with us a fundamental assumption that the immediate ionizing agent is in general an electron (+ or −).

There was one further early consideration that spoke strongly in favor of the photon theory of the nature of cosmic rays, namely, their banded structure, particularly the fact that the softest band, of absorption coefficient about 0.5 per meter of water, controls practically the whole ionization between the altitudes of 14,000 feet and, say, 50,000 feet. No incoming particle theory seemed able to account for the *exponential* rise of ionization between these limits. For Bowen, Epstein, and Langer all proved independently that incoming particle rays of a given energy coming in from all directions and ionizing uniformly down to the end of their range give rise to a linear increase in ionization with altitude starting from the depth corresponding to the range of the incoming particles. This was completely different from the observed depth-ionization curve, and furthermore, when one starts with this principle no summation of energies could be made to fit at all the observed depth-ionization curve. On the other hand, one single band of photons of uncertain energy, but practically certainly of energy less than 500 million electron-volts, would reproduce satisfactorily this portion of the depth-ionization curve.

II. THE LATITUDE EFFECT IN COSMIC RAYS

Whatever the nature of the primary cosmic rays themselves, it was to be expected that there would be some electronic, i.e., beta ray, component of the rays entering the earth from the cosmos. For, whether the rays are in process of being produced now and have always been in such process or whether they represent relics of a bygone condition of the universe in which atomic transformations no longer possible were once able to take place, as has been seriously suggested by some physicists of the expanding universe school, in any case no small part of those that strike the earth must have been shooting through space, like the photons that constitute light, for a billion years or more. Whenever in their travel through space they have gone through matter, even though in the creative act they were sent forth as pure photons, they have necessarily produced secondary electron rays such as we find them producing in our terrestrial electroscopes, so that there is no existing theory of the origin of the cosmic ray that does not require, once the existence of cosmic ray photons is postulated at all, that space be traversed both by these photons and by the secondary electrons or other particles which these photons have produced in their billions of years of journeyings through the universe. Also, since the observed uniformity of distribution of the rays means that if they originate as photons these must traverse space uniformly in all directions, it of course follows that the secondary electrons which they have produced must also traverse space uniformly in all directions. In other words, the rays which enter the earth must in any case be a mixture of photons and extra-terrestrial secondary electrons or else of electrons alone if the photon hypothesis is entirely barred out, and the only question that is left for experiment to settle is what fraction of the inflowing energy is carried by the photons and what by the electrons.

It was precisely these considerations that led to the first effort to detect the effect of the earth's magnetic field on the intensity of the cosmic rays. This was made in the summer of 1926 when Millikan and Cameron took three electroscopes on a voyage from Los Angeles to Mol-

lendo, Peru, taking continuous readings on two of them
(the third was too responsive to the ship's vibrations) in
the radio room of the S.S. "Mongolia" from Los Angelss
to Balboa, and then after transshipment to the S.S.
"Ebro" from Balboa to Mollendo, then up on Lake Titicaca and finally in Lake Maguilla near Caracolles, Bo-
livia.   The sea-level readings which were taken on the
two electroscopes during all of the twenty days of the
downward voyage and on one electroscope all the way
back showed, as we officially reported, "no variation with
latitude which was beyond the limits of our observational
uncertainty estimated at 6 per cent."   The observations
on Lake Titicaca (12,500 feet) and in Lake Maguilla
(15,000 feet) were reported as taken and are some 30 per
cent lower than those found by us at corresponding altitudes in California and Colorado.   But having found no
sea-level effects in the much more prolonged observations
taken under conditions entirely free from all uncertainties
as to possible electroscope-leaks and shielding from adjacent mountains—sources of error necessarily inherent in
our mountain work—and seeing no reason why we should
find in the equatorial belt altitude effects, if not sea-level
effects, we reported[1] that "we had found no latitude
effects *larger than our observational-uncertainties.*"

In the summer of 1928 I assisted Dr. Gish of the Carnegie Institution of Washington in calibrating electroscopes which he placed on the ill-fated iron-free ship
"Carnegie," and with which during a six-month's cruise
in the Pacific 507 observations were taken[1] between 20° S.
and 20° N., 509 between 20° S. and 60° S., and 262 be-

[1] See R. A. Millikan, *Phys. Rev.*, XLII (1933), 668, for more complete
discussion.

tween 20° N. and 60° N., with mean results which as reported to me by Dr. Gish seemed to show a reduction of intensity in the equatorial belt over that in the temperate belts of from 6 per cent to 7 per cent. Also Clay[1] in 1927 and 1928 in voyages from Amsterdam to Sumatra had reported at least a 15 per cent drop in sea-level cosmic-ray intensities in going from the temperate into the equatorial belt. Stimulated by these results, as well as by the foregoing theoretical considerations which seemed to demand that the earth's magnetic field produce some latitude effect on the observed cosmic-ray intensities, in the summer of 1930, having by this time devised and perfected pressure electroscopes much more sensitive and dependable than those used in any preceding geographical surveys, I made a long series of sea-level observations in the Pasadena area, then transported the same electroscope to Churchill, Canada (geographic latitude 59, magnetic latitude 70) and taking the mean of a week's continuous observations, night and day, at Churchill, *found the sea-level intensities the same in the two localities "within an error of not more than 1 per cent."*

Epstein[2] had by this time published his calculations on the effect of the earth's magnetic field on incoming electrons in which he found that even billion (10^9) volt electrons approaching the earth from outside could not get through this field at magnetic latitudes lower than 58° N., while from latitudes beginning somewhat above 58° they should come in with *uniform intensity*. According to his calculations the uniform polar cap of sea-level

[1] J. Clay, *Proc. Roy. Acad.* (Amsterdam), XXX (1927), 1115; XXXI (1928), 1090; XXXIII (1930), 711.

[2] P. S. Epstein, *Proc. Nat. Acad. Sci.*, XVI (1930), 658.

intensities which I had found extending clear down to the latitude of Pasadena meant that there could be no appreciable number of electrons accompanying the cosmic rays into the earth's magnetic field of energies between three billion volts, the energy which he assumed it was necessary for an electron to have *merely to get down to sea level through the resistance offered by the atmosphere,* and four billion volts, which represented the energy which he computed that it was necessary for an electron to have if it got through the blocking effect of the earth's magnetic field and struck the earth's surface in the latitude of Pasadena.

The non-existence of appreciable numbers of incoming electrons of energies between 4 and 3 billion volts was not disturbing from the standpoint of the observational data available in 1930, since I then estimated that no primary photons existed which could impart to electrons energies higher than 3 billion volts anyway, and according to Epstein electrons of energies lower than 3 billion volts could not get down to sea level.   All the secondary electrons of energies from 3 billion volts down coming into the earth's magnetic field with the cosmic rays should, however, be stopped somewhere in the upper atmosphere and produce therefore a latitude effect on cosmic-ray intensities which it should be possible to bring to light by making a high-altitude cosmic-ray survey in the northern part of the United States and Canada, or according to Epstein's calculations between magnetic latitudes 58 and 90 or geographic latitudes in North America between 47 and 79. Accordingly, in the fall of 1931, before any geographical cosmic-ray surveys other than those already noted, viz., those by the "Carnegie," Clay, and ourselves, had been made, or so far as I was aware projected, I made an application to the Carnegie Corporation of New York for funds for the development of specal instruments especially adapted for making an *accurate high-altitude airplane survey*.

The development of a vibration-free, self-recording cosmic-ray electroscope for accurate work on a moving platform and which was actually found to yield as good a

FIG. 79.—Dr. H. Victor Neher's vibration-free self-recording electroscope which wholly unattended by skilled observers of any sort has taken automatic records of cosmic-ray intensities in airplanes up to 29,000 feet, in stratosphere balloons up to 62,000 feet, and in world-encircling voyages of three months' duration on Dollar Line and Matson Line ships.

record in a diving airplane, a running automobile, a speeding train, or a tossing ship ás in the laboratory, was largely the work of Dr. H. Victor Neher. With the aid of this instrument (see Fig. 79) taken up to an altitude of 29,000 feet through the courtesy and generous assistance of Colonel Arnold and the other officers of the U.S. Air Corps at March Field, near Riverside, California, we first got the record of the cosmic rays' intensities at this latitude as shown in Figure 80.   The instructions to the pilot were to rise to an altitude of 15,000 feet, to then level off for an hour, then to rise to 22,000 feet and there level off for another hour, then to rise to 29,000 feet and repeat there the leveling-off operation.   In the summer of 1932, Bowen, Neher, and I took this instrument to Cormorant Lake in Northern Canada where the Royal Canadian Air

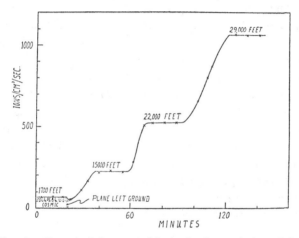

FIG. 80.—Record of the rates of ionization in an airplane flight at March Field (near Pasadena, California).

Force assisted us in making similar flights up to altitudes to 22,000 feet.   Dr. Neher and I then supervised similar most skilful flights made for us at Spokane by Captain Breene of the National Guard.   Dr. Neher then took the same instrument to Mollendo, Peru, and got, both coming and going, an accurate sea-level record as a check on our 1926 much less sensitive readings.   At Panama, through the orders of General Falois, Chief of the Army Corps, a flight to 22,000 feet similar to that already made at Cormorant Lake and Spokane was arranged for, and in Peru a ship of the Pan American Airways flying between Ariquipa and Lima flew at 19,000 feet both coming and going. In all these flights the true altitudes were recorded on our own barograph officially calibrated for us at the Bureau of Standards in Washington and checked by us in our laboratory at Pasadena.

The results of the high-altitude records are all graphed together in Figure 81. These graphs show the predicted high-altitude latitude effect between March Field and Spokane and a marked latitude effect between Panama and March Field (Pasadena) but no latitude effect between Panama and Peru and none north of Spokane. On the basis of these curves we shall divide the incoming cosmic rays into two parts, namely the *non-field-sensitive part* producing all the ionization beneath the lowest curve, and the *field-sensitive part* producing the ionization between the upper and the lower curves. This last is of course due to the direct and the indirect influence of incoming extra-terrestrial electrons.

III. RESISTANCE OF THE ATMOSPHERE TO AN INCOMING ELECTRON ABOUT SIX BILLION ELECTRON-VOLTS

The reason that no latitude effect whatever appears between Spokane and Cormorant Lake is presumably that the secondaries under 2 billion volts that we actually find in our energy-measuring magnetic field at Pasadena and that we therefore think must be present in the incoming rays in considerable numbers, cannot penetrate the atmosphere down to 22,000 feet, which is just under halfway through. If this is correct, a change in intensity should appear between these latitudes if flights were made

to sufficiently high altitudes.   According to our direct
measurements, to be presently detailed, there are more
secondary electrons formed of energies under 2 billion
volts than of energies above this value.   These latter
according to Epstein's computations as to the effect of the
earth's magnetic field—and these are essentially the same
as those made by LeMaitre and Vallerta[1]—are responsible
for all latitude effects found south of Spokane.   According to these computations, 2.4-billion-volt electrons will
just begin to get through the earth's magnetic field in the
latitude of Spokane (mag. lat. 54°).   The failure then to
find any latitude effect north of Spokane but a notable
effect south of it at a depth beneath the top of the atmosphere of $4\frac{1}{2}$ meters of water (22,000 feet), means that the
total resistance of the atmosphere (equivalent to 10 meters of water) to incoming electrons must be more than
$\frac{10}{4.5} \times 2.4 = 5.3$ billion electron-volts.   Indeed, if the computations of the foregoing theorists as to the blocking
effect of the earth's magnetic field are correct, then to account for the absence of a sea-level latitude effect between
Pasadena and Churchill requires the assumption that an
incoming electron must have an energy of about 6 billion
volts to get through the atmosphere.   Anything less than
6 would enable the upper-air latitude effect actually
shown in Figure 81 between Pasadena and Spokane to
manifest itself also at sea level.   Anything appreciably
more than 6 would prevent the upper-air latitude effect
shown in Figure 81 between Panama and March Field
from manifesting itself also at sea level; for the measurable decrease in sea-level intensity actually begins a very

[1] LeMaitre and Vallerta, *Phys. Rev.*, XLIII (1933), 91.

few miles south of Pasadena. Since the earth's magnetic field has been pretty well determined *the conclusion may be drawn then with some assurance from the constancy of the sea-level intensity between Pasadena and Churchill and its lack of constancy beginning just south of Pasadena that the total resistance of the atmosphere to the passage of high-energy electrons through it is about 6 billion electron-volts.*

From the same reasoning it would follow that the decrease in the high-altitude intensities found between Pasadena (March Field) and Panama, as shown in Figure 81, and the entire absence of any change in intensity in any altitude in the long stretch between Panama and Peru, means that enough electrons enter the earth's magnetic field of energies above 4 billion (the value at which they just begin to get through the blocking effect of the earth's field in magnetic latitude 41, but are unable to penetrate the atmosphere) to produce the excess of ionization shown in the March Field curve over that in the Panama curve, while between Panama and Peru there cannot be any considerable number of further field-sensitive particles coming in. Whether the large number of incoming cosmic rays that experiment reveals in these latitudes are non-field-sensitive because they are photons or because they are electrons of 10 billion volts of energy or more is to be determined on the basis of evidence to be later presented.

According to our assumption, north of Pasadena all vertically entering electrons of energies above 6 billion volts must get down to sea level, and in fact north of Pasadena there is as already indicated no sea-level change at all, while the means of the sea-level readings taken with our sensitive self-recording electrocope on the way to Mollendo and back gave a diminution of cosmic-

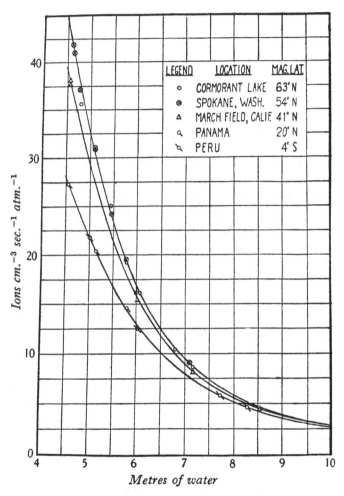

FIG. 81.—High-altitude-latitude effects

ray intensities setting in just at Pasadena and amounting to 7 per cent between Los Angeles and Panama, but no appreciable change below Panama. This, of course, meant that *about 7 per cent of the sea-level ionization in the temperate and polar latitudes above Los Angeles is due directly or indirectly to incoming electrons while some 93 per cent is due to some kind of non-field-sensitive rays.*

As a check on the important conclusion derived from the foregoing latitude survey that the resistance of the atmosphere to incoming electrons is 6 billion volts, we may bring forward Anderson and Neddermeyer's direct measurement of the loss in energy experienced by an electron of energy from a hundred million to five hundred million electron-volts in traversing a plate of lead a centimeter thick. Figures 82 and 83 show how satisfactorily this loss can be determined from the difference in curvature of the electron track above the lead and below it. The mean of a dozen or so plates like those shown in Figures 82 and 83 yielded 57 million electron-volts loss per cm. of lead. On the assumption that the ratio of the loss of energy of an electron in traversing water and lead remains the same for cosmic-ray electrons as for the highest energy electrons with respect to which this loss has been directly measured, namely about 1 to 9,[1] the resistance of the atmosphere (10 m. of water) comes out $57 \times 10^6 \times \frac{10.33}{9} = 6.5$ billion electron-volts, in reasonably good agreement with the results obtained by the two preceding modes of computation. This may be looked upon either as a third bit of evidence for the 6-billion-volt value of the resistance or as evidence that the nuclear absorption of electrons follows roughly the mass absorption law.

[1] Rutherford, *et al., Radiations* etc., 1930, p. 437. Also Bethe, *Zeit. f. Physik.,* LXXCVI (1932), 293.

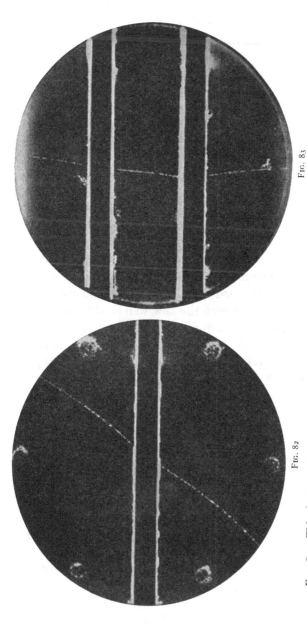

Fig. 82

Fig. 82.—This electron entered the lead from above with an energy of 113 million volts and emerged with an energy of 86 million.

Fig. 83

Fig. 83.—The electron entered the upper lead with an energy of 240 million volts and emerged from the lower plate with an energy of 160 million.

IV. THE EFFECTS OF NUCLEAR ENCOUNTERS

It is of much interest to inquire what fraction of this 6-billion-volt resistance to the passage of electrons through the atmosphere is due to nuclear encounters and what to extra-nuclear. This may be estimated as follows: Anderson and Neddermeyer have counted the mean number of ions formed along cosmic-ray tracks. This can be done quite accurately by choosing for the count electron shots that go through the chamber say a tenth of a second before the expansion so that the ions formed have a chance to diffuse before they are caught and fixed in position by the condensation of the water upon them. The droplets thus become separated so as to be accurately countable. The mean number of ions formed in air per cm. of path at standard pressure so determined comes out 31 which, combined with the now well-established value of the average energy required to detach an ion from the molecules constituting the atmosphere, namely 32 electron-volts,[1] leads to a resistance of close to 1,000 volts per cm. of air path at normal pressures, and this in turn leads to just under a billion volts for the resistance of the atmosphere to an electron entering vertically. This is to be regarded as a lower limit since any loss in the counting of droplets and any obliqueness in incidence would raise somewhat the estimate (the great bulk of the cosmic rays reaching sea level actually coming in within 45° of the vertical). Something between a billion and a billion and a half volts may be taken as a rough estimate of the primary ion-track resistance of the atmosphere to the average incoming cosmic-ray electron that reaches sea level.

But to this must be added the resistance due to such

[1] Eisl, *Ann. d. Phys.*, III (1929), 277.

close extra-nuclear encounters as give rise to secondary tracks, which are all of negative sign.   Doctors Anderson and Neddermeyer have studied these with great care experimentally, using the cloud-chamber technique; Drs. Carlson and Oppenheimer have analyzed them theoretically with results in reasonably satisfactory agreement. The net result is found in the conclusion that another billion and a half volts may disappear in this way as an incoming electron plunges through the atmosphere.   To a first approximation then we shall make the estimate that about half of the 6 billion volts of atmospheric resistance to the passage of an electron is due, directly or indirectly, to extra-nuclear ionization, the other half to nuclear encounters.

Three types of secondary rays may be expected to result from these nuclear encounters of electrons: (1) rays consisting of electrons (+ and −) directly emerging as a result of the encounter; (2) photons of the same nature as the general X-radiation emitted from the point of impact of the cathode ray beam on the anticathode; and (3) annihilation-ray photons of half-million-volt energy as first measured by Chao.   The existence of all three of these radiations has been directly brought to light in the Norman Bridge Laboratory.   The first is shown in Figure 62 (p. 347) in which a negative electron hits a nucleus and a positive electron emerges from the collision.   This is a very rare occurrence compared to a close encounter with an extra-nuclear electron—an event that of course gives rise to a new negative instead of a new positive.   The second is of the type shown in Figures 59, 60, and 61, though these "photon sprays" are here thought to result from the impact of photons upon a nucleus rather than the impact of an electron, though the latter event should produce similar effects, and Dr. Anderson has five or six cases in which he thinks he has direct evidence for it.

With the aid of these energy measurements of Anderson and Neddermeyer and the foregoing latitude survey we are able to draw some quite definite conclusions regarding the energy relations of the incoming or extra-terrestrial electrons and photons, and the secondary radiation produced within the atmosphere by them. These conclusions may be stated thus:

1. *Between 75 per cent and 85 per cent of all the ionization produced in the atmosphere by incoming electrons is due to secondary electron and photon rays produced within the atmosphere.* This result follows from the fact that the loss of energy *in the primary ion track* is only from 1 to $1\frac{1}{2}$ billion electron-volts while as above shown the actual resistance of the atmosphere to such an electron is from 4 to 6 times as great or about 6 billion volts.

2. The secondary electron and photon rays produced within the atmosphere *by incoming electrons* are *practically all low-energy rays.* This result follows from the fact that though the difference between the Panama and the Spokane curves shows that field-sensitive rays, i.e., incoming electrons, are doing much ionizing of the atmosphere above sea level, yet *the secondary rays produced by them are not sufficiently penetrating to extend their influence down to sea level* so as to cause any sea-level latitude effect at all between Pasadena and Spokane. In confirmation of this important conclusion, Anderson and Neddermeyer have measured directly a very large number of these secondary electron tracks that branch off in their cloud chamber from the tracks of high-energy electrons as

they traverse plates of lead, aluminum, and carbon. These secondary tracks are practically always negatives, and therefore arise almost wholly from extra-nuclear encounters and none of them have ever been found having energies higher than 100 million electron-volts.   Similarly, the *"bremsstrahlung"* photons exhibited in Figures 59, 60, and 61 have never shown energies higher than 50 million electron-volts.

3.   On the other hand the electrons released in the atmosphere by the nuclear encounters of incoming photons should have energies of the same order of magnitude as the photons themselves, no matter whether the encounters are of the photoelectric or the Compton sort.   Since then the great bulk of the cosmic-ray tracks measured by both Anderson and Kunze[1] have energies far above 100 million electron-volts and since, further, somewhat more than half of them are positives while secondaries produced by high-speed electrons are practically all negatives, it follows from statement 2 that these tracks are not the indirect effect of incoming electrons; and it follows from statement 1 (see also statement 4) that but a small percentage of them can be the incoming electrons themselves. *They must therefore in the main be secondaries produced by the nuclear encounters of incoming photons.*

4.   Since the total equatorial drop in sea-level cosmic-ray ionization going from Los Angeles to Mollendo is but 7 per cent and since according to statement 1 from 75 per cent to 85 per cent of this 7 per cent of field-sensitive ionization is due to secondaries produced within the atmosphere by incoming electrons, it follows that the num-

[1] P. Kunze, *Zeit. f. Phys.*, LXXIX (1932). 79; LXXX 1933). 550; LXXXIII (1933), 1.

ber of these incoming electrons found at sea level need be no more than 2 per cent or 3 per cent, at most, of the total number of ionizing particles that produce the cosmic-ray ionization found within an electroscope or a cloud chamber at sea level.

V. THE SIGNIFICANCE OF STRATOSPHERE FLIGHTS

The original observations of Kohlhörster[1] made in 1913 carried our knowledge of the cosmic-ray ionization of the atmosphere up to an altitude of about 9 kilometers or from about 29,000 to 30,000 feet. It is significant and highly creditable to this very early work that all the observations made twenty years later in a comparable magnetic latitude, namely, the observations of Mott-Smith and Howell[2] and of Bowen, Millikan, and Neher,[3] made in airplanes to about Kohlhörster's altitude, as well as those of Bowen and Millikan,[4] Regener,[5] and Piccard,[6] made in sounding or manned balloons to very much higher altitudes, have checked his values reasonably well up to the altitude to which he went.   Figure 81, however, shows at once how important an influence magnetic latitude may have on these intensities and warns against expecting agreement in different latitudes.

Kohlhörster computed from his high-altitude data an apparent absorption coefficient of 0.55 per meter of water,

[1] Kohlhörster, *Phys. Zeit.*, XIV (1913), 1066, 1153; and *Verh. d. Deut. Phys. Ges.*, XV (1914), 719.

[2] Mott-Smith and Howell, *Phys. Rev.*, XLIV (1933), 4.

[3] Bowen, Millikan, and Neher, *ibid.*, XLIV (1933), 246.

[4] Bowen and Millikan, *ibid.*, XLIII (1933), 695.

[5] Regener, *Naturwiss.*, XX (1932), 695; *Phys. Zeit.*, XXXV (1934), 782.

[6] Piccard and Cosyns, *Compt. Rend.*, CXCV (1932), 606.

and it was to see whether such a coefficient held to the top
of the atmosphere that Bowen and Millikan in 1922 made
their first stratosphere flight with sounding balloons.
Their balloons were sent up from near San Antonio, Texas
(mag. lat. 38°), and reached an altitude of 15.5 km. or
50,000 feet.   From their data Bowen and Millikan drew
the single conclusion that the apparent coefficient, 0.55,
could not possibly continue from 29,000 feet up to the top
of the atmosphere but must instead come back to very
much smaller values.   This conclusion has recently been
checked by the data obtained in the newer stratosphere
flights of Piccard, Regener, and Bowen and Millikan, all
of whom find the altitude-ionization curve even reversing
its curvature at very high altitudes.   The determination
of this shape near the top of the atmosphere, especially as
a function of latitude, is one of the significant objectives of
stratosphere flights.   All that has been done to date in
this direction is contained in Figure 84.

The most accurate of these curves is undoubtedly that
corresponding to the Fordney-Settle flight of November,
1933, for which we sent up one of our unshielded, vibration-free, self-recording, Neher electroscopes.   A portion
of the record thus obtained is shown in Figure 85.   In this
flight the balloonists remained near the 62,000-foot altitude for more than three hours so that the intensity at this
altitude is determined with quite a new precision.   As a
check, too, we obtained a good similar record from our
unshielded electroscope sent up in the Kepner-Stephens
flight of July, 1934.   These two flights were made in essentially the same magnetic latitude and in the latter
flight the balloonists remained for more than an hour at
the 40,000-foot altitude.   The intensity there recorded

falls very nicely on the Fordney-Settle curve shown in Figure 84. The three lower curves contain much greater uncertainties because of the continuous and more rapid

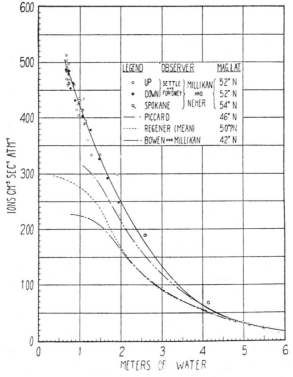

FIG. 84.—Cosmic-ray ionization of the atmosphere in different latitudes as a function of depth in equivalent meters of water beneath the top of the atmosphere. (Regener's new data [1934] fits his curve given above.)

rise and fall of the balloons and the less sensitive recording electroscopes; but the fact that all three of them show a reversal in curvature at the top, not shown in the Fordney-Settle curve, probably means a latitude difference and not merely observational uncertainties (see below).

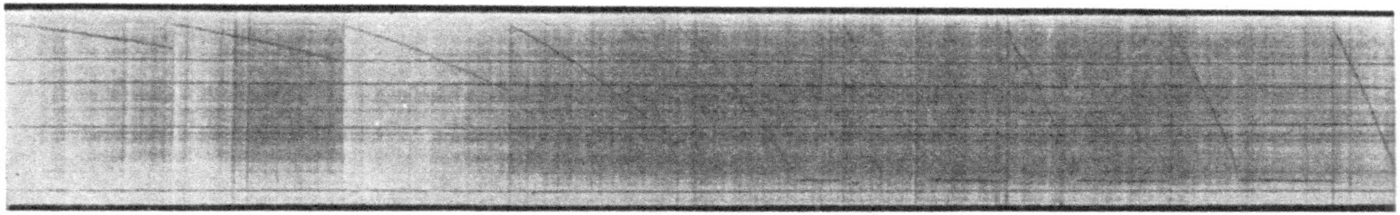

Fig. 85.—Section of film from the Neher automatic recording electroscope taken during the ascent to the stratosphere of the balloon in the Fordney-Settle flight of November 20, 1933. Each of the nine panels across which the sloping dark line moves from left to right corresponds to a five-minute time interval. The slope is proportional to the rate of discharge of the electroscope, which in turn is proportional to the intensity of the cosmic rays. In the first panel the balloon was at an elevation of about 20,000 feet, and in the next twenty minutes (four panels) while the balloon was rising to a height of 45,000 feet the slope, or cosmic-ray intensity, rose tenfold. From there on to the highest altitude attained, 62,000 feet, the rate of rise of the balloon was much less rapid. It remained at or near the top for three and a half hours, so that this film yielded accurate measurements of cosmic-ray intensities at the higher and lower altitudes, but not at the intermediate ones.

But before discussing further these differences it will be well to present the second significant objective of stratosphere flights, namely that of obtaining direct information as to the actual *penetrating power* of the rays existing at different altitudes.   This is done by taking an unshielded electroscope to a given altitude under such conditions that only cosmic rays affect it, observing its rate of discharge, then observing again the rate of discharge of the same electroscope in the same place when it is surrounded by a heavy shield of lead.

Figure 86 shows the result of such tests made at March Field (near Pasadena) through the co-operation of Col. Arnold and the other army officers stationed there. If the lead shield, here 10 cm. thick, acted simply as a resistance, cutting down the energy of the electrons incident upon its outer surface and supplying no new electrons whatever through the absorption of incident photons, then the ratio between the ionization found within it, first without screen, second with it, would reveal what fraction of the electrons existing at the given altitude possess sufficient energy to penetrate the given thickness of lead.   Since, however, it is known that many new secondary electrons are actually stimulated within the lead, and since all of these that get into the electroscope increase the actual ionization existing within it when shielded, it is clear that the foregoing ratio gives merely *the lower limit* to the fraction of electrons having energies under the value required to penetrate the lead.   Since the shield was 10 cm. thick,[1] Figure 86 shows that at 29,000

---

[1] I have used the actual thickness, 10 cm., rather than the average thickness given in Figure 86 since most of the rays come in from above, i.e., radially.

feet at least 70 per cent (the actual figure is probably over
90 per cent) of all the electron-rays existing at that altitude have energies under 570 million volts, this being,

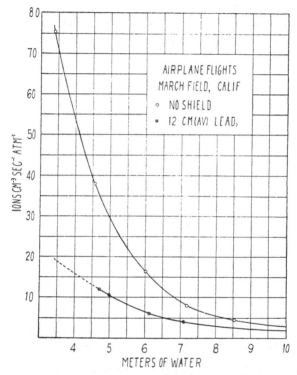

Fɪɢ. 86.—Comparison of ionizations found at different elevations up
to 29,000 feet in unshielded electroscope and ionizations in the same electroscope when surrounded with a 10 cm. shield of lead.

according to Anderson's measurements, the energy required to traverse 10 cm. of lead. *Practically none of
these electrons could have come in from outside the earth's
atmosphere, since it requires at this latitude at least 4-billion-*

volt electrons to get into the atmosphere through the earth's
magnetic field, and the atmosphere above 29,000 feet could
only have subtracted some 1.5 billion volts from this mini-
mum of 4 billion.

The argument is still more cogent when applied to
stratosphere flights, and Figure 87 gives the correspond-

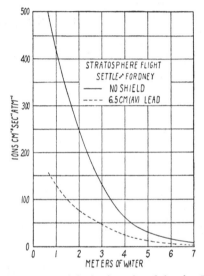

Fig. 87.—Comparison of ionizations found in the Fordney-Settle
flight in unshielded electroscope and in one shielded with 6 cm. of lead.

ing observations made in the same way in the Fordney-
Settle flight, the non-shielded readings being taken by us
with one of our Neher electroscopes, while the lower curve
gives the results obtained by Dr. A. H. Compton[1] with his
electroscope inside 6.0 cm. of lead. Figure 88 gives the
lower limits taken from Figure 87 to the fraction of the
electron-rays existing at each altitude the energy of which

[1] A. H. Compton and R. J. Stephenson, *Phys. Rev.*, XVL (1934), 441.

must lie under $57 \times 6 = 340$ million electron-volts. It shows that from an elevation corresponding to a depth beneath the top of the atmosphere of about 6 meters of water, i.e., from Pike's Peak up to a point but 60 cm. of water beneath the top, or 94 per cent of the total, not less than 70 per cent (probably more than 90 per cent) of the electron-rays have energies under 340 million volts and no one of these above 2 m. can possibly have come in from outside since at the least 2 billion volts are required

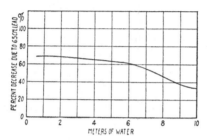

FIG. 88.—Showing the *lower limit* to the percentage of electrons existing at a given depth beneath the top of the atmosphere that have energies under 340 million electron-volts.

to get through the blocking effect of the earth's field in latitude 52. Since the figures apply to the whole atmosphere, i.e., both to field-sensitive and to non-field-sensitive parts of the ionization, we may conclude with certainty, as stated in 1932,[1] that *"all but a very small part of the ionization of the air is due to secondary rays produced within the atmosphere."*

All of the field-sensitive part of this ionization is, of course, produced either directly or indirectly by incoming or extra-terrestrial electrons, from 75 per cent to 85 per

[1] *Phys. Rev.*, XLIII (1933), 665.

cent of it (according to Section III) being indirect and due to the relatively non-penetrating secondaries produced by the incoming very high-energy electrons. Enough of these latter must be present of energies between 8 billion and 6 billion electron-volts to cause the 7 per cent sea-level latitude change found between Los Angeles and Panama, and enough of energies between 8 billion and 2 billion to produce the high-altitude-latitude effect shown in Figure 81 between Panama and Spokane, for it is only energies in this range that can get through the blocking effect of the earth's field.

We are now in position to understand, too, the probable cause of the difference between the shapes of the curves shown in Figure 84. We have seen in Section IV that the great bulk of the higher energy electrons measured by Anderson and Neddermeyer at sea level are secondaries produced by the collisions of incoming photons with atomic nuclei. The number of these electrons that have energies above 4 billion electron-volts is much less than the number having energies between 4 billion and 2 billion. A similar distribution must exist whether these same photons have produced their secondaries in the earth's atmosphere or in the matter through which they have passed before entering the earth's atmosphere. So that incoming electrons of energies between 4 and 2 billion volts are more numerous than those above 4 billion. Further, since 2-billion-volt electrons can only penetrate a third of the way through the atmosphere they must do all their ionizing above that level. This is why the ionization near the top of the curve representing the Fordney-Settle flight, made in magnetic latitude 52, is very much

larger than the ionization found in the two Bowen-Milli-
kan flights made at about magnetic latitude 42° where
no electrons under about 4 billion volts can get through
the blocking effect of the earth's field. There is still, as
the figure shows, some experimental uncertainty with respect to these lower curves of Figure 84, but not enough,
I think, to change the main conclusions here drawn.
Regener's and Piccard's magnetic latitudes, somewhat
uncertain, are at least intermediate between 42° and 52°
and their curves are also intermediate.

I interpret also the shapes of the upper part of the
curves obtained by Bowen and Millikan at latitude 42°
and also by Regener and Piccard as due to the effect of
the incoming electrons of energies between 4 billion and
8 billion electron-volts. For there are now no longer
present the incoming 2- to 4-billion-volt extra-terrestrial
secondaries to add to the ionization of the underlying
photons of energies under 340 million volts, such as are
needed to produce the bulk of the observed non-field-
sensitive ionization. Instead, the only incoming electrons that can get through the earth's field into magnetic
latitude 42° and produce the field-sensitive ionization
found between 42° and 20° (Panama) are from 4 to 8
billion ones, and these are approaching the energies at
which they penetrate the entire atmosphere and are therefore nearing the condition that gives rise to a linear law
of increase of ionization with altitude instead of an exponential one (see below). This, combined with the
effect of incoming photons of energy under 500 million
electron-volts, should give the reversals of curvature
shown in the lower curves of Figure 84.

Figure 89 shows the results of the Fordney-Settle flight plotted on a log scale. This displays nicely (1) how from sea level (10 m.) up to the height of Pike's Peak (6 m.) a sum of exponentials is needed to reproduce the observed ionization; (2) how a single absorption coefficient, $\mu = 0.55$, takes care of the ionization from 6 m. to 2.5 m. below the top despite the complex character of the causes at work in this region, i.e., relatively low-energy photons plus high-energy (8 billion to 2 billion

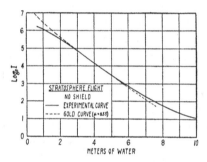

Fig. 89.—Showing the limits of validity in the upper atmosphere of the exponential rise of ionization, with coefficient $\mu = 0.55$.

volts) incoming electrons; (3) how, at a depth of about 2.5 m. below the top, the constancy of the exponential rise with the coefficient, $\mu = 0.55$, breaks down, presumably because of the effect of incoming extra-terrestrial secondary electrons.

A clearer notion, however, of the bearing upon the nature of cosmic rays of measurements like the foregoing on both ionization and penetration as a function of altitude can be gained by considering such measurements made in the equatorial belt, where no incoming electrons of energies under about 10 billion electron-volts can get through

the blocking effect of the earth's field.   As already indicated, Bowen, Millikan, and Neher[1] have made such accurate measurements in Panama up to 22,000 feet, and in
Peru up to 19,000 feet.   We found there an exponential
rise up to 22,000 feet but with a coefficient, toward the top
of the flight, of $\mu = 0.50$ instead of the coefficient 0.55 of
the Fordney-Settle flight.   *No such exponential rise is
possible as a result of the ionizing influence of incoming electrons.*   Indeed, as heretofore indicated, Bowen, Epstein,
and Langer have all shown independently that electrons
of a given energy coming in from all directions and ionizing
uniformly along their paths give rise to a *linear* rise in
ionization measured upward from the level corresponding
to the end of the range.

Now since the resistance of the atmosphere is about 6
billion volts, electrons of energies of 10 billion volts or
more will have a range corresponding to a depth of nearly
two atmospheres beneath the top of the atmosphere; and
since the secondaries produced by fast electrons have already been shown to possess energies that are practically
all under 100 million electron-volts, even these secondaries may be considered as being produced *uniformly
along the paths* of the hypothetical incoming 10-billion-
volt ones. So that a linear law of increase of ionization
from a depth of 8 or 10 meters of water below sea level up
to the top of the atmosphere is the type of ionization to
be expected from incoming electrons at the equator.
This would mean an increase in ionization in going from
sea level to 22,000 feet (4.5 meters) of some 60 per cent
instead of the observed increase of some 1,000 per cent.
This observed exponential rise constitutes one of the original

[1] Bowen, Millikan, and Neher, *Phys. Rev.*, XLIV (1933). 250.

and one of the most cogent arguments against the attempt to explain the non-field-sensitive portion of the cosmic-ray ionization as due in any major degree to incoming high-energy electrons.

The argument from the *penetrating powers* as directly observed in the equatorial belt is equally cogent. Thus if 10-billion-volt electrons produce in an unshielded electroscope at a given elevation, say 19,000 feet in Peru, a given amount of ionization, then when the electroscope is surrounded by a 10 cm. lead shield, even if no increase in secondary rays produced in the material of the shield entered the electroscope, since the lead would offer an insignificant resistance to the 10-billion-volt incoming electrons, the ionization within the electroscope should not be appreciably reduced by the presence of the shield. It is actually reduced to about one-third its unshielded value. These considerations speak quite as powerfully as those advanced in Section IV for the photon interpretation of the nature of the great bulk of the non-field-sensitive cosmic-ray effects. Nevertheless, the observations presented in the next section show that there are in fact a small number of incoming electrons of sufficient energy to reach the earth even in the equatorial belt. The first evidence that will be presented came from the discovery of the "longitude effect in cosmic rays."

VI. THE LONGITUDE AND THE EAST-WEST EFFECTS IN COSMIC RAYS

In view of the difference obtained by different observers in the measurement of the sea-level latitude effect, in January, 1934, we placed one of our self-recording electroscopes on Captain Hancock's yacht, the "Valero III,"

FIG. 90.—Automatic record taken by Neher electroscope on Captain Allan Hancock's yacht, the "Valero III"

and sent it with him on a three-month's cruise to Guayaquil and the islands off the coast of Equador. It required no attention save to have the clock wound once in two days. Because of the fact that Captain Hancock stopped for three separate weeks at different ports in the equatorial belt and also for nine days in the Los Angeles harbor, the mean intensity in each one of these points, as well as in intermediate latitudes where his three-day stops were made, was determined with much greater accuracy than had theretofore been attained. Figure 90 shows a bit of the developed film. The rate of discharge during an hour's run is given by the slope of each diagonal line of this film. The mean of 24 of these slopes gives the mean intensity for that day and the mean of $24 \times 7 = 168$ slopes, the mean for a week. In the upper curve of Figure 91 is shown the final graph of Captain Hancock's results. It is true that along this upper smooth curve are found also the readings obtained on other trips, but the findings on board the "Valero III," easily distinguishable, were given peculiar weight in the fixing of the upper curve. The readings in Los Angeles harbor were nearly a per cent lower (see Fig. 91) than the mean of the readings Dr. Neher and I took between Los Angeles and Seattle and also in New York; so that the total equatorial dip referred to Los Angeles is just over 7 per cent, while referred to the mean values farther north it is, as shown, very close to 8 per cent.

To see whether the sea-level variation between Mollendo, Peru, and the United States were different for an unshielded electroscope and one surrounded by a 10 cm. shield of lead, in May, 1934, Dr. Serge Korff, who had been making observations with our recording electroscopes in Peru, brought one of them unshielded from Mollendo to Los Angeles. The result is shown in the upper curve of Figure 92. The equatorial difference, referred again to the foregoing mean value, is here found to be 8.2 per cent. We thus have now the results of three different trips from Mollendo to Los Angeles and one more from New York through Panama to Los Angeles, and they all yield with great consistency and precision an equatorial dip in this region of 7 per cent $\pm$ 1 per cent referred to Los Angeles, 8 per cent referred to the mean of many values farther north. These values are lower than those obtained by the groups headed by Dr. A. H. Compton,[1] but the method used and the care we have taken in studying the films gives us great confidence in their accuracy. The films are all preserved at the Norman Bridge Laboratory so that other observers who wish to check for themselves the results need only measure up independently these films.

[1] A. H. Compton, *Phys. Rev.*, XLIII (1933), 387.

In a precisely similar way in August of 1933 Dr. Neher and I placed one of our self-recording instruments in the captain's room (inside its 10 cm. lead shield) on the Dollar Line ship, "President Garfield," and three months later, after her return to Los Angeles following her world-cruise to the China coast, Singapore, the Red Sea, through the Mediterranean and the Atlantic to New York, and back to Los Angeles via Panama, we removed and developed at Pasadena the film and found a good record until New York was reached.   This record revealed to us (January 10, 1934) the "longitude effect."[*]   We then in June, 1934, placed one shielded and one unshielded Neher electroscope in the room of First Officer Graham, of the S.S. "Monterey" of the Matson Line, and got an excellent record both going and coming between Los Angeles and Sydney, Australia, in the case of the shielded electroscope, and a good record in the unshielded one on the way down. To Captain Allan Hancock, owner and captain of the "Valero III," to First Officer Graham of the "Monterey" of the Matson Line, and to Captain Cullen of the "President Garfield" of the Dollar Line, all of whom kept our log as requested and wound our clocks with their chronometers, we owe the important discovery of the longitude effect in the cosmic rays.   This was first announced publicly at the meeting of the National Academy of Sciences in Washington on April 24, 1934.   All the results obtained from our records thus far taken are collected in Figures 91

[*] A letter from Prof. J. Clay of Amsterdam informs me that it was discovered independently by him on a voyage to Batavia, September, 1933 to January, 1934, very briefly referred to in *Physica*, March, 1934, and more fully in *Physica*, August, 1934.

and 92. It will be seen that the equatorial dip in intensity corresponding to the readings taken in the region of Sumatra is found to be 12 per cent instead of the 8 per cent dip found on the west coast of South America, while on the trip between Los Angeles and Sidney the equatorial dip has, quite consistently, the intermediate value

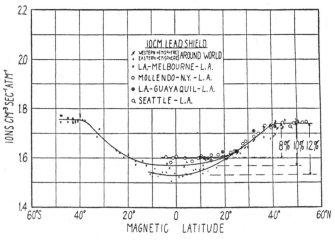

Fig. 91.—The longitude effect taken with a shielded electroscope

of 10 per cent. This lack of symmetry in the effect of the earth's magnetic field appears to reflect the dissymmetry in that field itself, for not only does the line connecting the north and south magnetic poles of the earth run closer to the Americas than to Asia, but the surface value of the horizontal component of the earth's magnetic field is 0.4 gauss near Batavia, while it is but 0.3 gauss in the vicinity of Equador and Peru. In a word, *the surface properties of the earth's magnetic field are here shown to be reflected in effects taking place thousands of miles above the surface, and we are now for the first time in possession of the means of studying the properties of the field far out in space.*

From the standpoint of the cosmic rays we interpret this longitude effect as meaning that there must exist some few incoming electrons which have energies enough to get through the earth's field at the magnetic equator on

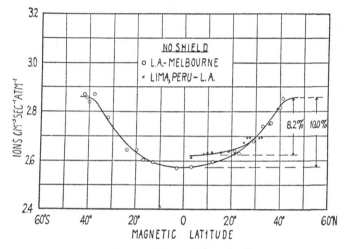

Fig. 92.—The longitude effect taken with an unshielded electroscope

the South American side but which are blocked off by the stronger field on the opposite side of the earth. This means that there are at least a few incoming electrons of energies above 10 billion volts.

Now it is precisely this small number of high-energy electrons coming into the equatorial belt that is needed to account for the so-called east-west effect first found by Johnson[1] and Alvarez and Compton[2] in Mexico City, stud-

[1] T. H. Johnson, *Phys. Rev.*, XLIII (1933), 834, and XLV (1934), 569.

[2] Alvarez and Compton, *ibid.*, XLIII (1933), 835.

ied by Rossi[1] and later found by Johnson to be as strong in Peru as in Mexico. The observations consist in attaching to a rigid system capable of rotating around a vertical axis two pairs of so-called counter tubes, one tube of each pair being set say a foot above the other. Suppose that the line connecting one pair of these tubes is set so as to make an angle of 45° with the vertical axis, and that the plane defined by this line and the vertical is an east-west plane, the line connecting the counters pointing upward to the west of the vertical. At the same time let the line connecting the second pair of counter tubes also lie in this plane and also make an angle of 45° with the vertical, but let it point upward to the east of the vertical. Such a pair of counter tubes will "respond" or produce a signal every time one single electron shoots through both tubes. Now the number of responses of the counter-pair that points upward to the west of the vertical is actually found at Mexico City to be about 9 per cent larger than the number of responses in the pair pointing upward to the east. A similar effect was found in Peru, but only a relatively minute effect in the United States. This west excess, amounting to a maximum of about 8 per cent (the maximum coming at 45°) at sea level in Peru, 16 per cent at an altitude of 14,350 feet, according to both Johnson's[1] and Korff's measurements,[2] is interpreted as showing that there are more incoming positive electrons than negatives in Peru, these positives being bent by the earth's field so that they strike the earth predominantly from the west.

Since we had already obtained the evidence, as set forth above, that some 90 per cent of all the cosmic-ray

[1] B. Rossi, summary at Int. Conf. on Physics, London, October, 1934.

[2] Serge Korff, *Phys. Rev.*, XLVI (1934), 74.

ionization of the atmosphere is due directly to secondary electrons within the atmosphere itself, we arranged with Dr. Korff to take to Peru in October, 1933, a new device for testing the east-west effect. The new element in this device was that it was capable of comparing the effect obtained when 30 cm. of lead was introduced above each pair of counters with the effect when the lead was not used. *The result was that the west-east ratio was uniformly reduced by the insertion of the lead in such amount as to show that the effect was predominantly due to secondary electrons formed within the atmosphere;* for, as already indicated, 30 cm. of lead could not have stopped appreciably any effects due directly to 10-billion-volt electrons, such as alone could have come from outside the earth into the equatorial belt.

The very small number of electrons of 10 billion volts or more required to account for the longitude effect is quite adequate to account also for this east-west effect. For, since the maximum equatorial dip found in the measurement of the latitude effect was 12 per cent, and since the ionizing effect of each incoming particle has already been shown to be multiplied five or six times by the secondaries it produces in the atmosphere, no more than about 2 per cent of the ionizing particles existing at sea level need have come in from outside to account for this 12 per cent equatorial dip.

The incoming electrons here in question may all be assumed to be positive, either arbitrarily, or better, from my point of view, because the highest energy photons which may be assumed to have produced them somewhere in their travel through space tend to be absorbed by the nuclei of atoms, and in all nuclei positive electrons of course predominate. Indeed, hydrogen nuclei contain no negatives at all, and hydrogen constitutes, according to Russell, more than 90 per cent of the universe. On the other hand, the lower energy photons will of course tend to be absorbed more in all substances by extra-nuclear electrons which are all negatives, for it is these extra-nuclear electrons that are responsible for all the ionization described by the Klein-Nishina formula. The earth's magnetic field then separates out the incoming electrons on the basis of energy, the lower energies, primarily negatives, coming in at the poles—the very highest energies, very few in number, being the only ones able to break through at the equator. In this connection it is interesting to note that in Anderson and Neddermeyer's most recent and most extended study of the distribution in energy and sign of the electrons passing through their cloud chamber, as presented at the London International Conference on Nuclear Physics in October, 1934,[1] positives and negatives are present in equal numbers in all ranges of energies from 300 million up to 3,000 million electron-volts, but above this range the positives seem to predominate. Even if one takes his stand upon the pair theory and therefore assumes equality in the number of positives and negatives "created," since the nucleus is positive it is natural to assume that the positive of the pair gets the bulk of the energy. Extra-nuclear negatives might make up the negative deficiency in the lower energy ranges, but not in the higher.

But whatever the cause, the east-west effect requires that the incoming 10-billion-volt electrons that reach the earth in the equatorial belt be predominantly positives.

[1] Anderson and Neddermeyer, *Proc. Int. Phys. Union.* Dec. 1934

With the introduction of the further consideration that
the secondaries, whether positives or negatives, produced
by each one of these in their journey through the atmosphere, must be distributed uniformly in direction about
the line of direction of the parent incoming electron, i.e.,
must follow in the mean that parent-electron's direction,
especially at these enormous energies, it will be seen that
the 2 per cent of incoming high-energy electrons mentioned above becomes, with its five-or six-fold progeny of
secondaries, 10 per cent or 12 per cent, so far as producing
a west excess is concerned.   Since the maximum west excess found at sea level in Peru is about 8 per cent, it is
clear that the electrons that produce the longitude effect
are entirely adequate to produce also the observed east-
west effect.   The above-mentioned tendency of the
secondaries to follow the direction of the primary as the
energy of the primary increases is nicely shown by a
comparison of Figs. 18, 19, 20 (p. 192), with C. T. R.
Wilson's photograph, opposite p. 200.

Further, it is precisely the above mentioned smallness
in the required number of incoming positives that is
needed to remove the contradiction heretofore existing
between the demands of the east-west effect for a large
excess of positives at sea level and Anderson's direct determination of the general equality at all energies up to 3
billion volts in the numbers of positives and negatives
passing through his cloud chamber as shown in Figure 93.
The from 1 to 3 per cent of ionization that must be attributed to the direct effect of incoming 10-billion-volt
positives to account for the from 6 to 18 per cent west
excess observed in Johnson's and Korff's Peru work with
counters involves an excess of positives wholly negligible

in Anderson's work. The remaining 98 per cent of Anderson's observed tracks may then all be due to the release from terrestrial atoms by incoming photons of electrons which with that origin may well be equally distributed between positives and negatives.

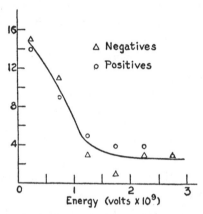

FIG. 93.—Distribution in energy of the positive and negative electrons. Each point on the curve represents the number of tracks in the corresponding half-billion-volt interval. The photographs were taken by the "random" method (not Geiger-Counter controlled).

This whole group of latitude, longitude and east-west effects here presented put certain fairly definite requirements upon the origin of the cosmic rays.

VII. THE INTERCONVERTIBILITY OF MASS AND ENERGY

The conclusion of the preceding sections that the non-field-sensitive part of the cosmic rays consists primarily of photons rests upon the two fundamental postulates, namely, (1) the validity of the analysis of Störmer, Epstein and LeMaitre-Vallerta as to the effect of the earth's magnetic field on incoming electrons; and (2) the validity of the evidence that, in general, the immediate ionizing agents in the production of cosmic-ray ionization are free positive and negative electrons, other ionizing agents not being necessarily absent entirely but at any rate exerting a negligible influence. The general *qualitative* validity of the first postulate will scarcely be questioned, and its *quantitative* validity is not of vital importance. For example, if the energies necessary to get through the earth's magnetic field were but half those computed by Epstein, the whole scale of electron energies would be reduced one-half without fundamentally modifying the reasoning or the conclusions.

The second postulate that the immediate ionizing agents are electrons (+ and −) rests upon the excellent evidence presented in Section I; but for energies above say 1.5 billion electron-volts even this distinction becomes unimportant, since in these ranges protons and electrons should be practically alike in ionizing power and not seriously different in deflectability. So that the conclusion seems inescapable that the whole non-field-sensitive portion of the ionization cannot possibly be due either to incoming electrons or protons of an energy of the order of 10 billion volts and more, and yet these are the only charged particles of even remotely possible properties which the earth's magnetic field permits to penetrate into the equatorial belt. The very fact, then, that we need incoming *photons* at all (and even the photographs seem to demand this much) to interpret cosmic-ray effects appears to eliminate the possibility that the source of the cosmic rays is to be found in cosmic electrical fields which impart these 10-billion-volt energies to electrons that fall through them.

So far, then, as we can now see, the only remaining possible source of these observed energies—the only other source that has even been suggested by anyone—lies in the theory of the convertibility of mass into radiant energy. The history of this theory is as follows. It was originally suggested by astronomers[1] because they could see no other way to account for the enormous quantities of heat that have been continuously pouring out from the stars in the billions of years during which we have evidence that they have been in existence. It was several years earlier than this, however, that the experimental physicist had begun to get direct laboratory evidence in his experiments on the pressure exerted by light, on the variability of mass with speed,[2] etc. (see p. 188) for such interconvertibility, and in 1905 Einstein deduced, as one of the most important consequences of his special theory of relativity, the equation which should govern such conversion, namely, $E = mc^2$.

This made it possible to compute from the known mass of the hydrogen atom and the known masses of the common elements like oxygen or silicon the amount of energy released when one of these common elements is produced by the union of the requisite number of hydrogen atoms, since, for example, 16 atoms of hydrogen actually weigh 0.8 per cent more than does an atom of oxygen. This sort of a hypothesis as to the origin of the heat of the stars was soon after suggested. It was elaborated at

[1] Jeans, *Nature*, LXX (1904), 101. See also "History of Research in Cosmic Rays, Millikan," *ibid.*, CXXVI (1930), 14.

[2] W. Kaufmann, 1901–06. See résumé *Ann. der Phys.*, XIX (1906), 487.

length by Harkins in 1915.[1]   A little later,[2] in order to permit of still greater stellar lifetimes, this hypothesis of the "partial transformation" of the mass of hydrogen atoms into radiant energy by the building up of the heavier atoms was extended so as to include the case of the complete annihilation of matter and the release in the interior of the star of the corresponding amount of radiant energy, and this theory of the transformation, partial or complete, of matter into radiant energy soon became recognized as orthodox astronomical theory.

Similarly, when the experiments of Cameron and myself convinced us of the enormous penetrability of individual cosmic rays, we could see no reasonable way of accounting for what we computed to be the corresponding energies except by this same process of the transformation of mass into radiant energy, and in particular we computed that the chief components of the cosmic rays had energies of about the right magnitude to correspond to the theory of the building up of the common elements —helium, oxygen, silicon, and iron—out of hydrogen; while the energy computed by means of the Einstein equation from the complete annihilation of any of the common elements, even of hydrogen (a billion electron-volts), was much too high to fit, in any event, our estimated energy of the least penetrating component, which actually produces 90 per cent of the ionization of the atmosphere.

In order to account for the uniformity of distribution of the cosmic rays over the celestial dome we tried to

[1] Harkins, *Phil. Mag.*, XXX (1915), 723.

[2] Eddington, *Nature*, XCIX (1917), 445.

devise a theory to explain how this building up of the common elements out of hydrogen might be taking place *in interstellar space*. At that time we thought it probable that the more penetrating components corresponded in the same way to the building up of heavier and rarer elements, the computed energies released in the building of which might rise in the case of uranium to the value of 1.8 billion electron-volts, and we saw no reason, then, for assuming that the act of complete annihilation ever took place in interstellar space, although we were quite willing to follow the astronomers in assuming that it might take place in the interior of the stars, and thus account for their long lifetimes.

The situation on January 1, 1935, is that our newer experiments detailed above have shown the invalidity of the theoretical Klein-Nishina formula on which we at first depended, but at the same time they have given us more direct and more certain knowledge than we had before of the *order of magnitude* of the energy contained in this chief component, though this knowledge is still far from quantitative, so that there is the same reason now for assuming this building up process that there has ever been.

It is important, too, to observe that in so far as this evidence is based on *energy* considerations it is quite independent of all questions as to *the nature of the radiations themselves*, i.e., as to whether they come in to us as photons or electrons. In fact, the whole evidence in favor of the hypothesis that the act of the building up of one or more of the common elements out of hydrogen in interstellar space may be summarized in the following statements, all of which are just as applicable to cosmic ray electrons as to photons:

1. Hydrogen actually constitutes a very large fraction of the universe.

2. The whole mass of isotope data constitutes excellent evidence that somewhere at some time or other the heavier elements have actually been built up out of hydrogen.

3. The energy so released computed from Einstein's equation has seemed before, and it seems now, to be of the right order of magnitude to account for the chief component of the observed rays.

4. The absence of any appreciable effect of the sun or the Milky Way, or any particular nebulae, points toward interstellar space as the most natural source of origin.

5. The extreme conditions of temperature and pressure existing in interstellar space ought to be conducive to more difficult *"condensation-processes"* than can be observed at the higher temperatures existing on the earth and in the stars.

Some have called this hypothetical act of the building up of the common elements out of hydrogen *"the creation of matter."* So far as I know I have never myself used that phrase, though it has a certain appropriateness since the possibility of the building up of new worlds of the type of ours is suggested.   In a foregoing chapter I have called it "the partial annihilation of matter," since from a thermodynamic standpoint this is precisely what it is.

But the newer experiments discussed above have also brought forth new, direct, and convincing evidence of the existence of cosmic rays of very much higher energies than those of this chief component. In the past we have only been able to make guesses, or at best very uncertain estimates, as to the values of the energies of these most penetrating components. Now, however, both Anderson and Neddermeyer, and Kunze have actually measured energies up to 6 billion volts, and the longitude and east-west effects described above have made it necessary to *assume* that some of the incoming rays have energies even higher than 10 billion electron-volts. Energies of this sort are far higher than those that can be obtained by the processes of atom-building even in the cases of the heaviest elements. Indeed, as indicated in chapter xiii, 10 billion electron-volts is an energy close to that which, in accordance with the Einstein equation, would be produced by the sudden complete transformation of the whole mass of a carbon atom into radiant energy. And if, as is at least possible, some of these enormously powerful rays have energies of 16 billion electron-volts this would correspond to the total energy that would be transformed if a complete atom of oxygen, far and away the most common of all the elements except hydrogen, were suddenly annihilated in the sense that its mass was completely transformed into a single photon or projected electron. Just such a transformation could not take place without violating the momentum principle (third law), but if we assume two 8-billion-volt photons (or electrons) going off in opposite directions we avoid this difficulty. In the same way we can imagine two 14-billion photons produced by the annihilation of silicon, another common element of atomic weight 28.

That such a complete annihilation of the whole of an

atom can take place at all will be doubted by many physicists and astronomers, and I am far from asserting that it does. Nevertheless, it is true that there is no other source now in sight that can yield the cosmic-ray energies that we actually observe. One can take this merely as evidence of the depth of our ignorance of the origin of these cosmic rays, or, if he is so inclined, he can take it as evidence that such annihilation processes actually take place, and I have in chapter xiii made some speculations as to how it might be possible both for the partial and complete annihilation processes to be taking place in interstellar space.

Some might even raise the question as to the legitimacy of using the Einstein equation as a basis for such energy computations, but it will have been observed that this is precisely what has been done in the discussion of the problem of the transmutation of the elements in the preceding chapter, and that the results based upon this assumption have there been rather remarkably checked experimentally. So that the validity of this equation for revealing the energy to be expected when matter is transformed into radiant energy is now more thoroughly established than it was a few years ago.

There is a further type of speculation in which the public seems to be much interested but which it is important to keep entirely distinct from the foregoing considerations, for all the processes of partial or complete mass-transformation thus far suggested are such as would be in line with the assumption of the universal validity of the second law of thermodynamics. This law asserts in substance that so far as our experience here on earth goes the energy of lifted weights tends to be transformed in general into heat and thus to escape from man's control through the radiation process. In other words, the potential energy of separated attracting bodies tends to go off into radiant energy when the two bodies come together. This "Second Law" is classically and simply stated in the Humpty-Dumpty rhyme. If this law, which represents merely man's experience here on earth, were applicable to the universe as a whole, then the universe would indeed be tending toward what has been called the "heat death," or the final extinction of activity of all sorts. To take a particular illustration, if the universe is now mostly hydrogen, and if the mass of hydrogen atoms is being all the time transformed into radiant energy, the time must sometime come when all the hydrogen has been destroyed. Some scientists and philosophers have then wished to raise the question of the possibility of the replenishment of this vanishing hydrogen by the retransformation somewhere of this radiant energy back into atoms of hydrogen. This is a question upon which there has never existed any experimental evidence of any sort, though we seem to be "getting a little warmer" now than we used to be. Thus, some may think it suggestive in this connection that, as I once pointed out, cosmic rays are apparently coming into the earth from all over the celestial dome, while radiant energy is continuously being lost from the earth to the celestial dome. Others, also, may find it suggestive that, as indicated in chapter xiv, the creation of a positive-negative electron pair out of a gamma ray is actually a case of the temporary transformation of radiant energy—gamma rays—back into the potential energy of separated positive and negative electrons, i.e., back into the form of ordinary mass. Since, however, in general, this positive electron seems to exist as such only for an infinitesimal time (a time long enough for it to lose most of its kinetic energy through its ionizing capacity, say 10^{-10} seconds), and then to have its mass re-transformed into the energy of a half-million-volt gamma ray, it is quite clear that on the average there is not in this process of pair formation *any permanent return* of energy to the potential form. On the other hand, the *denial of the possibility* of such transformation anywhere in the universe on the basis of our experience here on the earth's surface represents, from my point of view, a use of "the dogma of the heat death of the universe," which is quite akin to the use of the worst form of ecclesiastical dogma, and I have myself not hesitated to express disapproval of both kinds of dogmatism. Thus, the dogma of the heat death rests squarely on the assumption that we infinitesimal mites on a speck of a world know all about how the universe behaves in all its parts, or more specifically, that the radiation laws that seem to hold here cannot possibly have any exceptions anywhere, even though that is the sort of sweeping generalization that has led physicists into error half a dozen times during the past thirty years, and also though we know quite well that conditions prevail outside our planet which we cannot here duplicate or even approach. This is why I think the dogma of the heat death has always been treated with reserve by the most thoughtful of scientists. No more crisp nor more cogent statement of what seems to me to be the correct position of science in this regard has come to my attention than is found in the following quotation from G. N. Lewis: "Thermodynamics gives no support to the assumption that the universe is running down.

Gain in entropy always means loss of information and nothing more."

In any case, the cosmic rays up the present have no significant bearing upon the question as to whether the universe is ultimately running down or whether there are processes which keep it in equilibrium. These are questions which at present must be left to the philosopher and metaphysician.

APPENDIX A

ne FROM MOBILITIES AND DIFFUSION COEFFICIENTS

If we assume that gaseous ions, which are merely charged molecules or clusters of molecules, act exactly like the uncharged molecules about them, they will tend to diffuse just as other molecules do and will exert

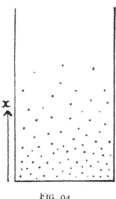

FIG. 94

a partial gas pressure of exactly the same amount as would an equal number of molecules of any gas. Imagine then the lower part of the vessel of Fig. 94 to be filled with gas through which ions are distributed and imagine that these ions are slowly diffusing upward. Let n' be the ionic concentration, i.e., the number of ions per cubic centimeter at any distance x from the bottom of the vessel. Then the number N of ions which pass per second through 1 sq. cm. taken perpendicular to x at a distance x from the bottom must be directly proportional to the concentration gradient $\frac{dn'}{dx}$ and the factor of proportionality in a given gas is by definition the diffusion coefficient D of the ions through this gas, i.e.,

$$N = D\frac{dn'}{dx} \ldots\ldots\ldots\ldots\ldots\ldots (42)$$

But since N is also equal to the product of the average velocity V with which the ions are streaming upward at

x by the number of ions per cubic centimeter at x, i.e., since $N = n'V$, we have from equation (42)

$$V = \frac{D}{n'}\frac{dn'}{dx}.$$

The force which is acting on these n'-ions to cause this upward motion is the difference in the partial pressure of the ions at the top and bottom of a centimeter cube at the point x. It is, therefore, equal to $\frac{dp}{dx}$ dynes, and the ratio between the force acting and the velocity produced by it is

$$\frac{\frac{dp}{dx}}{\frac{D}{n'}\frac{dn'}{dx}}.$$

Now this ratio must be independent of the particular type of force which is causing the motion. Imagine then the same n'-ions set in motion, not by the process of diffusion, but by an electric field of strength F. The total force acting on the n'-ions would then be Fen', and if we take v as the velocity produced, then the ratio between the force acting and the velocity produced will now be $\frac{Fen'}{v}$. By virtue then of the fact that this ratio is constant, whatever kind of force it be which is causing the motion, we have

$$\frac{Fen'}{v} = \frac{\frac{dp}{dx}}{\frac{D}{n'}\frac{dn'}{dx}} \quad \dots\dots\dots\dots\dots (43)$$

Now if v_0 denote the velocity in unit field, a quantity which is technically called the "ionic mobility," $\frac{v}{F} = v_0$. Again since the partial pressure p is proportional to n', i.e., since $p = Kn'$, it follows that $\frac{dp}{p} = \frac{dn'}{n'}$. Hence equation (43) reduces to

$$\frac{en'}{v_0} = \frac{1}{\dfrac{D}{p}}$$

or

$$v_0 = De\frac{n'}{p} \dotfill (44)$$

But if we assume that, so far as all pressure relations are concerned, the ions act like uncharged molecules (this was perhaps an uncertain assumption at the time, though it has since been shown to be correct), we have $\frac{n'}{p} = \frac{n}{P}$ in which n is the number of molecules per cubic centimeter in the air and P is the pressure produced by them, i.e., P is atmospheric pressure. We have then from equation (44)

$$ne = \frac{v_0 P}{D} \dotfill (45)$$

APPENDIX B

TOWNSEND'S FIRST ATTEMPT AT A DETER-
MINATION OF e

Figure 95 shows the arrangement of apparatus used. The oxygen rising from the electrode E is first bubbled through potassium iodide in A to remove ozone, then through water in B to enable the ions to form a cloud. This cloud-laden air then passes through a channel in an electrical insulator—a paraffin block P—into the tubes

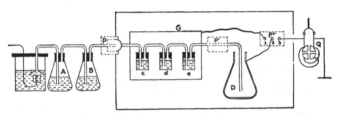

FIG. 95

c, d, e, which contain concentrated sulphuric acid. These drying tubes remove all the moisture from the air and also such part of the charge as is held on ions which in the process of bubbling through c, d, e have actually touched the sulphuric acid. The dry air containing the rest of the charge passes out through a channel in the paraffin block P' into the flask D. (If the gas being studied was lighter than air, e.g., hydrogen, D was of course inverted.) The outside of D is covered with tin foil which is connected to one of the three mercury cups held by the paraffin block P''. If the air in D contained

at first no charge, then an electrical charge exactly
equal to the quantity of electricity which enters the flask
D will appear by induction on the tin-foil coating which
covers this flask and this quantity q_1 can be measured by
connecting the mercury cup 2 to cup 3 which is connected
to the quadrant electrometer Q, and observing the deflection per minute. Precisely similarly the total quantity
of electricity which is left per minute in the drying tubes
c, d, e is exactly equal to the quantity which appears by
induction on the outer walls of the hollow metal vessel G,
which surrounds the tubes c, d, e. This quantity q_2 can
be measured by connecting mercury cup 1 to cup 3 and
observing the deflection per minute of the quadrant
electrometer. The number of cubic centimeters of gas
which pass through the apparatus per minute is easily
found from the number of amperes of current which are
used in the electrolysis apparatus E and the electro-
chemical equivalent of the gas. By dividing the quantities of electricity appearing per minute in D and G by
the number of cubic centimeters of gas generated per
minute we obtain the total charge per cubic centimeter
carried by the cloud.

The increase in weight of the drying tubes c, d, e per
cubic centimeter of gas passing, minus the weight per
cubic centimeter of saturated water vapor, gives the
weight of the cloud per cubic centimeter. This completes
the measurements involved in (2) and (3), p. 47.

As to (4), p. 48, the average size of the droplets
of water Townsend found by passing the cloud emerging
from B into a flask and observing how long it took for
the top of the cloud to settle a measured number of
centimeters. The radius of the drops could then be

obtained from a purely theoretical investigation made by Sir George Stokes,[1] according to which the velocity v_1 of fall of a spherical droplet through a gas whose co-efficient of viscosity was η is given by

$$v_1 = \frac{2}{9} \frac{g a^2}{\eta} \sigma$$

in which σ is the density of the droplet. From this Townsend got the average radius a of the droplets and computed their average weight m by the familiar formula $m = \frac{4}{3} \pi a^3 \sigma$. He was then ready to proceed as in (5). see p. 48.

[1] Lamb, *Hydrodynamics*, 1895, p. 533.

APPENDIX C

THE BROWNIAN-MOVEMENT EQUATION

A very simple derivation of this equation of Einstein has been given by Langevin of Paris[1] essentially as follows: From the kinetic theory of gases we have $PV = RT = \frac{1}{3}Nm\overline{c^2}$ in which $\overline{c^2}$ is the average of the squares of the velocities of the molecules, N the number of molecules in a gram molecule, and m the mass of each. Hence the *mean* kinetic energy of agitation E of each molecule is given by $E = \frac{1}{2}m\overline{c^2} = \frac{3}{2}\frac{RT}{N}$.

Since in observations on Brownian movements we record only motions along one axis, we shall divide the total energy of agitation into three parts, each part corresponding to motion along one of the three axes, and, placing the velocity along the x-axis equal to $\frac{dx}{dt}$, we have

$$\frac{E}{3} = \frac{1}{2}m\left(\overline{\frac{dx}{dt}}\right)^2 = \frac{1}{2}\frac{RT}{N} \quad \dots\dots\dots\dots (46)$$

Every Brownian particle is then moving about, according to Einstein's assumption, with a mean energy of motion along each axis equal to $\frac{1}{2}\frac{RT}{N}$. This motion is due to molecular bombardment, and in order to write an equation for the motion at any instant of a particle subjected to such forces we need only to know (1) the

[1] *Comptes Rendus*, CXLVI (1908), 530.

value X of the x-component of all the blows struck by the molecules at that instant, and (2) the resistance offered by the medium to the motion of the particle through it. This last quantity we have set equal to Kv and have found that in the case of the motion of oil droplets through a gas K has the value $6\pi\eta a\left(1+A\dfrac{l}{a}\right)^{-1}$. We may then write the equation of motion of the particle at any instant under molecular bombardment in the form

$$m\frac{d^2x}{dt^2} = -K\frac{dx}{dt} + X \ldots\ldots\ldots\ldots(47)$$

Since in the Brownian movements we are interested only in the absolute values of displacements without regard to their sign, it is desirable to change the form of this equation so as to involve x^2 and $\left(\dfrac{dx}{dt}\right)^2$. This can be done by multiplying through by x. We thus obtain, after substituting for $x\dfrac{d^2x}{dt^2}$ its value $\frac{1}{2}\dfrac{d^2(x^2)}{dt^2} - \left(\dfrac{dx}{dt}\right)^2$,

$$\frac{m}{2}\frac{d^2(x^2)}{dt^2} - m\left(\frac{dx}{dt}\right)^2 = -\frac{K}{2}\frac{d(x^2)}{dt} + Xx \ldots\ldots\ldots(48)$$

Langevin now considers the *mean* result arising from applying this equation at a given instant to a large number of different particles all just alike.

Writing then s for $\dfrac{d(\overline{x^2})}{dt}$ in which $\overline{x^2}$ denotes the mean of all the large number of different values of x^2, he gets after substituting $\dfrac{RT}{N}$ for $m\left(\dfrac{dx}{dt}\right)^2$, and remembering that in taking the mean, since the X in the last term is as likely to be positive as negative and hence that $\overline{Xx}=0$,

$$\frac{m}{2}\frac{dz}{dt} - \frac{RT}{N} = -\frac{Kz}{2}.$$

Separating the variables this becomes

$$\frac{dz}{\left(z-\frac{2RT}{NK}\right)} = -\frac{K}{m}dt,$$

which yields upon integration between the limits o and τ

$$z = \frac{2RT}{NK} + Ce^{-\frac{K}{m}\tau} \ldots \ldots \ldots \ldots \ldots (49)$$

For any interval of time τ long enough to measure this takes the value of the first term. For when Brownian movements are at all observable, a is 10^{-4} cm. or less, and since K is roughly equal to $6\pi\eta a$ we see that, taking the density of the particle equal to unity,

$$\frac{m}{K} = \frac{\frac{4}{3}\pi(10^{-4})^3}{6\pi.00018\times10^{-4}} = 10^{-8}.$$

Hence when τ is taken greater than about 10^{-8} seconds, $e^{-\frac{K}{m}\tau}$ rapidly approaches zero, so that for any measurable time intervals

$$z = \frac{2RT}{NK}$$

or

$$\frac{d(\overline{x^2})}{dt} = \frac{2RT}{NK}$$

and, letting $\overline{\Delta x^2}$ represent the change in $\overline{x^2}$ in the time τ

$$\overline{\Delta x^2} = \frac{2RT}{NK}\tau \quad \dots \dots \dots \dots \dots (50)$$

This equation means that if we could observe a large number n of exactly similar particles through a time τ, square the displacement which each undergoes along the x-axis in that time, and average all these squared displacements, we should get the quantity $\frac{2RT}{NK}\tau$. But we must obviously obtain the same result if we observe the same identical particle through n-intervals each of length τ and average these n-displacements. The latter procedure is evidently the more reliable, since the former must assume the exact identity of the particles.

APPENDIX D

THE INERTIA OR MASS OF AN ELECTRICAL CHARGE ON A SPHERE OF RADIUS a

If Fig. 96 represents a magnet of pole area A, whose two poles are d cm. apart, and have a total magnetization M, a density of magnetization σ, and a field strength between them of H, then the work necessary to carry a unit pole from M to M' is Hd, and the work necessary to create the poles M and M', i.e., to carry M units of magnetism across against a mean field strength $\frac{H}{2}$ is $\frac{HMd}{2}$. Hence the total energy E_1 of the magnetic field is given by

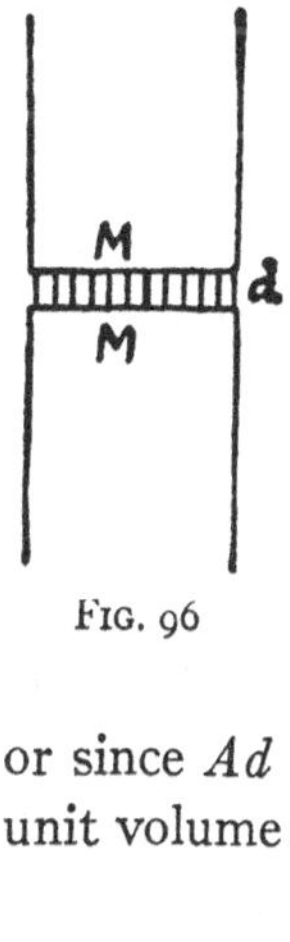

$$E_1 = \frac{HMd}{2} = \frac{HA\sigma d}{2},$$

but since $H = 4\pi\sigma$

$$E_1 = \frac{H^2 A d}{8\pi},$$

Fig. 96

or since Ad is the volume of the field the energy E per unit volume of the magnetic field is given by

$$E = \frac{H^2}{8\pi}. \dots\dots\dots\dots\dots\dots (51)$$

Now the strength of the magnetic field at a distance r from a moving charge in the plane of the charge is $\frac{ev}{r^2}$, if e is the charge and v its speed. Also the magnetic field strength at a point distant $r\theta$ from the charge, θ being the angle between r and the direction of motion, is given by

$$H = \frac{ev}{r^2}\sin\theta.$$

Hence the total energy of the magnetic field created by the moving charge is

$$\int E d\tau = \int \frac{H^2}{8\pi} d\tau$$

in which τ is an element of volume and the integration is extended over all space. But in terms of v, θ, and ϕ.

$$d\tau = r d\theta, dr, r \sin\theta d\phi$$

$\therefore$ Total energy=

$$\frac{e^2v^2}{8\pi}\int \frac{\sin^2\theta}{r^4} d\tau = \frac{e^2v^2}{8\pi}\int_a^\infty \frac{dr}{r^2}\int_0^{2\pi} d\phi \int_0^\pi \sin^3\theta d\theta = \frac{e^2v^2}{3a}.$$

Since kinetic energy $=\frac{1}{2}mv^2$, the mass-equivalent m of the moving charge is given by setting

$$\frac{1}{2}mv^2 = \frac{e^2v^2}{3a}$$

$$\therefore \quad m = \frac{2}{3}\frac{e^2}{a} \dots\dots\dots\dots\dots (52)$$

The radius of the spherical charge which would have a mass equal to the observed mass of the negative electron is found by inserting in the last equation $e = 4.770 \times 10^{-10}$

electrostatic units $= 1.591 \times 10^{-20}$ electromagnetic units and $\frac{e}{m} = 1.757 \times 10^7$ electromagnetic units. This gives $a = 1.9 \times 10^{-13}$ cm.

The expression just obtained for m obviously holds only so long as the magnetic field is symmetrically distributed about the moving charge, as assumed in the integration, that is, so long as v is small compared with the velocity of light. When v exceeds .1 the speed of light c, the mass of the charge begins to increase measurably and becomes infinite at the speed of light. According to the theory developed by Lorentz, if the mass for slow speeds is called m_0 and the mass at any speed v is called m, then

$$\frac{m}{m_0} = \frac{1}{\sqrt{1 - \frac{v^2}{c^2}}} \dots\dots\dots\dots\dots (53)$$

This was the formula which Bucherer found to hold accurately for the masses of negative electrons whose speeds ranged from .3 to .8 that of light.

APPENDIX E

MOLECULAR CROSS-SECTION AND MEAN
FREE PATH

If there is one single molecule at rest in a cubical space 1 cm. on a side, the chance that another molecule which is shot through the cube will impinge upon the one contained is clearly $\frac{\pi d^2}{1}$ in which d is the mean diameter of the two molecules. If there are n contained molecules the chance is multiplied by n, that is, it becomes $\frac{n\pi d^2}{1}$. But on the average the chance of an impact in going a centimeter is the number of impacts actually made in traversing this distance. The mean free path l is the distance traversed divided by the number of impacts made in going that distance. Hence

$$l = \frac{1}{n\pi d^2} \quad \dots\dots\dots\dots\dots\dots (54)$$

This would be the correct expression for the mean free path of a molecule which is moving through a group of molecules at rest. If, however, the molecules are all in motion they will sometimes move into a collision which would otherwise be avoided, so that the collisions will be more numerous when the molecules are in motion than when at rest—how much more numerous will depend upon the law of distribution of the speeds of the molecules. It is through a consideration of the Maxwell distribution law that the factor $\sqrt{2}$ is introduced into the denominator (see Jeans, *Dynamical Theory of Gases*) so that equation (54) becomes

$$l = \frac{1}{\sqrt{2}\,n\pi d^2} \quad \dots\dots\dots\dots\dots\dots (55)$$

APPENDIX F

NUMBER OF FREE POSITIVE ELECTRONS IN THE NUCLEUS OF AN ATOM BY RUTHERFORD'S METHOD

If N represents the number of free positive electrons in the nucleus, e the electronic charge, E the known charge on the a-particle, namely $2e$, and $\frac{1}{2}mV^2$ the known kinetic energy of the a-particle, then, since the inertias of the negative electrons are quite negligible in comparison with that of the a-particle, if the latter suffers an appreciable change in direction in passing through an atom it will be due to the action of the nuclear charge. If b represents the closest possible approach of the a-particle to the center of the nucleus, namely, that occurring when the collision is "head on," and the a-particle is thrown straight back upon its course, then the original kinetic energy $\frac{1}{2}mV^2$ must equal the work done against the electric field in approaching to the distance b, i.e.,

$$\tfrac{1}{2}mV^2 = \frac{NeE}{b} \dotfill (56)$$

Suppose, however, that the collision is not "head on," but that the original direction of the a-particle is such that, if its direction were maintained, its nearest distance of approach to the nucleus would be p (Fig. 97). The deflection of the a-particle will now be, not $180°$, as before, but some other angle ϕ. It follows simply from the geometrical properties of the hyperbola and the elementary principles of mechanics that

$$p = \frac{b}{2} \cot \frac{\phi}{2} \quad \ldots \ldots \ldots \ldots \ldots \ldots (57)$$

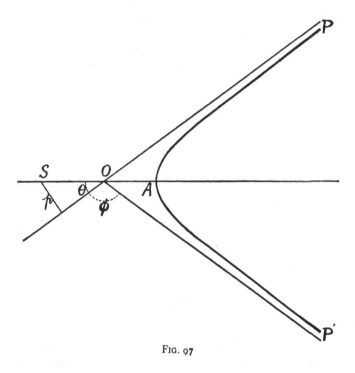

FIG. 97

For let PAP' represent the path of the particle and let $POA = \theta$. Also let V = velocity of the particle on entering the atom and v its velocity at A. Then from the conservation of angular momentum

$$pV = SA \cdot v \ldots \ldots \ldots \ldots \ldots \ldots (58)$$

and from conservation of energy

$$\tfrac{1}{2}mV^2 = \tfrac{1}{2}mv^2 + \frac{NeE}{SA}$$

$$\therefore \quad v^2 = V^2\left(1 - \frac{b}{SA}\right) \dots\dots\dots\dots (59)$$

Since the eccentricity $\epsilon = \sec\theta$, and for any conic the focal distance is the eccentricity times one-half the major axis, i.e., $SO = OA \cdot \epsilon$, it follows that

$$SA = SO + OA = SO\left(1 + \frac{1}{\epsilon}\right) = p\ csc\ \theta(1 + \cos\theta) = p\cot\frac{\theta}{2}.$$

But from equations (58) and (59)

$$p^2 = SA(SA - b) = p\cot\frac{\theta}{2}\left(p\cot\frac{\theta}{2} - b\right)$$

$$\therefore \quad b = 2p\cot\theta \dots\dots\dots\dots\dots (60)$$

and since the angle of deviation ϕ is $\pi - 2\theta$, it follows that

$$\cot\frac{\phi}{2} = \frac{2p}{b} \dots\dots\dots (61) \quad \text{Q.E.D.}$$

Now it is evident from the method used in Appendix E that if there are n atoms per cubic centimeter of a metal foil of thickness t, and if each atom has a radius R, then the probability M that a particle of size small in comparison with R will pass through one of these atoms in shooting through the foil is given by

$$M = \pi R^2 nt.$$

Similarly the probability m that it will pass within a distance p of the center of an atom is

$$m = \pi p^2 nt.$$

If this probability is small in comparison with unity, it represents the fraction ρ of any given number of particles shooting through the foil which will actually come within a distance p of the nucleus of an atom of the foil.

The fraction of the total number which will strike within radii p and $p+dp$ is given by differentiation as

$$dm = 2\pi pnt \cdot dp$$

but from equation (57)

$$dp = -\frac{b}{2}\tfrac{1}{2}csc^2\frac{\phi}{2}d\phi$$

$$\therefore \quad dm = -\frac{\pi}{4}ntb^2 \cot\frac{\phi}{2}\,csc^2\frac{\phi}{2}d\phi.$$

Therefore the fraction ρ which is deflected between the angles ϕ_1 and ϕ_2 is given by integration as

$$\rho = \frac{\pi}{4}ntb^2\left(\cot^2\frac{\phi_1}{2} - \cot^2\frac{\phi_2}{2}\right)$$

It was this fraction of a given number of α-particles shot into the foil which Geiger and Marsden found by direct count by the scintillation method to be deflected through the angles included between any assigned limits ϕ_1 and ϕ_2. Since n and t are known, b could be at once obtained. It was found to vary with the nature of the atom, being larger for the heavy atoms than for the lighter ones, and having a value for gold of 3.4×10^{-12} cm. This is then an upper limit for the size of the nucleus of the gold atom.

As soon as b has thus been found for any atom, equation (56) can be solved for N, since E, e, and $\frac{1}{2}mV^2$ are all known. It is thus that the number of free positive electrons in the nucleus is found to be roughly half the atomic weight of the atom, and that the size of the nucleus is found to be very minute in comparison with the size of the atom.

APPENDIX G

BOHR'S THEORETICAL DERIVATION OF THE VALUE OF THE RYDBERG CONSTANT

The Newtonian equation of a circular orbit of an electron e rotating about a central attracting charge E, at a distance a, with a rotational frequency n, is

$$\frac{eE}{a^2} = (2\pi n)^2 ma \quad \ldots\ldots\ldots\ldots\ldots (62)$$

The kinetic energy of the electron is $\frac{1}{2}m(2\pi na)^2 = \frac{1}{2}\frac{eE}{a}$. The work required to move the electron from its orbit to a position at rest at infinity is $\frac{eE}{a} - \frac{1}{2}m(2\pi na)^2 = \frac{1}{2}\frac{eE}{a}$. If we denote this quantity of energy by T, it is seen at once that

$$2a = \frac{eE}{T}$$

and $\qquad\qquad\qquad\qquad\qquad\qquad \left.\begin{array}{c}\\ \\ \\ \\ \end{array}\right\} \quad\ldots\ldots\ldots (63)$

$$n = \frac{\sqrt{2}T^{\frac{3}{2}}}{\pi eE\sqrt{m}} \ \ldots\ldots$$

If we combine this with (37), p. 213, there results at once

$$T = \frac{2\pi^2 me^2 E^2}{\tau^2 h^2} \qquad 2a = \frac{\tau^2 h^2}{2\pi^2 meE} \qquad n = \frac{4\pi^2 me^2 E^2}{\tau^3 h^3} \ldots (64)$$

Upon change in orbit the radiated energy must be

$$T_{\tau_1} - T_{\tau_2} = \frac{2\pi^2 me^2 E^2}{h^2}\left(\frac{1}{\tau_1^2} - \frac{1}{\tau_2^2}\right),$$

477

and, if we place this equal to $h\nu$, there results the Balmer formula (34), p. 210,

$$\nu = N\left(\frac{1}{\tau_1^2} - \frac{1}{\tau_2^2}\right)$$

in which

$$N = 2\pi^2 e^2 E^2 \frac{m}{h^3}$$

Since for hydrogen $E = e$, we have

$$N = \frac{2\pi^2 m e^4}{h^3}$$

and from (64)

$$a = \frac{\tau^2 h^2}{4\pi^2 m e^2}.$$

APPENDIX H

A. H. COMPTON'S THEORETICAL DERIVATION OF THE CHANGE IN THE WAVE-LENGTH OF ETHER-WAVES BECAUSE OF SCATTERING BY FREE ELECTRONS

Imagine, as in Fig. $42A$, that an X-ray quantum of frequency ν_0 is scattered by an electron of mass m. The momentum of the incident ray will be $h\nu_0/c$, where c is the velocity of light and h is Planck's constant, and that

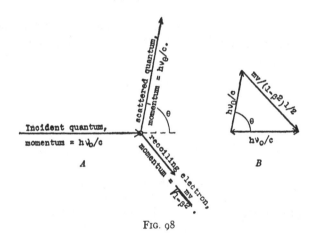

F$_{IG}$. 98

of the scattered ray is $h\nu_0/c$ at an angle θ with the initial momentum. The principle of the conservation of momentum accordingly demands that the momentum of recoil of the scattering electron shall equal the vector difference between the momenta of these two rays, as in

Fig. 42B.   The momentum of the electron, $m\beta c/\sqrt{1-\beta^2}$, is thus given by the relation

$$\left(\frac{m\beta c}{\sqrt{1-\beta^2}}\right)^2 = \left(\frac{h\nu_0}{c}\right)^2 + \left(\frac{h\nu_\theta}{c}\right)^2 - 2\frac{h\nu_0}{c}\cdot\frac{h\nu_\theta}{c}\cos\theta\, ,\ .\,.(65)$$

where β is the ratio of the velocity of recoil of the electron to the velocity of light.   But the energy $h\nu_\theta$ in the scattered quantum is equal to that of the incident quantum $h\nu_0$ less the kinetic energy of recoil of the scattering electron, i.e.,

$$h\nu_\theta = h\nu_0 - mc^2\left(\frac{1}{\sqrt{1-\beta^2}}-1\right)\ldots\ldots\ldots(66)$$

We thus have two independent equations containing the two unknown quantities β and ν_θ.   On solving the equations we find

$$\nu_\theta = \nu_0/(1+2a\sin^2\tfrac{1}{2}\theta),\ldots\ldots\ldots\ldots(67)$$

where

$$a = h\nu_0/mc^2 = h/mc\lambda_0\ldots\ldots\ldots\ldots\ldots(68)$$

or, in terms of wave-length instead of frequency,

$$\lambda_\theta = \lambda_0 + (2h/mc)\sin^2\tfrac{1}{2}\theta\ldots\ldots\ldots\ldots(69)$$

Substituting the accepted values of h, m, and c,

$$\lambda_\theta - \lambda_0 = \Delta\lambda = 0.0484\sin^2\tfrac{1}{2}\theta\ldots\ldots\ldots\ldots(70)$$

APPENDIX I

THE ELEMENTS, THEIR ATOMIC NUMBERS, ATOMIC WEIGHTS, AND CHEMICAL POSITIONS

1 H
1.008

0	I	II	III	IV	V	VI	VII	VIII		
2 He 3.99	3 Li 6.94	4 Be 9.1	5 B 11.0	6 C 12.00	7 N 14.01	8 O 16.00	9 F 19.0			
10 Ne 20.2	11 Na 23.00	12 Mg 24.32	13 Al 27.1	14 Si 28.3	15 P 31.04	16 S 32.06	17 Cl 35.46			
18 A 39.88	19 K 39.10	20 Ca 40.07	21 Sc 44.1	22 Ti 48.1	23 V 51.0	24 Cr 52.0	25 Mn 54.93	26 Fe 55.84	27 Co 58.97	28 Ni 58.68
....	29 Cu 63.57	30 Zn 65.37	31 Ga 69.9	32 Ge 72.5	33 As 74.96	34 Se 79.2	35 Br 79.92			
36 Kr 82.92	37 Rb 85.45	38 Sr 87.63	39 Y 88.7	40 Zr 90.6	41 Nb 93.5	42 Mo 96.0	**43 Ma**	44 Ru 101.7	45 Rh 102.9	46 Pd 106.7
....	47 Ag 107.88	48 Cd 112.40	49 In 114.8	50 Sn 118.7	51 Sb 120.2	52 Te 127.5	53 J 126.92			
54 X 130.2	55 Cs 132.81	56 Ba 137.37	57 La 139.0	58 Ce 140.25 59 Pr 140.6 60 Nd 144.3 61 Il 62 Sm 150.4 63 Eu 152 64 Gd 157.3 65 Tb 159.2 66 Ds 162.5						
67 Ho 163.5 68 Ev 167.7 69 Tu 168.5 70 Yb 173.5 71 Lu 175.0 72 Hf							73 Ta 181.5	74 W 184.0	**75 Re**	76 Os 190.9 77 Ir 193.1 78 Pt 195.2
....	79 Au 197.2	80 Hg 200.6	81 Tl 204.0	82 Pb 207.20	83 Bi 208.0	84 Po (210.0)	85 —			
86 Em (222.0)	87 —	88 Ra 226.0	89 Ac (227)	90 Th 232.15	UrX2 (234)	92 Ur 238.2	....			

Elements, the atomic numbers of which are not in the order of atomic weights, are in italics. The numbers corresponding to missing elements are in bold-faced type.

1 Hydrogen	24 Chromium	47 Silver	70 Ytterbium
2 Helium	25 Manganese	48 Cadmium	71 Lutecium
3 Lithium	26 Iron	49 Indium	72 Hafnium
4 Beryllium	27 Cobalt	50 Tin	73 Tantalum
5 Boron	28 Nickel	51 Antimony	74 Tungsten
6 Carbon	29 Copper	52 Tellurium	75 Rhenium
7 Nitrogen	30 Zinc	53 Iodine	76 Osmium
8 Oxygen	31 Gallium	54 Xenon	77 Iridium
9 Fluorine	32 Germanium	55 Caesium	78 Platinum
10 Neon	33 Arsenic	56 Barium	79 Gold
11 Sodium	34 Selenium	57 Lanthanum	80 Mercury
12 Magnesium	35 Bromine	58 Cerium	81 Thallium
13 Aluminium	36 Krypton	59 Praseodymium	82 Lead
14 Silicon	37 Rubidium	60 Neodymium	83 Bismuth
15 Phosphorus	38 Strontium	61 Illinium	84 Polonium
16 Sulphur	39 Yttrium	62 Samarium	85 ———
17 Chlorine	40 Zirconium	63 Europium	86 Niton
18 Argon	41 Niobium	64 Gadolinium	87 Ekacaesium
19 Potassium	42 Molybdenum	65 Terbium	88 Radium
20 Calcium	43 Mazurium	66 Dyprosium	89 Actinium
21 Scandium	44 Rhuthenium	67 Holmium	90 Thorium
22 Titanium	45 Rhodium	68 Erbium	91 Uranium X.
23 Vanadium	46 Paladium	69 Thulium	92 Uranium

APPENDIX J

The following are the most important physical constants, the values of which it has become possible to fix,[1] within about the limits indicated, through the isolation and measurement of the electron.

The electron........................ $e = (4.770 \pm 0.005) \times 10^{-10}$

The Avogadro constant.............. $N = (6.064 \pm 0.006) \times 10^{23}$

Number of gas molecules per cc. at

0° C. 76 cm..................... $n = (2.706 \pm 0.003) \times 10^{19}$

Boltzmann constant $\left(\dfrac{R_0}{N} = \dfrac{8.3136 \times 10^7}{6.064 \times 10^{23}} \right) = k = (1.371 \pm 0.001) \times 10^{-16}$

Mass of atom of unit atomic weight.. $m_0 = (1.649 \pm 0.0016) \times 10^{-24}$

Mass of an atom of hydrogen in grams. $m = (1.6618 \pm 0.0017) \times 10^{-24}$

Planck's element of action.......... $h = (6.55 \pm 0.01) \times 10^{-27}$

Wien constant of spectral radiation.... $c_2 = 1.432 \pm 0.0030$

Stefan-Boltzmann constant of total

radiation....................... $\sigma = (5.72 \times 0.034) \times 10^{-12}$

Grating spacing in calcite........... $d = 3.028 \pm 0.001 \text{Å}$

[1] See *Phys. Rev.*, XXXV (1930) 1231; also Raymond T. Birge, *Phys. Rev.*, Supplement I, July, 1929.

INDEX OF AUTHORS

Aepinus, 12, 14

Alvarez, 441

Ampere, 20, 21

Anderson, Carl D., chap. xiv

Aristotle, 9

Arnold, 95 ff., 124

Arnold, Col., 412, 428

Avogadro, 30, 183, 261

Baade, 315

Bacon, 8, 9

Balmer, 210, 478

Bär, 173, 181

Barkla, 196, 199

Barnett, 212

Becker, 259 f., 361, 368, 379, 397

Becquerel, 301, 350

Begeman, 57

Bennett R. D., 362, 405

Birge, R. T., 482

Blake, 214

Bleakney, W., 144

Blockett, 335, 337, 339, 343, 398, 401

Bodoszewski, 145, 147

Bohr, 210, 211, 213 f., 221, 223, 244, 248, chaps. ix, xi, and xii

Boltwood, 160

Boltzmann, 81, 261

Bothe, 361, 368, 369, 370, 379, 397, 418

Bowen, 143, 203, 220, 248, chap. xii, 304, 305, 309, 355, chap. xvi

Bragg, 139, 189, 197

Breene, Capt. 413

Brown, 145

Bucherer, 188, 214, 469

Burton, 302

Cameron, G. H., 304, 307, 316, 355, 408, 449

Campbell, 27

Carbonelle, 145

Cario, 256

Cassen, 362

Chadwick, chaps. xiv and xv

Chao, 341, 356, 357, 358, 397, 398, 399, 401, 421

Clark, 246, 259

Clausius, 8

Clay, 410, 411, 439

Cockroft, 365, 388

Compton, A. H., 245, 256, 257 ff., 260, 264, 340, 356, 370, 401, 405, 423, 430, 438, 441

Cook, 302

Corlin, 310

Coster, 221, 225

Cosyns, 424

Coulomb, 32

Crane, H. R., 351–57 and chap. xv

Crookes, Sir William, 24, 41

Cullen, Capt. 439

Cunningham, 90, 163, 167

Curie, Madam, 301

Curie-Joliot, 335, 336, 337

Dalton, 2

Darrow, Carl, 387

Davidson, 262

Davis, 247

D'Albe, Fournier, 41

DeBroglie, Louis, 148 f., 150, 152, 161 f., 198, 199, 200, 231, 246

DeBroglie, Maurice, 260, 268

Dee, 366, 367

DeHaas, 212

Delsasso, 389
Delsaulx, 145
Democritus, 2, 6, 8, 9, 15
Derieux, 135, 178
DeWatteville, 148, 161
Dirac, 264, 339, 349, 389
Drude, 251
Duane, 214, 246, 259, 260
Dufay, 11
Du Mond, 259, 260
Dunning, J. R., 387
Dunnington, 42

Eddington, 449
Ehrenfest, 260
Ehrenhaft, 148, 162 f., 164, 166 ff., 168 f., 170, 171 ff., 175 f., 180, 181
Einstein, 146, 147, 155, 175, 180, 188, 212, 237 f., 244 f., 247 f., 253, 256, chaps. xiii to xvi, 463
Eisl, 420
Ellis, C. D., 246, 361
Enright, 46
Epicurus, 6
Epstein, 212, 219 220, 256, 260, 410, 415, 417, 435, 446
Erikson, 38
Eyring, 261

Faraday, 12, 15, 16, 17, 19, 24, 25, 27, 28, 29, 58, 158, 237 f., 249
Fletcher, 129, 151 f., 153 ff., 164 168, 169, 172
Foote, 247
Fordney-Settle, 426, 430, 434
Foulois, Gen., 413
Fowler, 142, 389
Franck, 38, 39, 40, 126 f., 134, 247, 256
Franklin, Benjamin, 11 ff., 14, 15, 20, 24
Friedrick, 263, 264

Gaertner, 117
Geiger, 158, 159, 180, 195, 196, 475

Gerlach, 135 f., 295
Germer, 262
Gibson, 94
Gilbert, 11
Gilchrist, 93, 94
Gish, 409, 410
Gockel, 303
Goldhaber, 396
Goucher, 247
Goudsmit, 292
Gouy, 145 f.
Graham, First Officer, 439
Gray, 341, 356, 358
Grindley, 94

Hadamard, 88
Hancock, Capt. Allan, 436, 437
Hapfield, 341
Harkins, 449
Harper, 352
Harrington, 94, 120
Heisenberg, 264, 268, 293, 300
Helmholtz, 22, 23, 24
Hemsalech, 148, 161
Henderson, 354
Hertz, 17, 247
Hess, 304, 305, 310, 311
Hevesy, 225
Hoffmann, 310, 405
Hogg, 94
Houston, W. A., 42, 274, 292
Howell, 425
Hughes, 245
Hund, 293
Huygens, 233

Ishida, Y., 178

Jeans, 398, 448, 471
Joffé, 135
Johnson, T. H., 441, 443, 445
Joliot, M. F., 351, 358, 398
Joule, 7

Karpowicz, 171
Kaufmann, 448
Kelly, 136, 142
Kelvin, Lord, 4, 23, 24
Kikuchi, 264
Kinsey, 385, 392, 394, 395
Kirchner, 366
Kirkpatrick, 259, 260
Klein, 255
Klein-Nishina, 398, 401, 402, 444, 450
Knipping, 263, 264
Kolhörster, 304, 308, 309, 424
Konstaninowsky, 166
Korff, Serge, 438, 442, 443, 445
Kossell, 231
Kramers, 221
Kunze, 423, 452
Kurz, 303

Ladenburg, 97, 98, 245
Lamb, 267, 462
Langer, 407, 435
Langevin, 35, 134, 147, 463
Laplace, 46
Laue, 197
Laurence, 354, 363, 368
Lauritsen, 261, 351–57, and chap. xv
Lavoisier, 46
Lee, J. Y., 178
Le Maitre, 415, 446
Lenard, 236, 251
Leucippus, 2
Lewis, G. N., 368, 445
Lindholm, 310, 405
Livingston, 354, 368
Lodge, Sir Oliver, 17, 46
Loeb, 37, 38
Lorentz, 21, 274
Lucretius, 2, 6
Lunn, 89
Lyman, 204, 210, 213, 283

McLennan, 302
MacMillan, 402
Main-Smith, 296
Marsden, 195, 196, 475
Maxwell, Clerk-, 8, 17, 19, 20, 24, 77, 242, 470
Meitner, L., 335, 337, 341, 398
Mendeleéff, 202
Meyer, 32, 135 f.
Millikan, 57, 120, 129, Fig. 36, 142, 150, 184, 203, 209, 214, 248, 254, 260
Millikan and Bowen, 203
Mohler, 247
Moseley, 196, 198, 200, 202, 203, 204 f., 216
Mott-Smith, 424

Neddermeyer, Seth, 321, 339, 340, 352, 398, 401, 418, 421, 422, 432, 444, 452
Neher, H. V., 309, 313, 406, 424, 435, 437–39
Nernst, 27
Newton, 211, 233, 256
Nichols, 248
Nordlund, 157, 181

Occhialini, 335, 337, 339, 343, 398, 401
Oliphant, 385, 392, 394–96
Oppenheimer, 261, 264, 339, 341, 342, 357, 399, 402
Ostwald, 10, 157
Otis, R., 304, 307

Paschen, 204, 210, 274
Pauli, 293, 295
Peltier, 20, 21
Perrin, 147, 170
Phillip, 335, 337, 398
Piccard, A., 309, 424, 433
Pierson, 136
Planck, 117, 212, 213, 237 f., 242, 244, 253 f., 261

Plato, 9
Plesset, 339, 399
Pohl, 166, 245
Prenin, 27
Pringsheim, 245
Prout, 205, 206
Przibram, 156, 166
Pythagoras, 4

Quincke, 52

Ramsay, 103
Rapp, 93, 94
Rasetti, 379
Rayleigh, Lord, 252
Regener, 159, 160, 180, 309, 316, 398, 404, 424, 433
Richardson, 27, 245, 303
Ritz, 210
Roetgen, 301
Ross, 258, 259
Rosseland, 255
Rossi, 442
Rowland, 185
Russell, A. H., 293, 297, 298, 313, 444
Rutherford, Sir Ernest, 27, 33, 35, 50, 158, 159, 160, 180, 195, 209, 252, 301, 327, 360, 361, 385, 392, 394, 395, 418, 472
Rydberg, 210, 477

Sadler, 199
Saunders, 293
Sawyer, 283
Schidlof, 171
Schmid, 172
Schreidenger, 264
Schuster, Sir Arthur, 42
v. Schweidler, 304, 305
Siegbahn, 198, 200
Skobelzyn, 329
Smoluchowski, 148, 154
Smythe, 259 f.

Soddy, 147, 202, 301
Soltan, 354, 368
Sommerfeld, 211, 218, 271, 275, 278, chap. xii
Spencer, 103
Stearns, 405
Stefan-Boltzmann, 117, 261
Steinke, 310
Stephenson, R. J., 430
Stern, Otto, 263, 295
Stokes, 48, 90, 91, 97, 124, 462
Stoner, 296
Stoney, G. Johnstone, 21, 25, 26, 31
Störmer, 446
Sutherland, 86
Svedberg, 150
Symmer, 13

Tarrant, 341, 356, 358
Tear, 248
Thales, 1, 6, 205
Thibaud, 358, 398
Thirion, 145
Thompson, G. P., 262
Thomson, Sir J. J., 14, 27, 33, 34, 42, 43, 49, 50, 51, 52, 53, 54, 55, 57, 58, 70, 143, 144, 186, 235, 237, 238, 248, 249, 253, 320
Tomlinson, 94
Townsend, 36, 39, 40, 45, 46, 47, 50 f., 53, 54, 55, 58, 70, 125, 126 f., 134, 265 f., 460
Tuve, M. A., 363
Tyndall, 8, 9

Uhlenbeck, 292

Vallerta, 415, 446
van de Graaff, 363
Varley, 24
Volta, 309

Wahlin, 38
Walton, 365, 366, 367, 388

Warburg, 32
Watson, 246
Weber, 20, 21
Webster, 246
Weiss, 153, 156, 166
Wellish, 37
Westgren, 157, 181
Westphal, 38, 39, 40, 126 f., 134
Wien, 43, 117
Wigand, 305

Wilson, C. T. R., 48, 51, 52, 134 f., 138, 139 f., 141, 445
Wilson, H. A., 52, 54, 55 f., 57, 63, 70, 161 f.
Wood, 247

Zeeman, 43
Zeleny, 35
Zerner, 166 f., 168 ff.
Zwicky, F., 315

INDEX OF SUBJECTS

Absorption frequencies, 199 ff.

Absorption spectra, 200 ff.

Alkalin metals, spectra of, 275

Alpha particles: charge of, 158 f., 180; deflection of, 192; penetration by, 190, 194; range of, 190

Amperian current, 21

Angular momentum, atomicity of 216

Annihilation of matter, 453

Annihilation rays, 354, 399

Arequipa, 414

Aristotelian philosophy, 9

Arrowhead Lake, 307, 316

Artificial radioactivity, 350

Atmospheric resistance to electrons, 6 billion volts, 414

Atom, 26; the Bohr, 209; constituents of, 41, 189; in helium, 141; hydrogen, 26, 27, 42, 215; impenetrable portion, 194; loose structure of, 139, 194; miniature stellar system, 141; multiply charged, 143; nucleus of, 193; size of, 183; stripped, 282; structure of, 182; a system, 184

Atom-ion theory, 37

Atomic numbers of elements, 481

Atomic structures, 222, 224, 229, 230, 231; and spectral lines, 205

Atomic system, 193

Atomic theories: of matter, 6, 7, 8, 10, 15, 77, 157; of electricity, 15, 23 f., 66, 163; and strain theory, 18

Atomic weight, 196, 202, 206, 207

Atomic weight table, 481

Atoms, number of, 195

Avogadro's constant, 30, 261; rule, 183

Azimuthal quantum number, 272

Balance, electrical, 103; quartz, 103

Balanced-drop method, 57 ff., 66

Balmer formula, 478

Balmer-Ritz equation, 210, 215, 216

Balmer series, 204, 213, 214 f.; constant of, 214, 273

Bell Laboratories, 262

Beryllium rays, 361

Black-body radiation 232 236, 237

Bohr atom, 209, 211 ff.; orbits for hydrogen, 215; theory, 210, 214, 230, 231; quantum principle, 218

Bohr-Sommerfeld model of hydrogen atom, 219

Bohr's: derivation of Rydberg constant, 282; periodic table, 226

Boltzmann constant, 482

Bremsstrahlung photons, 423

Brownian movements, 10, 131, 136, 145 ff., 147, 164, 167, 170, 171, 175, 268; determining e from, 167; Einstein-, equation, 151, 156, 175, 180; equation, 268; experiments with dust-free air, 165; fluctuations of, 171; in gases, 145, 149, 150; in liquids, 156; 463

"Carnegie," the, 409

Cathode rays, 23 f.; charge on, 80; $\frac{e}{m}$ for, 42; frequencies of X-ray spectra, 197 f.

Cavendish Laboratory, 33, 35, 44, 48, 54

Charge: of alpha particle, 158 f.; change of, in drop, 77, 135, 249; constant molecular, 19; electrical, 22, 28, 30, 40, 46; free positive, on atoms, 196; frictional, 72, 73, 76; granular structure of, 77; ionic, 47, 76; multiple, 143; positive and negative, 12, 27; ratio of, to mass, 42; single, 134, value of, 59

Churchill, Manitoba, 410

Cluster-ion theory, 37

Cold emission, 261, 365

Compton effect, 256, 257 f., 259, 260, Fig. 36

Compton wave-length displacement formula, 479

Cormorant Lake, 413, 417

Corpuscular theory, 233

Cosmic rays, banded structure, 316; discovery of, chap. xiii; east-west effect, 436; energy of, chap. xiv; hypothesis of origin, 314; latitude effect in, 407; longitude, 436; nature of, chap. xvi; penetrating powers of, 436; place of origin, 310; showers, 341; spectral distribution of, 317

Coulomb, 28

Creation of matter, 451

Democritus, principles of, 9

Deutrons, 368

Diffraction patterns, 263, 264, 265

Diffusion coefficients of ions, 457

Discontinuity, of radiation, 244, 254

Discovery of the neutron, 368

Dollar Line, 439

Drop: density of, 114; law of motion of, 124; rigid, 87; velocity of, 86; weighing the, 102

Duane's effect, 246

e: constant value of, 169, 177; essential elements of measurement, 124; exact evaluation of,

90, 105, 114 f.; final value of, 26, 120; fundamental physical and chemical constant, 17; H. A. Wilson's work on, 54; method of obtaining, 158, 160, 162 f.; Sir J. J. Thomson's work on, 49, 53; Townsend's work on, 45 ff.; variation in, 169, 171, 172

East-west effect, 436

Einstein-Bohr equation, 248

Einstein's equation, 146 f., 149, 156, 157, 238, 239, 242, 245, 247, 250, 266; history of, 244 ff.

Einstein's quantum theory, 237, 256

Electricity: absolute unit of, 29, 31; atomic theory of, 6, 10, 21, 66; early views of, 6 f.; Franklin's theory of, 14, 20, 24; growth of theories of, 10 f.; ion of, 28; and light, 252; and matter, 186; proof of atomic nature of, 66, 131; structure of, 3, 4; two-fluid theory, 13

Electrolysis, 27, 28, 43

Electrolytic laws, 25

Electromagnetic theory of mass, 20, 185, 187; conflict with, 212

Electron, 4, 26, 77, 261; basis of all static charges, 72; early values of, 27–63; energy of, 239, 250; mass of, 27; origin of the word, 25; positive and negative equal, 82, 85, 182, 208; radius of, 185 f., 188; speed of, 191; theory of, 11, 21, 24

Electronic energy, 245

Electronic orbits of atoms, 221, 222

Electrons: emission of, 235, 236; number of, in atom, 195; number of free positive, in nucleus, 277

Elementary processes, 266

Elements, the, 92; with atomic numbers and weights, and chemical positions, 481

$\frac{e}{m}$: for cathode rays, 42; value of, in electrolysis, 27, 30; value of, in exhausted tubes, 43; for negative and positive ions, 43; Bucherer's value of, 214

Elliptical orbits, 279

Emission energy of electron, 236, 250 f.; explosive, 253; from light, 251

Energies released in atom building, 314; in atom annihilation, 314

Energy, total in cosmic rays, 309

Epstein theory of orbit, 219, 220

Ether, 16, 26; theory, 232, 235

Ether-stress theory, 17 f.; objections to, 248

Ether-string theory, Thomson's, 236, 237, 248, 249

Faraday constant, 120

Faraday's laws, 15, 19, 22, 25

Faraday lines, 237

Faraday-Maxwell theory, 17

Field currents, 261

Fordney-Settle flights, 426, 430, 434

Franklin theory, 20

Gamma rays, 232, 252; in transmutation processes, 379

Gaseous conduction, nature of, 32

Gases: electrical properties of, 46; ionization of, 125; gases, quantitative measurements in, 147

Geiger-Müller counters, 335, 343 345, 368, 379, 446

Gem Lake (Calif.), 317

Gram molecule, volume of in gases, 30, 31

Grating: molecular, 197; crystal, 264; spacing, 261

Greek philosophy, 1, 2, 9 f., 232

h, value of, 117, 212, 213, 242. 245, 261

High-altitude latitude effects, 417

Hipp chronoscope, 74

"Hot spark," vacuum spectrometer, 282

Inner quantum number, 282

Interconvertibility of mass and energy, 446

Interpenetration ideas, 279

Ion, 28, 29; diffusion coefficient of, 36; gaseous and electrolytic 34; isolation of, 67; mobility of, 35, 36 f., 39, 262; positive and negative, 38, 48; univalent, in electrolysis, 34, 39

Ionic charge, 29, 39, 45, 47, 176; elementary, 76; mobility, 264

Ionization: by α-rays, 139 f., 144; by β-rays, 138, 144; by ether waves, 134, 144; gaseous and electrolytic, 39; mechanism of gaseous, 125; oil-drop experiments in, 127, 141; by X-rays and radium rays, 125, 134

Irregular doublet law, 280

Kinetic energy: of atom, 194, 273; of light, 239; of translation of molecule, 261

Kinetic theory, 8, 31, 49, 157; of gases, 145, 268

Klein-Nishina formula, 398, 401, 402

L-Doublet, 275

Lake Maguilla, 409

Lake Titicaca, 409

Latitude effect in cosmic rays, 407

Lauritsen million-volt tube, 364

Lenard's trigger theory, 251

Light-quants, 244, 256

Lima, Peru, 414, 441

Lithium spectrum, 276

Longitude effect, 436

Magnetic quantum number, 295

March Field, 417, 428

Mass, electrical theory of, 185, 188; of hydrogen atom, 261; variation of with speed, 189

Mass of charge on sphere of radius, *a*, 467

Mass of electrical charge, 467

Matson Line, 439

Maxwell-Boltzmann law, 81

Maxwell distribution law, 151, 275 f.; theory, 24

Mean free path of a gas molecule, 8, 183, 470

Mean free path of a negative electron, 191

Measurements, exact, 58, 65

Melbourne, 440, 441

Mobilities of ions, 457

Molecular cross-section, 470

Molecule: diameter of, 215; of gas, 261; kinetic energy of agitation of, 80, 81, 194

"Monterey," the, 439

Moseley's discovery, 196, 203; atomic numbers, 198, 200, 203; atomic weights, 202; X-ray frequencies equal, 246

Moseley's law, 204 f.; in optics, 284; inexactness of, 216

Mount Whitney, 307

Muir Lake, 307, 316

Ne, value of, in electrolysis, 31; in gases 35, 262

Ne, discrepancies, 168 ff.; in gases, 34, 125, 166 f.; value of, for electrolysis, 30, 31; value of, for negative and positive ions, 125

Neher electroscope, 412

Neutron, the, chap. xv

Norman Bridge Laboratory, 4

Nuclear electronic encounters, effects of, 420

Nucleus: of atom, 193; charge on, 195; number of electrons in, 207, 472, 476

Panama, 414

p₁p₂ orbits, 282

Peltier effects, 20

Penetrating properties of cosmic rays, 436

Perrin's value of *N*, 170

Peru, 417

Photo-electric effect, 236, 244, 246, 255

Photon sprays, 421

Photons, 260

Physical constants, table of, 482

Pikes Peak, 431

Planck's *h*, 117, 212, 213, 242, 245, 261

Pointing vector, 266

Positive and negative electrical charges, 11, 12

Positrons due to γ-rays, 335

Positrons, chap. xiv

Potential barrier, 262, 365

"President Garfield," the, 439

Quantitative measurements in gases, 147

Quantization rules, 272

Quantization, spacial, 295

Quantum theory of radiation, Einstein's, 237 f.; Planck's, 253

Radiant energy, 242, 245; nature of, 232

Radiation, black-body, 237; discontinuous, 244; quantum theory of, 237, 253; theories of, 232; Thomson's ether-string theory, 236, 237

Radio quantum number, 272

Radium, conductivity in air due to, 33; as ionizing agent, 53, 57, 79; mechanism of ionization by, 125; structure of atom of, 230

Regular doublet law, 280

Relativity formula, 275

Relativistic fine structure, 271

Resistance to motion independent of charge, 86

Rutherford's method, 277

Rydberg constant, 210, 477; value of, 477; Bohr's derivation of, 477

Ryerson Laboratory, 4, 37, 93, 94, 95, 150, 178, 209, 239

Singapore, 439

Spectropic rules, 293

Spinning electron, 270

Spokane, 417

Special relativity, 273

Spectroscopics, new, 293

Stark effect, 219

Statistical behavior, 268

Stefan-Boltzmann constant σ, 117, 261

Stokes's law, 48, 49, 52, 55, 63, 66, 167; correction of, 98 f., 101, 113, 162; failure of, 90 ff., 93, 98, 163; limits of validity of, 95 ff.

Stratisphere flights, 424, 426, 430, 434

Sub-electron, 161 f., 166, 175, 181

Sydney, Australia, 440

Thomson-Einstein hypothesis, 253

Three-electron system, 286

Transmutation of the elements, chap. xv

Two-electron system, flag of, 286

Uncertainty principle, 300

Valency in gaseous ionization, 127 f.

"Valero III," 436, 437

Velocity of agitation of molecules, 8; of emission of electrons, under monochromatic light, 245

Velocity of drop, 86, 101

Vienna Academy, 153, 175

Viscosity of air, 93

Wave-length, electron waves, 261

Wave theory, 233, 234

Wien constant c_2, 117, 261

Wilson-Sommerfeld rules, 271

X-rays, 23, 33, 39, 40, 46, 50, 57, 192, 204, 235; absorption of, 199; Duane's method, 214; frequencies, 202, 246; ionizing agent, 79, 126, 134; K and L series, 204; and light, 248; mechanism of ionization by, 125; monochromatic, 246; photographs of spectra, Fig. 21; spectra, 196, 200; spectrometer, 197; wave-lengths of metals, 197

X-ray doublets, regular, 281; irregular, 281; separation of, 275